# Advanced Sensing and Image Processing Techniques for Healthcare Applications

# Advanced Sensing and Image Processing Techniques for Healthcare Applications

Editors

**Vahid Abolghasemi**
**Hossein Anisi**
**Saideh Ferdowsi**

MDPI • Basel • Beijing • Wuhan • Barcelona • Belgrade • Manchester • Tokyo • Cluj • Tianjin

*Editors*

| Vahid Abolghasemi | Hossein Anisi | Saideh Ferdowsi |
| School of Computer Science | School of Computer Science | School of Computer Science |
| Electronic Engineering | Electronic Engineering | Electronic Engineering |
| University of Essex | University of Essex | University of Essex |
| Colchester | Colcehster | Colchester |
| United Kingdom | United Kingdom | United Kingdom |


*Editorial Office*
MDPI
St. Alban-Anlage 66
4052 Basel, Switzerland


This is a reprint of articles from the Special Issue published online in the open access journal *Sensors* (ISSN 1424-8220) (available at: www.mdpi.com/journal/sensors/special_issues/adv_sens_image_process_tech_healthcare_appl).

For citation purposes, cite each article independently as indicated on the article page online and as indicated below:

LastName, A.A.; LastName, B.B.; LastName, C.C. Article Title. *Journal Name* **Year**, *Volume Number*, Page Range.

**ISBN 978-3-0365-4032-0 (Hbk)**
**ISBN 978-3-0365-4031-3 (PDF)**

# Contents

# About the Editors

**Vahid Abolghasemi**

Dr Abolghasemi is an Assistant Professor at the School of Computer Science and Electronic Engineering, University of Essex. His primary research focus is on compressive sensing, sparse representation, dictionary learning, image and signal processing, and computer vision. Most of his research findings and proposed solutions, e.g., in medical applications, biometrics, and computer vision, have been published in high-quality journals and conferences. Currently, he is working toward expanding the usability of his theoretical knowledge and expertise in new industrial and practical real-life applications such as compressive sensing for smart and low-power sensing and communication technologies, artificial intelligence for healthcare and global environmental challenges, etc.

**Hossein Anisi**

Dr Anisi is an Associate Professor at the School of Computer Science and Electronic Engineering, University of Essex, and head of Internet of Everything (IoE) Laboratory. Prior to this, he worked as a Senior Research Associate at University of East Anglia, UK, and Senior Lecturer at University of Malaya, Malaysia, where he received the 'Excellent Service Award'for his achievements.

His research has focused specifically on real-world application domains such as energy management, agriculture, transportation, healthcare and other potential life domains. As a computer scientist, he has designed and developed novel architectures and routing protocols for Internet of Things (IoT), enabling technologies including wireless sensor and actuator networks, vehicular networks, heterogeneous networks, and body area networks, and his research results have directly contributed to the technology industry. He has collaborated on many projects in industry and has been working with several companies in the UK with a focus on monitoring and automating IoT-based systems capable of reliable and seamless generation, transmission, processing and the demonstration of data.

He has published more than 100 articles in high-quality journals and several conference papers and won two medals for his innovations from PECIPTA 2015 and IIDEX 2016 expositions. He has received several international and national funding awards for his fundamental and practical research as PI and Co-I.

**Saideh Ferdowsi**

Dr. Ferdowsi is a Senior Research Officer at School of Computer Science and Electronic Engineering, University of Essex. She received her PhD from the University of Surrey in Biomedical signal and Image processing. Her main research interests are biomedical signal and image processing, data fusion, blind source separation and machine/deep learning for EEG, fMRI and ECG. She is currently working on a project which promotes social interaction through emotional body odours. Her focus in the project is on developing Bayesian computational models of multi-modal social interaction. The designed model will be used to cover the role of human chemosignal perception in social interactions. Neurophysiological signals such as EEG, behavioural signals such as f-EMG and peripheral physiological activation such as ECG, RESP, and EDA will be used to identify the model.

*sensors*

MDPI

*Article*

# Intraoperative Hypotension Prediction Model Based on Systematic Feature Engineering and Machine Learning

Subin Lee [1], Misoon Lee [2], Sang-Hyun Kim [2] and Jiyoung Woo [1,*]

[1] Bigdata Engineering Department, SCH Media Labs, Soonchunhyang University, Asan 31538, Korea; lsb102030@naver.com
[2] Department of Anesthesiology and Pain Medicine, Soonchunhyang University Bucheon Hospital, Soonchunhyang University College of Medicine, Bucheon 14584, Korea; misoonlee@schmc.ac.kr (M.L.); skim@schmc.ac.kr (S.-H.K.)
* Correspondence: jywoo@sch.ac.kr

**Abstract:** Arterial hypotension is associated with incidence of postoperative complications, such as myocardial infarction or acute kidney injury. Little research has been conducted for the real-time prediction of hypotension, even though many studies have been performed to investigate the factors which affect hypotension events. This forecasting problem is quite challenging compared to diagnosis that detects high-risk patients at current. The forecasting problem that specifies when events occur is more challenging than the forecasting problem that does not specify the event time. In this work, we challenge the forecasting problem in 5 min advance. For that, we aim to build a systematic feature engineering method that is applicable regardless of vital sign species, as well as a machine learning model based on these features for real-time predictions 5 min before hypotension. The proposed feature extraction model includes statistical analysis, peak analysis, change analysis, and frequency analysis. After applying feature engineering on invasive blood pressure (IBP), we build a random forest model to differentiate a hypotension event from other normal samples. Our model yields an accuracy of 0.974, a precision of 0.904, and a recall of 0.511 for predicting hypotensive events.

**Keywords:** machine learning; vital sign; invasive blood pressure; feature engineering; hypotension; arterial hypotension

**Citation:** Lee, S.; Lee, M.; Kim, S.-H.; Woo, J. Intraoperative Hypotension Prediction Model Based on Systematic Feature Engineering and Machine Learning. *Sensors* **2022**, *22*, 3108. https://doi.org/10.3390/s22093108

Academic Editors: Hossein Anisi, Vahid Abolghasemi and Saideh Ferdowsi

Received: 12 December 2021
Accepted: 11 April 2022
Published: 19 April 2022

**Publisher's Note:** MDPI stays neutral with regard to jurisdictional claims in published maps and institutional affiliations.

## 1. Introduction

Arterial hypotension that occurs during anesthesia may increase the incidence of postoperative complications, such as myocardial infarction or acute kidney injury [1]. Careful monitoring of the patient's hemodynamic changes is required during anesthesia, and when hypotension is detected, immediate treatment is provided to maintain hemodynamic stability. If the patient's hemodynamic changes are predicted in advance, it will be possible to provide safer anesthesia to the patient by maintaining hemodynamic stability. Most patient monitor devices that monitor a patient's vital signs store the data for a short time [2], and the data are mostly deleted without being utilized for other purposes.

However, these vital sign data can be useful in developing a tool which can predict a patient's hemodynamic changes.

While research on hypotension in operation room mostly focuses on investigating the factors affecting a hypotension event, not much research has been performed on real-time prediction of hypotension. The advanced warning that hypotension is imminent at least 5 min ahead enables clinicians to take proper measures to reduce the impact of hypotension. This forecasting problem is quite challenging compared to diagnosis that detects high-risk patients at current. The forecasting problem that does not specify when the event occurs is easier than the forecasting problem that specifies the event time. Furthermore, it is very difficult to advance the predictable time compared to the event occurrence time. In this work, we will challenge the forecasting problem in 5 min advance.

Previous works on hypotension prediction have proposed various indices that originate from the waveform of vital signs. Recently, machine learning algorithms have replaced the scoring system, identified significant factors, and measured their effect on an event automatically.

In this study, we propose a systematic feature engineering that is applicable to any kind of vital signs and build a machine learning model that predicts hypotension in advance. We aim to build a simple model that does not require many vital signs and only requires invasive blood pressure (IBP). Instead of hand-crafted features on IBP, we propose a common feature extraction model that can be applicable to various kinds of vital signs. The feature extraction model includes the statistical analysis, peak analysis, change analysis, and frequency analysis. We build an ensemble model using a random forest model to handle numerous features in heterogenous samples.

## 2. Related Works

Many studies using vital signs have been performed in the intensive care unit (ICU); however, there is little research for the operation room where vitality is relatively constant compared to ICU [3–9].

Recently, studies that predict hypotension, depth of anesthesia, hypothermia, etc., have been conducted in the operating room. Topics of the studies using vital signals during surgery encompass estimation of the depth of anesthesia, estimation of blood pressure, event prediction regarding blood pressure, and heart failure. The former models were designed to predict whether a patient would suffer an event or not at the initial stage of operation [10–14]. These works can inform high-risk patients, but are limited in alerting an alarm for real-time treatment for an event. The recent prediction models are developed into real-time prediction models and the number of works is limited. We briefly reviewed real-time models in terms of classification and regression.

### 2.1. Real-Time Event Detection

Yang et al. [15] reported a convolutional neural network (CNN)-based deep learning model that predicts the stroke volume with a 20 s arterial blood pressure waveform. Lee et al. [16] created a CNN-based deep learning model to predict hypotension before 5 min, 10 min, and 15 min, respectively, using IBP, electrocardiography (ECG), photoplethysmography (PPG), and capnography ($CO_2$). They demonstrated that the precision and recall were higher than our research, but their experimental setting was different from ours. They included only the period where non-hypotension lasted for 20 min only. Their environment was less realistic because their model did not work on samples that included any data below the criteria. In addition, it is not sure that they focused on predictions for the very first time point of hypotension. As hypotension occurred, an alarm given in a timely manner was required in the first place. Chen and Qi [17] proposed a feature-based model. They predicted heart failure by statistical features; textualization; and imaging using HR, SBP, DBP, $SpO_2$, and pulse pressure difference (PP). Among the statistical feature models, the gradient boosting tree model had the highest accuracy of 84%, while textualization and imaging models had accuracies of 81% and 83% for the logistic regression and convolution neural network models, respectively. Furthermore, in predicting heart failure, the statistical feature-based model gave the best results. The statistical features used in this study included the mean; variance; minimum; maximum; 25%, 50%, and 75% quantiles; skewness; kurtosis; and first-order difference of each feature.

These real-time detection models suffer from the class imbalance problem and rarely achieve good performance. Most works set up an artificial environment to make the models work.

### 2.2. Real-Time Regression

The following works have been proposed to real-time regression for blood pressure or depth of anesthesia. The real-time regression model showed better performance compared to the event detection model because regression models are free from the class imbalance problem that the event detection model suffers from. This imbalance problem makes the model difficult to generalize. The models adopted in previous works were developed from machine learning models incorporated with feature engineering to the deep learning model. RNN-based models suitable for time sequence were adopted, and CNN models suitable for imaging were also adopted after the vital sign transformed into an image.

Regarding the model adopted machine learning with feature engineering, Jeong et al. [18] developed a blood pressure prediction model by applying the deep learning model to non-invasive blood pressure and other vital signs. This work proposed a concise model using derived variables rather than the original waveform data.

Gopalswamy et al. [10] proposed a long short-term memory (LSTM) model to predict intraoperative blood pressure and length of stay (LOS) using temperature, respiratory rate (RR), heart rate (HR), diastolic blood pressure (DBP), systolic blood pressure (SBP), fraction of inspired $O_2$ ($FiO_2$), and end-tidal $CO_{2n}$ ($EtCO_2$). Sadrawin et al. [1] reported artificial neural networks (ANNs) which can predict the depth of anesthesia using electroencephalography (EEG), electromyography (EMG), HR, pulse, SBP, DBP, and signal quality index (SQI). Regarding CNN models, Liu et al. [19] presented a CNN model that can predict the depth of anesthesia by transforming the EEG signal into a spectral image through modified short-time Fourier transform (STFT) transformation. Chowdhury et al. [20] demonstrated that a deep learning model can predict the depth of anesthesia by imaging the ECG and PPG signals as a heat map.

### 2.3. Research Gaps

From the literature review, we found several research gaps:

- Little research has been conducted using the vital signs collected in the operation room, while plenty of research has been carried out in ICU.
- Previous works focusing on the vital signs in the operation room deal with the depth of anesthesia. Rare events such as hypotension are important for patient health.
- Most studies focus on diagnoses that can identify high-risk patients who will suffer an event rather than prognosis. To react to the event in a preventive way, a real-time prediction model is required.
- Light-weight real-time prediction models are more effective for instant answering. However, existing works used many kinds of vital sign [14,15,18,19,21].

## 3. Materials and Methods

### 3.1. Patient Population

The data used in this paper were collected in Soonchunhyang University Bucheon Hospital through the Vital Recorder [21] program, which used the Bx50 monitor for patients whose blood pressure was measured with intra-arterial catheters (ART) during operations. These data were based on the continuous monitoring of blood pressure as IBP and were collected from 30 December 2019 to 30 October 2020 using an IBP time series of 888 patients. IBP data were recorded in units of 100 Hz.

### 3.2. Preprocessing

A moving average was widely used to smooth data and remove short-term fluctuations to highlight the patterns embedded in time sequences. High-resolution data naturally exhibit fluctuations, making patterns distorted and feature extraction difficult.

To derive samples from waveform IBP, we set the specific feature observation period, delay period, and event observation period, accordingly. The feature observation period refers to the period where features are extracted, the delay period refers to how far into the future the forecasting targets, and the event observation period is when the event is observed.

For our model, the observation period was set as 20 s, the delay period was set as 5 min to provide enough time for medical staff to react, and the event observation period was set as 1 min. To differentiate the samples related to hypotension from normal samples, the observation period was kept as short as possible. However, the frequency-based features required many time points. Thus, we compromised these two contradictions and set up the observation period as 20 s. In other works, the observation period was set to 30 s. We aimed to vary the observation period up to 30 s and check the performance. The class information was retrieved over a 1 min observation period. A class observation period was set up instead of picking a point, though this was not due to difficulties in characterizing a certain point. The class observation period was long enough to generate more samples for the hypotension event. In our future work, we aim to perform various experiments with varying observation and class observation periods.

A hypotension sample is defined as a case where the maximum value of a 2 s moving average of IBP during the class observation period falls short of 65 mmHg. A normal sample is defined as a case where the minimum value of a 2 s moving average of IBP during 1 min exceeds 65 mmHg. Blood pressure data were used for feature extraction during the observation period. We excluded samples associated with hypotensive events which occur during the observation window or the delay period; otherwise, it would be unnecessary to make the prediction.

Any sample that satisfied the hypotension event during the data observation and delay periods were also excluded. In addition, if the hypotensive event occurred consecutively, only the first event needed to be considered. This specifically relates to cases with a maximum value of the 2 s moving average of the data combined with the observation section, whereby the delay section is <65 was excluded. This aimed to make a prediction at least 5 min in advance, except for cases where hypotension was predicted in a situation with hypotension. The results of preprocessing are shown in Table 1.

**Table 1.** The number of normal samples and hypotension samples with different windowing interval.

| Interval | Normal Samples | Hypotension Samples |
|---|---|---|
| 30 s | 240,314 | 11,956 |
| 10 s | 721,020 | 35,887 |

For real-time forecasting, data samples were continuously generated through windowing, as shown in Figure 1. We attempted two choices for the length of windowing interval, 30 and 10 s, and compared their respective prediction results. As the windowing interval decreased, more samples were generated, which helped to examine the data in a fine grain.

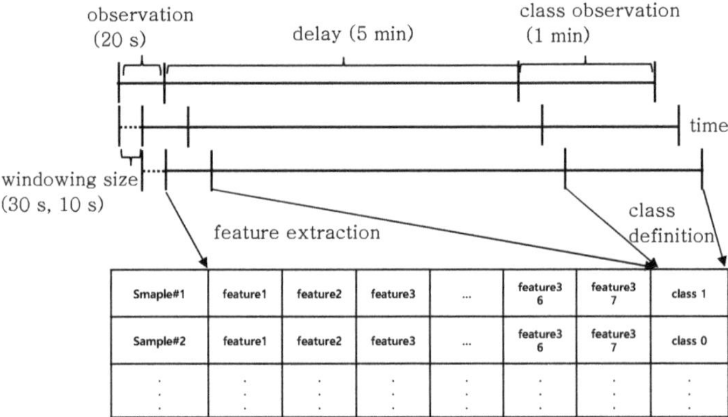


**Figure 1.** Sample generation process with an observation, delay, and class observation period. The observation period is how long feature generation can be observed, the delay is how far future events can be predicted, and the class observation is how long the event can be recognized.

Vital signs, as a form of time series through continuous monitoring, may display artifacts and noises due to electronic device errors, intraoperative events, or external pressure, as shown in Figure 2. To exclude artifacts and noise, we developed a criteria and excluded the samples that can satisfy various conditions. For example, the feature observation period and the class observation period, of which the maximum value exceeds 200 and the minimum value is under 20, were excluded. The case where the difference between the maximum and minimum during the feature observation or class observation is <30 conformed to an artifact. The difference between continuous values of 30 or less also conformed to artifacts. These slight variations for IBP occurred when the external pressure was applied to patients, usually to measure non-invasive blood pressure (NIBP) with cuffs.

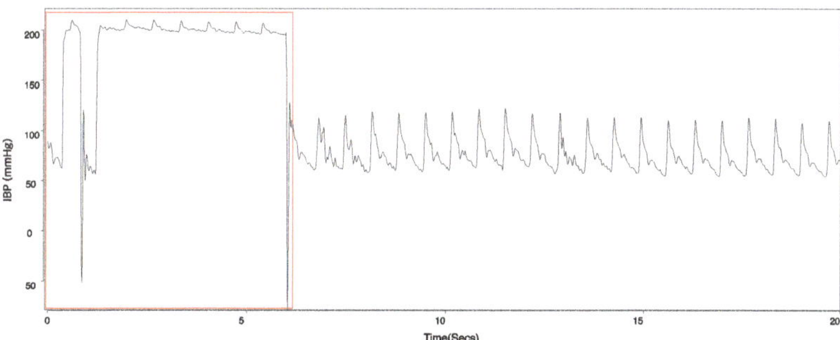

**Figure 2.** Artifacts marked in red on the time series of IBP.

## 4. Methodology

### 4.1. Feature Engineering

We proposed a systematic feature engineering process using domain knowledge. The proposed feature engineering process is not specific to only one vital sign, but can generally be applied to any vital sign signal.

The feature can be extracted in terms of the time domain and the frequency domain. The extensive feature engineering on the data observation period provides a hint for future events. To extract abnormality in values and their distribution, descriptive statistical analysis and peak analysis were both applied accordingly. The abrupt changes through change analysis were also captured.

### 4.2. Descriptive Analysis

Through descriptive statistical analysis, the representative values were selected through the mean, minimum, and maximum. The dispersion metrics describe the size of the distribution of values. The dispersion metrics include the range, variance, standard deviation, and inter-quartile range (IQR). To explain the shape and symmetry of data distribution, skewness and kurtosis were used as representative metrics. Skewness is a statistic which can indicate the degree of asymmetry of a distribution. If the distribution is symmetrical, such as a normal distribution or a T distribution, the skewness is 0. The skewness of a distribution with a long tail to the right and that to the left denote positivity and negativity, respectively. The kurtosis describes the weight of the tails of data distribution compared to standard normal distribution. The root sum square (RSS) was adopted by taking the square root of the sum of the squares of all the data points. RMS takes the square root of the arithmetic mean square of data points. These metrics represent the data as representative values. RSS implies the signal strength, while the RMS indicates the average of RSS.

### 4.3. Peak Analysis

The peak analysis aims to find the location of the local maxima or the minimum of a signal, and sorts the peaks by height, width, or prominence. Since our goal was to detect hypotension event, we defined the peak as the downward-sloping portion below 65, as marked in red in Figure 3. The statistical features on the peak detection results can be derived by the number of peaks, the mean, the standard deviation for the peak interval, the mean, the maximum, the minimum, the standard deviation for the peak value, and the crest factor. The crest factor shows the ratio of peak values to other values and represents the degree to which the peak is abnormal.

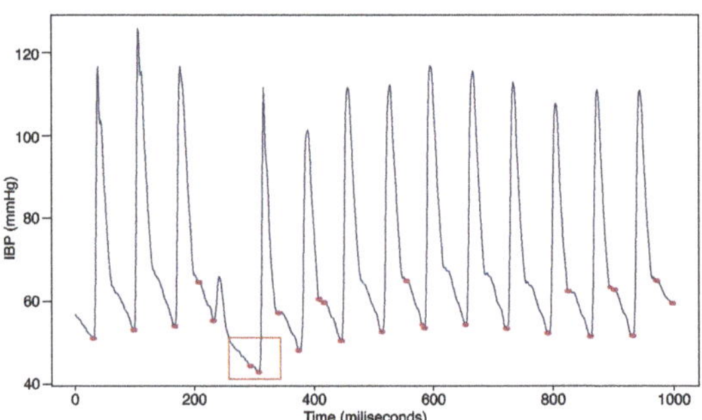

**Figure 3.** Downward peaks detected with a certain threshold.

Figure 3 demonstrates that the peaks can characterize the cyclic patterns, even though the patterns seem apparently similar to each other. The bounding box area shows different patterns with a low peak and a downward peak as well. Two peak points in the downward peak appear consecutively, as marked in red in Figure 3.

As demonstrated in Figure 3, peaks are useful to characterize cyclic patterns, even when they appear similar to one another. The bounding box area in red in Figure 3 shows different patterns from other time points. Two peak points in the downward peak appeared consecutively compared to other peak points.

### 4.4. Change Analysis

In the change analysis, the changes in mean and variance were detected. The change detection algorithm partitions a signal into adjacent segments where a statistic, such as the mean and the variance, is constant within a segment. To be more specific, the algorithm partitions the data into two parts and calculates the sum of the residual error of each part from its local mean. After detecting change points, the statistics, such as the number of changes in the mean, variance, and mean variance of blood pressure values, were accordingly derived. The red line in Figure 4 depicts the time point at which the mean changes (Figure 4a) and the time points at which the variance changes (Figure 4b).

### 4.5. Frequency Analysis

The waveform data recorded in the time domain can be transformed into the frequency domain, as shown in Figure 5. The frequency analysis extracts major frequencies in forming the time series. The frequency analysis was divided into Fourier transform and wavelet transform. The spectrum through the Fourier transform, displaying the power, indicates how much a given frequency contributes to the signal. We used the fundamental frequency with the highest power and other frequencies which follow the fundamental frequency. The frequencies with the top three powers were used as features.

In the wavelet transform, a wavelet, i.e., an oscillation form, was convolved with time-series data by scaling the wavelet and shifting into timelines.

Wavelet families include various mother wavelets that can be applied differently depending on domains. The Morlet parent function can identify oscillated patterns.

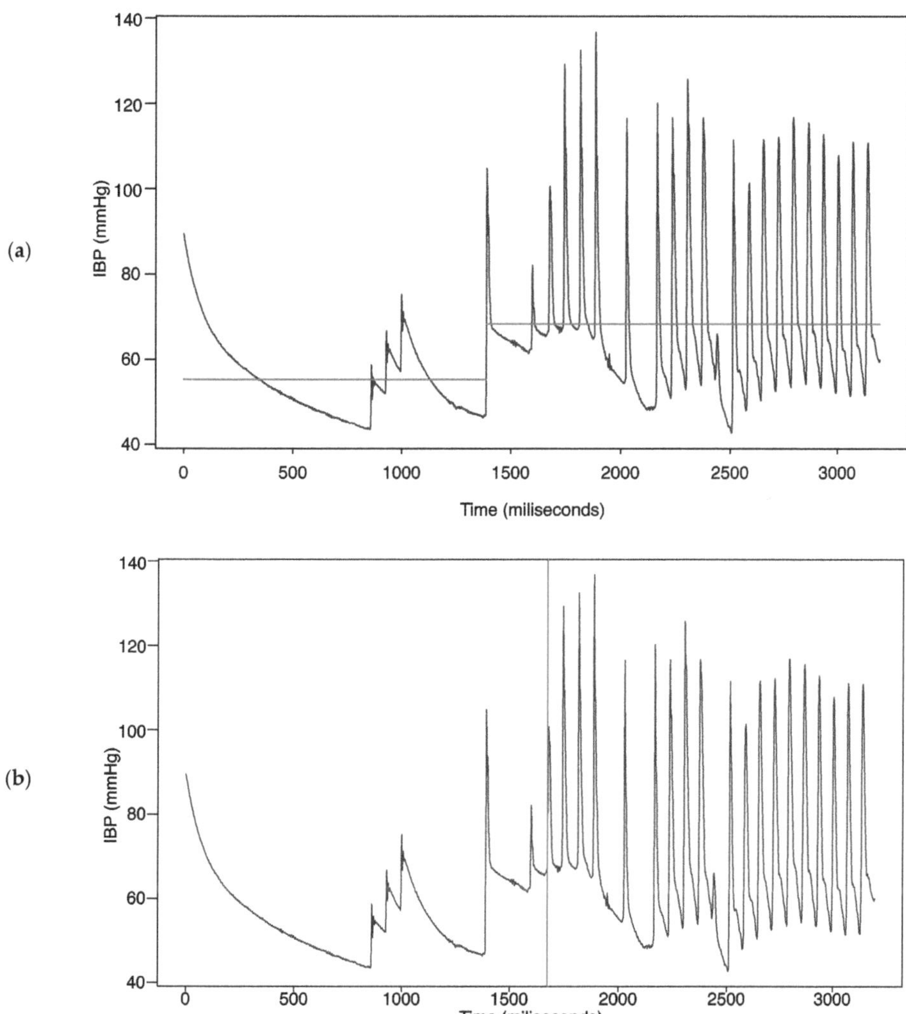

**Figure 4.** Change points analysis results. (**a**) Mean change; (**b**) variance change.

Wavelet transform is a form of time–frequency representation. It gives the coefficients of scaling and shifting coefficients. The baseline of the signal's scalogram is extracted through continuous wavelet transform. The scalogram value represents how much a wavelet scaled by a scale contributes to a signal at a certain time. We derived 10 scale values with the top scalogram values as features. The transformation of the time domain data into the frequency domain is shown in Figure 5. At the right upper panel in Figure 5, the periodogram from FFT shows the fundamental frequencies that lay at 0.02 and its multiples in terms of the relative frequency.

The scalogram at the right bottom panel indicates the absolute value of the continuous wavelet transform of an IBP time series, plotted as a function of scale and power. Wavelet algorithm changes the wavelet scale and checks how much the scaled wavelet fits to the signal. It gives the contribution of each scale to the total energy of the signal.

The 36 aforementioned features are listed in Table 2 below.

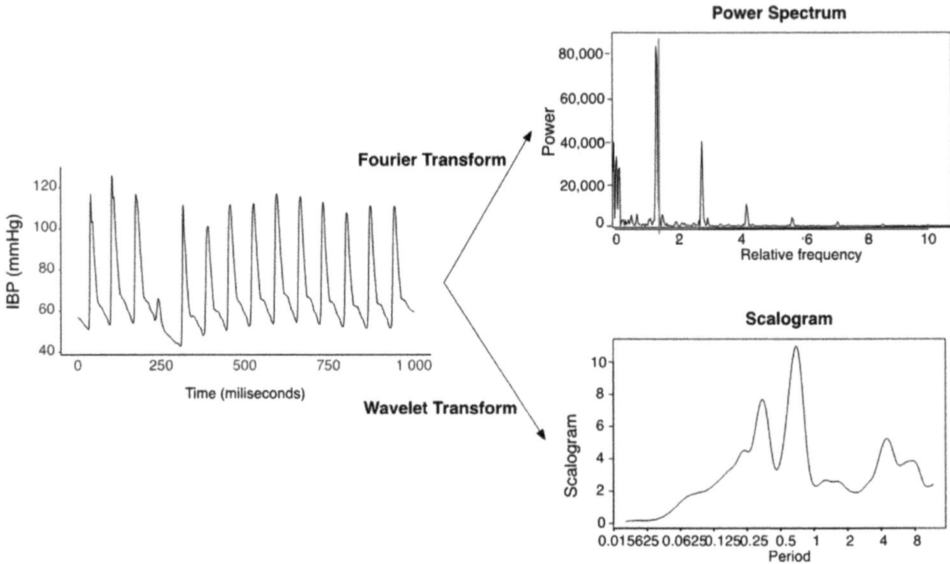



**Figure 5.** Fourier transform and wavelet transformation of the time domain signal.

**Table 2.** Details of the feature set.

| Category | Features | Number of Features |
|---|---|---|
| Statistics | Mean of blood pressure<br>Max of blood pressure<br>Min of blood pressure<br>Sd of blood pressure<br>Skewness of blood pressure<br>Kurtosis of blood pressure<br>RMS of blood pressure<br>RSS of blood pressure<br>IQR of blood pressure | 9 |
| Peak | Number of peak<br>Mean of peak interval<br>Sd of peak interval<br>Mean of peak value<br>Max of peak value<br>Min of peak value<br>Sd of peak value<br>Crest factor<br>The number of changes in mean<br>The number of changes in var<br>The number of changes in mean-var | 11 |
| Fourier | Top 3 power<br>Frequency of top 3 power | 6 |
| Wavelet | Top 10 scales with high scalog values | 10 |
| Total | | 36 |

*4.6. Model*

We then applied machine learning to extract the features. We adopted the sophisticated model on account of numerous features. Random forest is a machine learning technique proposed by [22] and is one of the ensemble learning methods used for classification and regression analysis. In a random forest model, several decision trees are constructed, and each tree individually learns the sampled data using bagging with different sets of features. Bagging is a method used to sample datasets by allowing duplicates. Then, the results of classification are voted on, and the result that receives the most is determined as the final classification result. This is effective for large data processing and has the advantage of improving model accuracy by avoiding the overfitting problem. A random forest was constructed for each extracted feature combination. The number of decision trees of random forest was designated as 100. Figure 6 presents the overall framework of our model.

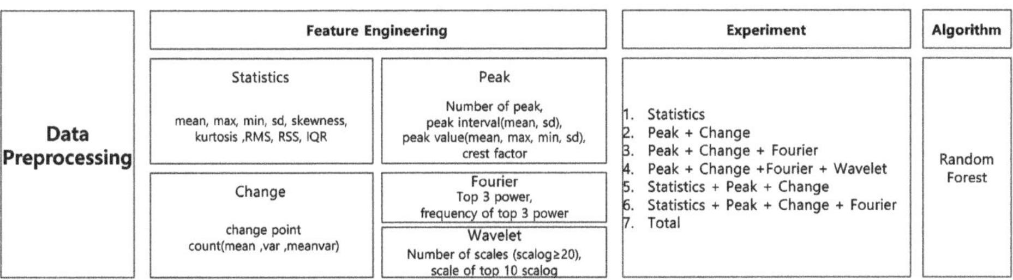

**Figure 6.** Research framework to build the hypotension prediction model.

## 5. Experiment and Results

### 5.1. Data Collection

Our dataset collected vital signs, EMR, and anesthesia record from adult patients (age ≥18 years) who underwent laparoscopic cholecystectomy under general anesthesia at Soonchunhyang University Bucheon Hospital, Bucheon City, Republic of Korea between 30 December 2019 and 31 October 2000. The vital signs were collected using the vital recorder [21]. Data collection was approved by the Soonchunhyang Bucheon hospital review board (approval No. SCHDB_IRB_2011-11-015). Informed consent was obtained from all subjects or their legal guardians. All methods were performed in accordance with the relevant guidelines and regulations.

### 5.2. Experiment Results

The hypotension prediction model was built under a different feature set, as shown in Figure 6. Our dataset had an imbalance problem with far less hypotensive samples than normal samples. To resolve the class imbalance problem, the most widely used methods are up-sampling and down-sampling. Up-sampling upsizes the small class at random, while down-sampling downsizes the large class at random.

To overcome this imbalance, the data for the minor class were augmented by up-sampling the training dataset. Up-sampling copies the data from the low-quantity class as much as the data from the high-quantity class to make the distribution of the classes the same. Up-sampling was performed by merely copying the hypotension samples for as many normal samples, as shown in Figure 7. Up-sampling processing was only performed in training data, but the validation dataset was kept as original.

Stratified k-fold cross validation was performed to evaluate the model. In k-fold cross validation, the data were divided into k splits, k-1 splits were used as the train set, and the remaining one split was used as the test set. k-fold is used when the data are independent and have the same distribution. For the data in this study, stratified k-fold was used instead of k-fold, because the distribution of each class was not the same. Stratified k-fold cross validation performs k-fold while maintaining the distribution of classes, as shown in Figure 8. In this study, we set k to 5.

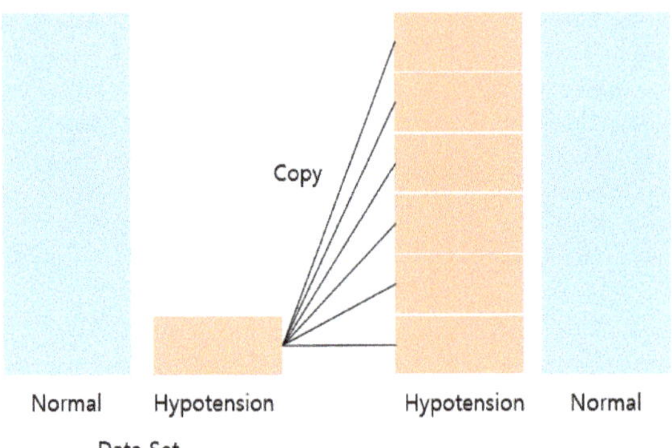



**Figure 7.** Up-sampling of hypotension class to resolve the class imbalance.

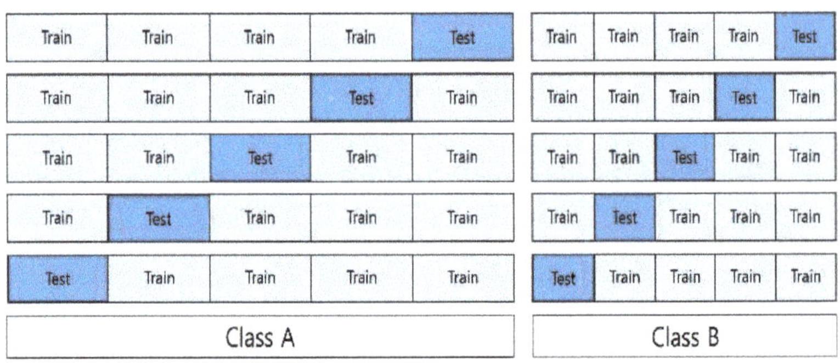

| Train | Train | Train | Train | Test | Train | Train | Train | Train | Test |
|---|---|---|---|---|---|---|---|---|---|
| Train | Train | Train | Test | Train | Train | Train | Train | Test | Train |
| Train | Train | Test | Train | Train | Train | Train | Test | Train | Train |
| Train | Test | Train | Train | Train | Train | Test | Train | Train | Train |
| Test | Train | Train | Train | Train | Test | Train | Train | Train | Train |
| Class A | | | | | Class B | | | | |

**Figure 8.** Visual representation of cross validation.

Accuracy for all classes and precision and recall for the hypotension class were used as the model performance indicators. Each expression is as follows. The hypotension is the same metric of sensitivity.

$$Accuracy = \frac{TP + TN}{TP + FN + FP + TN} \tag{1}$$

$$Precision = \frac{TP}{TP + FP} \tag{2}$$

$$Recall\ (Senstivity) = \frac{TP}{TP + FN} \tag{3}$$

*TP, TN, FP,* and *FN* represent the true positives, true negatives, false positives, and false negatives, respectively. As indicated by the performance, precision and recall of hypotension class were primarily used. Due to the sample imbalance, the metric should focus on the minor class. The precision and recall of hypotension class were presented first and the accuracy was presented as an overall metric.

The performances according to the feature sets are listed in Table 3. The results in Table 3 show that the fundamental frequencies and the Morlet wavelet, which captures the oscillation patterns, are both effective in improving characterization between the hypotension and normal class. The accuracy was as high as 0.974, but precision and recall for the positive class (hypotension) were rather low. This shows that the model works better with the normal class than with the hypotension class. The model was trained to precisely detect the hypotension and, as a result, it misses a significant portion of the hypotension, consequently yielding a low recall.

**Table 3.** Prediction results according to different windowing intervals and different feature sets.

| Feature Set | Windowing | | | | | |
| --- | --- | --- | --- | --- | --- | --- |
| | 30 s | | | 10 s | | |
| | Accuracy | Precision | Recall | Accuracy | Precision | Recall |
| Statistics | 0.961 | 0.679 | 0.316 | 0.967 | 0.792 | 0.403 |
| Peak + Change | 0.958 | 0.646 | 0.269 | 0.963 | 0.750 | 0.338 |
| Peak + Change + Fourier | 0.963 | 0.759 | 0.335 | 0.971 | 0.873 | 0.451 |
| Peak + Change + Fourier + Wavelet | 0.964 | 0.780 | 0.345 | 0.972 | 0.891 | 0.468 |
| Statistics + Peak + Change | 0.963 | 0.745 | 0.339 | 0.970 | 0.861 | 0.444 |
| Statistics + Peak + Change + Fourier | 0.965 | 0.775 | 0.372 | 0.973 | 0.887 | 0.491 |
| Total | 0.966 | 0.970 | 0.379 | 0.974 | 0.904 | 0.511 |

To improve the performance, we modified the machine learning algorithm by adding different class weights to the cost function of the algorithm. Various methods were used to assign the weight onto the class as shown in Table 4. The balanced method involves adding weight in reverse order to the class distribution. The balanced subsample calculates weights which are inversely proportional to the class frequency based on bootstrap samples. We could improve the recall when adjusting the weight assigned to each class, but should compromise the precision metric. Thus, we kept the original normal without the weight assignment.

**Table 4.** Prediction results according to different weights on the classes.

| Class Weight | Windowing | | | | | |
| --- | --- | --- | --- | --- | --- | --- |
| | 30 s | | | 10 s | | |
| | Accuracy | Precision | Recall | Accuracy | Precision | Recall |
| balanced | 0.964 | 0.795 | 0.326 | 0.972 | 0.914 | 0.461 |
| balanced subsample | 0.964 | 0.800 | 0.326 | 0.972 | 0.914 | 0.462 |
| 6:4 | 0.966 | 0.789 | 0.388 | 0.974 | 0.901 | 0.515 |
| 7:3 | 0.965 | 0.764 | 0.393 | 0.974 | 0.893 | 0.519 |
| 8:2 | 0.965 | 0.757 | 0.400 | 0.974 | 0.884 | 0.523 |
| 9:1 | 0.965 | 0.739 | 0.408 | 0.974 | 0.867 | 0.526 |

We also checked the receiver operating characteristic (ROC) curves for the best performed model in Figure 9. The ROC curves are consistently close to the ideal point which is (0, 1) for all cross-validation sets. As shown in Figure 9, the specificity relating to the normal class, calculated as the 1-$x$ axis value (False Positive Rate), is very close to 1. This is because most of the samples are normal and the algorithm works well for this major class.

To build an explainable machine learning algorithm, we assessed the impact of any given variable on the performance using feature importance. Feature importance is computed based on how important any given feature is to aid in the classification process when the classifier is built, determined by its effect on the performance measures. Gini importance is computed from the random forest structure. As shown in Figure 10, the most important features are listed as mean, RSS, RMS, and min of IBP. In terms of feature groups, the statistical feature set, the peak analysis feature set, the frequency analysis feature set, and the change analysis feature set were found to be important in that order.

*5.3. Exploratory Analysis*

Table 5 lists the vital signs according to hypotension and non-hypotension. All vital signs, except for the number of changes in the mean, were found to be significantly different. Overall, the IBP of the hypotension class is lower than that of the non-hypotension class. However, its skewness is higher. The IBPs of hypotension patients reach higher peaks than the non-hypotension class, and the peak values of the hypotension class are lower than those of the non-hypotension class. In addition, the peak values of the hypotension class have rather larger deviation than the normal class. The frequency of the hypotension class is higher than the non-hypotension class. This implies that IBP right before hypotension exhibits high vibration. The wavelet's scales of the hypotension class are lower than those of the non-hypotension class. This implies that more sharp oscillations occur in the hypotension class compared to the non-hypotension class.

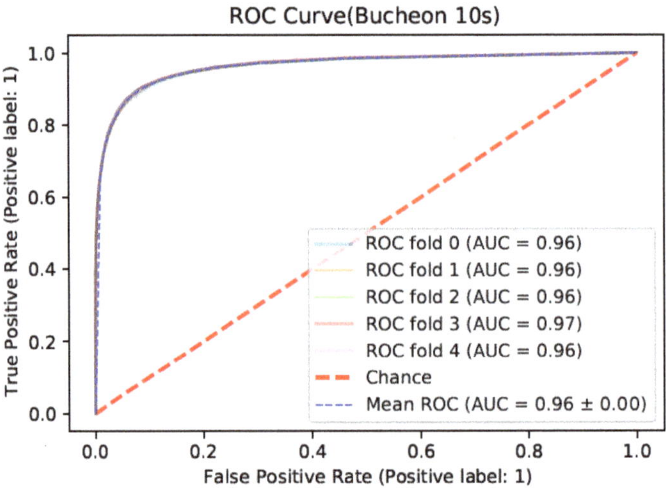


**Figure 9.** Receiver operating characteristic curves for each fold.

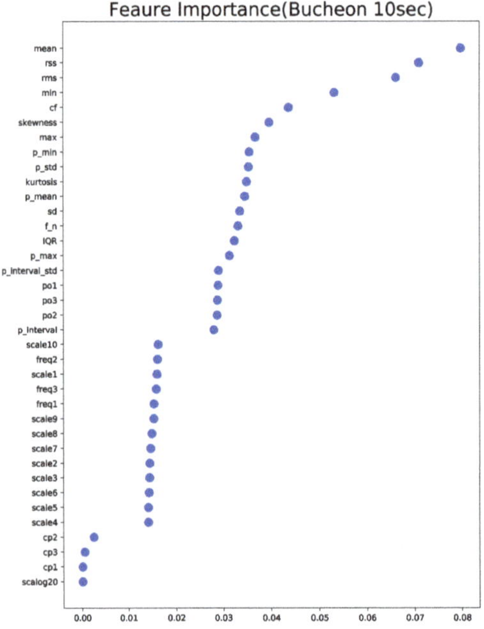


**Figure 10.** Feature importance plot.

**Table 5.** Clinical patient characteristics in terms of vital signs.

| Features | Non-Hypotension | | Hypotension | | *p*-Value |
|---|---|---|---|---|---|
| | Mean | Standard Deviation | Mean | Standard Deviation | |
| mean | 84.403 | 12.810 | 65.480 | 10.956 | <0.001 |
| min | 60.737 | 10.545 | 44.821 | 9.254 | <0.001 |
| max | 124.119 | 18.869 | 105.325 | 17.668 | <0.001 |
| sd | 17.528 | 4.873 | 16.007 | 4.485 | <0.001 |
| skewness | 0.717 | 0.250 | 0.917 | 0.319 | <0.001 |
| rms | 86.299 | 13.085 | 67.514 | 11.215 | <0.001 |
| rss | 172,598.647 | 26,169.658 | 135,027.662 | 22,429.642 | <0.001 |
| IQR | 28.128 | 10.679 | 22.535 | 8.861 | <0.001 |
| kurtosis | −0.678 | 0.841 | −0.252 | 1.130 | <0.001 |
| p_n: number of peaks | 30.221 | 23.457 | 52.457 | 30.431 | <0.001 |
| p_interval:means of peaks interval | 81.605 | 67.791 | 45.741 | 30.487 | <0.001 |
| p_interval_std: standard deviation of peak intervals | 47.246 | 79.895 | 25.111 | 29.733 | <0.001 |
| p_mean: mean of peaks | 58.673 | 4.928 | 49.040 | 10.455 | <0.001 |
| p_max: maximum of peaks | 62.625 | 3.514 | 56.908 | 11.508 | <0.001 |
| p_min: minimum of peaks | 55.299 | 6.916 | 42.889 | 10.630 | <0.001 |
| p_std: standard deviation of peaks | 2.052 | 1.766 | 3.927 | 2.327 | <0.001 |
| Cf: crest factor | 1.442 | 0.111 | 1.566 | 0.154 | <0.001 |
| cp1: number of mean changes | 1.000 | 0.002 | 1.000 | 0 | 0.699 |
| cp2: number of variance changes | 0.257 | 0.437 | 0.282 | 0.450 | <0.001 |
| cp3: number of mean and variance changes | 0.991 | 0.095 | 0.987 | 0.112 | <0.001 |
| freq1: first strongest frequency | 1.270 | 0.350 | 1.321 | 0.471 | <0.001 |
| freq2: second strongest frequency | 1.271 | 0.349 | 1.323 | 0.469 | <0.001 |
| freq3: third strongest frequency | 1.273 | 0.350 | 1.327 | 0.467 | <0.001 |
| po1: power of freq1 | 44,614.235 | 29,670.556 | 34,609.313 | 24,780.054 | <0.001 |
| po2: power of freq2 | 44,543.150 | 29,603.424 | 34,514.360 | 24,627.371 | <0.001 |
| po3: power of freq3 | 44,427.961 | 29,481.031 | 34,375.202 | 24,425.994 | <0.001 |
| scale1: the first largest scales of wavelet | 0.785 | 0.288 | 0.775 | 0.426 | <0.001 |
| scale2: the second largest scales of wavelet | 0.787 | 0.290 | 0.777 | 0.431 | <0.001 |

*5.4. Comparison Analysis with Another Dataset*

We performed an extra experiment with another dataset to verify the universality of our model. The public data from Seoul National University Hospital include all 6388 cases published in VitalDB whereby arterial pressure waveform monitoring was performed under general anesthesia. Those who are under the age of 18, weigh less than 30 kg or more than 140 kg, or who are less than 135 cm or more than 200 cm in height were excluded. In addition, the data cover 3278 files, excluding cases of transplant surgery, heart surgery, and vascular surgery. Like the data from Bucheon Hospital, it is recorded in units of 100 Hz. In this paper, only 983 were used for comparison, i.e., 30% of the data from Seoul National University Hospital. We found that the performance of our model for this dataset decreased, especially in terms of precision, as shown in Table 6.

**Table 6.** Verification results of the proposed model for other dataset.

| Windowing | | | | | |
|---|---|---|---|---|---|
| 30 s | | | 10 s | | |
| Accuracy | Precision | Recall | Accuracy | Precision | Recall |
| 0.989 | 0.652 | 0.441 | 0.992 | 0.764 | 0.540 |

The data from Seoul National University Hospital performed worse than the data from Soonchunhyang University Bucheon Hospital. This appears to be due to the difference in the type of surgery between the two datasets. In the case of the Seoul National University Hospital data, all surgeries, except for transplant surgery, heart surgery, and vascular surgery, were included, whereas the data from Soonchunhyang University Bucheon Hospital were only for laparoscopic cholecystectomy. In addition, although the Seoul National University Hospital data had a larger number of samples than the Bucheon Hospital data, the event imbalance was more severe. Based on 30 s, the ratio of hypotension samples over entire samples for Bucheon hospital was 4.7% (11,956/240,314) and the ratio for Seoul hospital was 1.3% (3559/260,683).

## 6. Discussion and Conclusions

Currently, several studies that predict the amount of stroke, heart failure, and hypotension using vital signs during surgery have been published [15,16,18,19]. In the near future, results of these studies may be adopted as useful diagnostic tools, enabling an immediate reaction to hemodynamic changes and improving perioperative prognosis.

The present authors conducted a study to predict the occurrence of hypotension 5 min in advance using vital signs. For that, we proposed a systematic feature engineering to build a real-time prediction model for hypotension in the operation room. This forecasting problem is quite challenging compared to diagnosis that detects high-risk patients at current. In particular, the forecasting problem that specifies the event occurrent time is very difficult to advance the predictable time. In this work, we challenged this problem through a systematic feature engineering and machine learning algorithm.

To process this problem, we tried to set up more a realistic condition than previous works. We included any hypotension, while previous works included the hypotension events that last for long time. One-off occurrences are more difficult to detect because there may be less precursor symptoms. In addition, we doubted whether previous works focus on the first point rather than following points during the hypotension. Any samples that embed hypotension during the observation and the delay should be deleted because they may give hints.

For more information, we performed the comparison between the patients who suffer hypotension or not. Appendix A Table A1 lists the clinical characteristics of patients, including electronic medical record and laboratory data. The only age among demographics and anesthesia time, operation time, crystal fluid amount, blood loss, and anesthesia method among operation-related variables recorded in EMR differed significantly between hypotension and normal groups. Among the preoperative test results, most variables such as Hb, Hct, Plt, PT, INR, aPTT, AST, ALT, Alb, Na, K, and Cl have significantly lower values of hypotensive patients than those of normal patients. Glc, BUN, and Cr did not differ significantly and had no clinical implication. Among preoperative laboratory test results, chloride concentration differed significantly between the groups. Among past disease records, valvular heart disease, Diabete smellitus, HbA1c, and cerebrovascular disease showed a significant difference between normal and hypotension groups. The presence of this disease is found to significantly increase the risk of hypotension.

From the current experiment, we could identify several future research directions.

Our problem is highly imbalanced for the hypotension class; thus, the model tends to be fitted to the normal class. As a consequence, it is hard to achieve good performance for the hypotension class. More specifically, our model does not cover hypotension samples, resulting in low recall. The low recall indicates that many patients who suffer hypotension later show no difference 5 min later compared to normal patients. This arguably suggests that the 5 min delay was too long, or that our feature engineering was insufficient. In future work, we will compromise the delay by checking the time point when differences between hypotension class and normal class are maximized.

From the feature importance, we found that the IBP values themselves were lower in hypotension than in the normal class. From this observation, more sophisticated statistical features can improve the performance. *p*-Values corresponding to a certain one-side test statistic will tell the difference in the distributions of IBP in normal and hypotension classes. These *p*-values indicate how a large portion of the data is lower than the threshold. We aim to vary the threshold to improve the performance.

Data were generated with windows of 30 s and 10 s, and features were extracted accordingly. The shorter the windowing interval, the better the performance. Furthermore, the model using all the features among the feature combinations showed the best performance. For future work, we will generate samples with the windowing interval in small units, such as 1 s. Furthermore, we will vary the observation and class observation period and check the performance. The best combination will be derived through the experiment.

Lastly, we will also apply other algorithms, such as deep learning on raw data, or other assemble methods, such as XGboost or stacking based on the same feature sets.

**Author Contributions:** J.W. and S.-H.K. designed the research; S.L., M.L. and J.W. wrote the manuscript; S.-H.K. and M.L collected data; S.L. and J.W analyzed the data and drafted the article; S.-H.K. edited the manuscript. All authors have read and agreed to the published version of the manuscript.

**Funding:** This research was supported by the Soonchunhyang University Research and also partially supported by the Basic Science Research Program through the National Research Foundation of Korea, funded by the Ministry of Education under Grant NRF-2020R1I1A3056858.

**Institutional Review Board Statement:** Data collection was approved by the Soonchunhyang Bucheon hospital review board (approval No. SCHDB_IRB_2011-11-015).

**Informed Consent Statement:** Informed consent was obtained from all subjects. All methods were performed in accordance with the relevant guidelines and regulations.

**Data Availability Statement:** Sample data is available on http://aibig.sch.ac.kr/data/listPageDATAHealthcare. do (accessed on 10 April 2022).



## Appendix A

**Table A1.** Clinical patient characteristics in terms of demographics, operation records, preoperative test results, and past disease history (% for discrete variable or standard deviation for continuous variable).

| Characteristics | No Hypotension ($n$ = 347) | Hypotension ($n$ = 541) | $p$ |
|---|---|---|---|
| | Demographics | | |
| Sex (Male) | 182 (52.4%) | 304 (56.2%) | 0.306 |
| Age | 58.75 (15.04) | 64.23 (15.15) | <0.001 * |
| Wt | 62.34 (16.71) | 61.53 (18.32) | 0.507 |
| Ht | 154.31 (36.22) | 152.71 (39.43) | 0.541 |
| BMI | 23.00 (6.81) | 22.59 (7.46) | 0.404 |
| ANE.Time | 246.48 (133.48) | 270.44 (174.82) | 0.030 * |
| Operation Time | 177.87 (119.72) | 203.90 (178.96) | 0.017 * |
| Crystal (mL) | 894.81 (742.36) | 1184.80 (1187.60) | <0.001 * |
| Colloid(mL) | 239.39 (287.54) | 274.62 (295.75) | 0.080 |
| Blood loss | 201.06 (296.78) | 373.76 (732.39) | <0.001 * |
| Urine. output | 473.90 (968.22) | 403.19 (548.59) | 0.166 |

Table A1. Cont.

| Characteristics | | No Hypotension ($n$ = 347) | Hypotension ($n$ = 541) | $p$ |
|---|---|---|---|---|
| ASA | 1 | 94 (27.1%) | 88 (16.3%) | |
| | 2 | 129 (37.2%) | 192 (35.5%) | |
| | 3 | 112 (32.3%) | 201 (37.2%) | <0.001 * |
| | 4 | 12 (3.5%) | 54 (10.0%) | |
| | 5 | 0 (0.0%) | 3 (0.6%) | |
| | 6 | 0 (0.0%) | 3 (0.6%) | |
| EM | | 71 (20.5%) | 175 (32.3%) | <0.001 * |
| Anesthesia Method | BPB | 0 (0.0%) | 1 (0.2%) | |
| | BPB_Volatile | 0 (0.0%) | 1 (0.2%) | |
| | CSE | 0 (0.0%) | 1 (0.2%) | |
| | MAC | 0 (0.0%) | 2 (0.4%) | <0.001 * |
| | spinal | 5 (1.4%) | 8 (1.5%) | |
| | TIVA | 195 (56.2%) | 171 (31.6%) | |
| | volatile | 147 (42.4%) | 357 (66.0%) | |
| Preoperative Test | | | | |
| Hb | | 10.03 (5.52) | 8.22 (5.92) | <0.001 * |
| Hct | | 29.83 (16.32) | 24.50 (17.63) | <0.001 * |
| Plt | | 197.13 (134.62) | 164.36 (140.36) | 0.001 * |
| PT | | 10.25 (5.40) | 8.99 (6.39) | 0.002 * |
| INR | | 0.79 (0.42) | 0.69 (0.50) | 0.004 * |
| aPTT | | 27.18 (14.72) | 23.91 (17.26) | 0.004 * |
| AST | | 21.12 (17.61) | 18.01 (17.89) | 0.011 * |
| ALT | | 19.10 (19.65) | 14.93 (17.92) | 0.001 * |
| Alb | | 3.00 (1.80) | 2.47 (1.92) | <0.001 * |
| Glc | | 71.67 (62.64) | 64.42 (71.44) | 0.122 |
| BUN | | 13.40 (10.95) | 12.19 (12.15) | 0.134 |
| Cr | | 1.97 (9.75) | 1.43 (7.69) | 0.357 |
| Na | | 110.41 (56.63) | 93.87 (65.06) | <0.001 * |
| K | | 3.27 (1.73) | 2.80 (1.99) | <0.001 * |
| Cl | | 82.88 (42.61) | 70.80 (49.13) | <0.001 * |
| History of Diseases | | | | |
| HBsAg | | 17 (6.2%) | 10 (2.7%) | 0.050 * |
| RPR | | 4 (1.5%) | 2 (0.5%) | 0.441 |
| Hypertension | | 150 (43.2%) | 271 (50.1%) | 0.054 |
| Atrialfibrillation | | 16 (4.6%) | 29 (5.4%) | 0.734 |
| Coronary artery disease | | 18 (5.2%) | 38 (7.0%) | 0.338 |
| Angina pectoris | | 11 (3.2%) | 18 (3.3%) | 1.000 |
| Myocardial infarction | | 3 (0.9%) | 8 (1.5%) | 0.620 |
| Congestive heart failure | | 5 (1.4%) | 16 (3.0%) | 0.221 |
| Valvular heart disease | | 3 (0.9%) | 16 (3.0%) | 0.062 * |
| Asthma | | 15 (4.3%) | 36 (6.7%) | 0.190 |

**Table A1.** *Cont.*

| Characteristics | No Hypotension ($n$ = 347) | Hypotension ($n$ = 541) | $p$ |
|---|---|---|---|
| COPD | 8 (2.3%) | 20 (3.7%) | 0.337 |
| Interstitial lung disease | 1 (0.3%) | 1 (0.2%) | 1.000 |
| Hepatitis | 8 (2.3%) | 6 (1.1%) | 0.263 |
| Liver cirrhosis | 9 (2.6%) | 12 (2.2%) | 0.894 |
| Viral carrier | 4 (1.2%) | 3 (0.6%) | 0.552 |
| Fatty liver | 1 (0.3%) | 0 (0.0%) | 0.823 |
| HBV | 12 (3.5%) | 8 (1.5%) | 0.088 |
| HCV | 4 (1.2%) | 4 (0.7%) | 0.786 |
| Alcoholic | 4 (1.2%) | 5 (0.9%) | 1.000 |
| Autoimmune | 0 (0.0%) | 1 (0.2%) | 1.000 |
| Acute kidney injury | 4 (1.2%) | 7 (1.3%) | 1.000 |
| Chronic kidney injury | 19 (5.6%) | 26 (4.8%) | 0.237 |
| End stage renal disease | 16 (4.6%) | 30 (5.5%) | 0.647 |
| Diabetes mellitus | 77 (22.2%) | 178 (32.9%) | 0.001 * |
| HbA1c | 1.42 (2.80) | 2.04 (3.20) | 0.003 * |
| Thyroid disease | 17 (4.9%) | 18 (3.5%) | 0.452 |
| Myasthenia gravis | 0 (0.0%) | 1 (0.2%) | 1.000 |
| Morbid obesity | 2 (0.6%) | 2 (0.4%) | 1.000 |
| Epilepsy | 2 (0.6%) | 2 (0.4%) | 1.000 |
| Cerebrovascular disease | 14 (4.0%) | 42 (7.8%) | 0.037 * |
| Cerebral aneurysm | 8 (2.3%) | 11 (2.0%) | 0.971 |
| Dementia | 6 (1.7%) | 15 (2.8%) | 0.440 |

Note: * $p < 0.05$.

## References


1. Salmasi, V.; Maheshwari, K.; Yang, D.; Mascha, E.J.; Singh, A.; Sessler, D.I.; Kurz, A. Relationship between intraoperative hypotension, defined by either reduction from baseline or absolute thresholds, and acute kidney and myocardial injury after noncardiac surgery: A retrospective cohort analysis. *Anesthesiology* **2017**, *126*, 47–65. [CrossRef] [PubMed]
2. Santos, R.J.; Bernardino, J.; Henriques, J. The HTP tool: Monitoring, detecting and predicting hypotensive episodes in critical care. In Proceedings of the IEEE EUROCON 2011—International Conference on Computer as a Tool, Lisbon, Portugal, 27–29 April 2011, pp. 1 4.
3. Barrett, L.A.; Payrovnaziri, S.N.; Bian, J.; He, Z. Building computational models to predict one-year mortality in ICU patients with acute myocardial infarction and post myocardial infarction syndrome. *AMIA Summits Transl. Sci. Proc.* **2019**, *2019*, 407. [PubMed]
4. Champion, S.; Lefort, Y.; Gaüzère, B.-A.; Drouet, D.; Bouchet, B.J.; Bossard, G.; Djouhri, S.; Vandroux, D.; Mayaram, K.; Mégarbane, B. CHADS2 and CHA2DS2-VASc scores can predict thromboembolic events after supraventricular arrhythmia in the critically ill patients. *J. Crit. Care* **2014**, *29*, 854–858. [CrossRef] [PubMed]
5. Dervishi, A. A deep learning backcasting approach to the electrolyte, metabolite, and acid-base parameters that predict risk in ICU patients. *PLoS ONE* **2020**, *15*, e0242878. [CrossRef] [PubMed]
6. Kong, G.; Lin, K.; Hu, Y. Using machine learning methods to predict in-hospital mortality of sepsis patients in the ICU. *BMC Med. Inform. Decis. Mak.* **2020**, *20*, 1–10. [CrossRef] [PubMed]
7. Moghadam, M.C.; Abad, E.M.K.; Bagherzadeh, N.; Ramsingh, D.; Li, G.-P.; Kain, Z.N. A machine-learning approach to predicting hypotensive events in ICU settings. *Comput. Biol. Med.* **2020**, *118*, 103626. [CrossRef] [PubMed]
8. Qin, K.; Xu, G.; Huang, J. Blood Pressure Prediction by Exploiting Informative Features from ICU Patients' ECG and PPG Signals under a Heterogeneous Ensemble Learning Framework. 2020. Available online: https://www.semanticscholar.org/paper/Blood-pressure-prediction-by-exploiting-informative-Qin-Xu/d591097e8e71ef1258c0bc28318a2a476ae80fd8 (accessed on 10 April 2022).
9. Zhang, P.; Roberts, T.; Richards, B.; Haseler, L.J. Utilizing heart rate variability to predict ICU patient outcome in traumatic brain injury. *BMC Bioinform.* **2020**, *21*, 1–11. [CrossRef] [PubMed]

10.  Gopalswamy, S.; Tighe, P.J.; Rashidi, P. Deep recurrent neural networks for predicting intraoperative and postoperative outcomes and trends. In Proceedings of the 2017 IEEE EMBS International Conference on Biomedical & Health Informatics (BHI), Orlando, FL, USA, 16–19 February 2017; pp. 361–364.
11.  Kang, A.R.; Lee, J.; Jung, W.; Lee, M.; Park, S.Y.; Woo, J.; Kim, S.H. Development of a prediction model for hypotension after induction of anesthesia using machine learning. *PLoS ONE* **2020**, *15*, e0231172. [CrossRef]
12.  Kim, H.; Jeong, Y.-S.; Kang, A.R.; Jung, W.; Chung, Y.H.; Koo, B.S.; Kim, S.H. Prediction of post-intubation tachycardia using machine-learning models. *Appl. Sci.* **2020**, *10*, 1151. [CrossRef]
13.  Lee, J.; Woo, J.; Kang, A.R.; Jeong, Y.-S.; Jung, W.; Lee, M.; Kim, S.H. Comparative analysis on machine learning and deep learning to predict post-induction hypotension. *Sensors* **2020**, *20*, 4575. [CrossRef] [PubMed]
14.  Jeong, Y.-S.; Kim, J.; Kim, D.; Woo, J.; Kim, M.G.; Choi, H.W.; Kang, A.R.; Park, S.Y. Prediction of postoperative complications for patients of end stage renal disease. *Sensors* **2021**, *21*, 544. [CrossRef] [PubMed]
15.  Yang, H.-L.; Lee, H.-C.; Jung, C.-W.; Kim, M.-S. A Deep Learning Method for Intraoperative Age-agnostic and Disease-specific Cardiac Output Monitoring from Arterial Blood Pressure. In Proceedings of the 2020 IEEE 20th International Conference on Bioinformatics and Bioengineering (BIBE), Cincinnati, OH, USA, 26–28 October 2020; pp. 662–666.
16.  Lee, S.; Lee, H.-C.; Chu, Y.S.; Song, S.W.; Ahn, G.J.; Lee, H.; Yang, S.; Koh, S.B. Deep learning models for the prediction of intraoperative hypotension. *Br. J. Anaesth.* **2021**, *126*, 808–817. [CrossRef] [PubMed]
17.  Chen, Y.; Qi, B. Representation learning in intraoperative vital signs for heart failure risk prediction. *BMC Med. Inform. Decis. Mak.* **2019**, *19*, 1–15. [CrossRef]
18.  Jeong, Y.-S.; Kang, A.R.; Jung, W.; Lee, S.J.; Lee, S.; Lee, M.; Chung, Y.H.; Koo, B.S.; Kim, S.H. Prediction of blood pressure after induction of anesthesia using deep learning: A feasibility study. *Appl. Sci.* **2019**, *9*, 5135. [CrossRef]
19.  Liu, Q.; Cai, J.; Fan, S.-Z.; Abbod, M.F.; Shieh, J.-S.; Kung, Y.; Lin, L. Spectrum analysis of EEG signals using CNN to model patient's consciousness level based on anesthesiologists' experience. *IEEE Access* **2019**, *7*, 53731–53742. [CrossRef]
20.  Chowdhury, M.R.; Madanu, R.; Abbod, M.F.; Fan, S.-Z.; Shieh, J.-S. Deep learning via ECG and PPG signals for prediction of depth of anesthesia. *Biomed. Signal Process. Control* **2021**, *68*, 102663. [CrossRef]
21.  Lee, H.-C.; Jung, C.-W. Vital Recorder—A free research tool for automatic recording of high-resolution time-synchronised physiological data from multiple anaesthesia devices. *Sci. Rep.* **2018**, *8*, 1–8. [CrossRef] [PubMed]
22.  Breiman, L. Random forests. *Mach. Learn.* **2001**, *45*, 5–32. [CrossRef]

*Article*

# Simulation of 3D Body Shapes for Pregnant and Postpartum Women

Chanjira Sinthanayothin [1,*], Piyanut Xuto [2], Wisarut Bholsithi [1], Duangrat Gansawat [1], Nonlapas Wongwaen [1], Nantaporn Ratisoontorn [1], Parut Bunporn [1] and Supiya Charoensiriwath [1]

[1] National Electronics and Computer Technology Center, National Science and Technology Development Agency, Pathum Thani 12120, Thailand; wisarut.bholsithi@nectec.or.th (W.B.); duangrat.gansawat@nectec.or.th (D.G.); nonlapas.wongwaen@nectec.or.th (N.W.); nantaporn.ratisoontorn@nectec.or.th (N.R.); parut.bunporn@nectec.or.th (P.B.); supiya.charoensiriwath@nectec.or.th (S.C.)
[2] Faculty of Nursing, Chiang Mai University, Chiang Mai 50200, Thailand; piyanut.x@cmu.ac.th
* Correspondence: chanjira.sinthanayothin@nectec.or.th

**Abstract:** Several studies have reported that pre-pregnant women's body mass index (BMI) affects women's weight gain with complications during pregnancy and the postpartum weight retention. It is important to control the BMI before, during and after pregnancy. Our objectives are to develop a technique that can compute and visualize 3D body shapes of women during pregnancy and postpartum in various gestational ages, BMI, and postpartum durations. Body changes data from 98 pregnant and 83 postpartum women were collected, tracked for six months, and analyzed to create 3D model shapes. This study allows users to simulate their 3D body shapes in real-time and online, based on weight, height, and gestational age, using multiple linear regression and morphing techniques. To evaluate the results, precision tests were performed on simulated 3D pregnant and postpartum women's shapes. Additionally, a satisfaction test on the application was conducted on new 149 mothers. The accuracy of the simulation was tested on 75 pregnant and 74 postpartum volunteers in terms of relationships between statistical calculation, simulated 3D models and actual tape measurement of chest, waist, hip, and inseam. Our results can predict accurately the body proportions of pregnant and postpartum women.

**Keywords:** 3D body shapes; body weights and measures; postpartum period; pregnancy period; anthropometry

## 1. Introduction

Obesity during pregnancy is a serious health problem for women. Worldwide, obstetricians and midwives have confronted increasing obesity among pregnant women [1,2]. Reports [3–5] showed that women with a pre-pregnant body mass index (BMI) of either overweight or obese levels are at risk of developing diabetes during pregnancy compared to women with normal pre-pregnant BMI, even after taking the weight gains during a normal pregnancy into account. Women with diabetes during pregnancy tend to have high blood pressure, which can lead to abdominal surgery and premature birth [6]. Therefore, it is important for all women who are planning to become pregnant to control a proper weight before and during pregnancy using multi-faceted interventions throughout the reproductive years as a part of a long-term follow up and behavioral interventions to minimize pregnancy weight gain [7]. The BMI before pregnancy affects not only the weight gain during pregnancy, but also the postpartum weight retention [8,9]. It was reported that if a woman in the postpartum period was unable to regulate her weight to her pre-pregnant weight within six months, postpartum weight retention could predict future weight gain and long-term obesity [10]. Another study suggested that the BMI of a woman of more than six months postpartum would indicate the retaining of extra body fluids produced during

**Citation:** Sinthanayothin, C.; Xuto, P.; Bholsithi, W.; Gansawat, D.; Wongwaen, N.; Ratisoontorn, N.; Bunporn, P.; Charoensiriwath, S. Simulation of 3D Body Shapes for Pregnant and Postpartum Women. *Sensors* **2022**, *22*, 2036. https://doi.org/10.3390/s22052036

Academic Editors: Hossein Anisi, Vahid Abolghasemi and Saideh Ferdowsi

Received: 29 December 2021
Accepted: 1 March 2022
Published: 5 March 2022

**Publisher's Note:** MDPI stays neutral with regard to jurisdictional claims in published maps and institutional affiliations.

pregnancy, as well as extra fat during the first six months postpartum [11]. Sui Z. et al. [12] reported a statistically significant indication that women with a high degree of body image dissatisfaction were more likely to have higher gestational weight gain. Hill B. et al. [13] also reported that the timing of pregnancy and body attitudes could predict gestational weight gain (GWG). The findings suggested that lower attractiveness in early-to-middle pregnancy was associated with higher GWG [13]. Therefore, in this work, the 3D body shape simulation of women before, during, and after pregnancy has been developed with an expectation that women will not misestimate their BMI to prevent being overweight during pregnancy and the postpartum period. A 3D body simulation may encourage women to prevent overweight behaviors before and throughout pregnancy to maintain good health in the long run. Having 3D models in three stages of before, during, and after pregnancy to compare, is one of the ways to motivate women to develop long-term healthy behaviors.

There are some research reports that look at correlations between body shapes and BMI for women [14–16]. Nevertheless, there is no physical simulation for the shape proportion of pregnant and postpartum woman subjects, as they are a population group vulnerable to the use of the 3D body scanner for data collection. The word "vulnerable" is in the context of human research protections. Pregnant women are considered vulnerable due to the involvement of the fetus that may be affected by the research and the fetus cannot give consent [17]. It was not possible to collect the body shape data using the 3D body scanner in the study. The reason for this is that most pregnant women are concerned about the safety of 3D body scanners and have questions about the potential consequences of the use of the scanner at all stages of pregnancy [18]. It is difficult to obtain an approval for applications for research projects involving human subjects from the Institutional Review Board (IRB), and especially when asking for consent forms for research subjects [17]. We also needed pregnant and postpartum women to measure their own body circumferences at home every four weeks during pregnancy and postpartum.

Therefore, it was necessary to collect data of various shapes using a tape measurement of pregnant and postpartum women by forward tracking six months before and after giving birth. Then data were analyzed and processed to create a simulation of 3D modeling of pregnant and postpartum women based on data of non-pregnant female shapes from SizeThailand [19,20]. Although there is a web application with a non-pregnant female simulation, including Body Visualizer [21] developed by Black and Broscaru [22] from Max Planck Gesellschaft, which published as a part of US Patent Application [23] and became US Patent [24]. The process mentioned in the patent has been in use for studying female sensitivity to changes in their perceived weight by altering the body mass index (BMI) of the participants' personalized avatars to deal with body perception [25]. Furthermore, the process has also been in use for body size estimation in females varying in BMI as a measure to deal with rising cases of Anorexia patients [26]. In addition, the process uses the virtual caliper for the accurate 3D body measurements [27].

Body Visualizer has used the dataset based on SizeUSA, American and European Surface Anthropometry Resource (CAESAR) as the basis for creating 3D body shapes visualization [28,29]. It is a visualization tool for a parametric 3D body model that provides metrically accurate anthropomorphic measurements based on laser scans of thousands of people from different ethnicities. However, it is still lacking the 3D body models for pregnant and postpartum women, especially for Asian women. Therefore, in this study we focused on simulating the 3D shape of pregnant and postpartum women for Thais. In this article, 3D body shape simulation of non-pregnant, pregnant, and postpartum womens' shape with body proportions are predicted in real-time and online from weight, height and gestational age. A real-time prediction in our study is a service that provides the predictions via an HTTP call to simulate the 3D shape of pregnant and postpartum women via the web browser after the users input their data. The body shape proportion of pregnancy and postpartum women were analyzed using the linear regression of 587 pregnancy data and 503 postpartum data. The simulations of pregnant and postpartum women were further modified based on non-pregnant simulations from our previous study, which was analyzed from the SizeThailand database with 6767 females' data [19,20]. SizeThailand is a national sizing surveys project that includes 13,442 adults, both males, and females across Thailand at various ages [30].

Application Z-Size Ladies [31], described in this paper, was intended to collect user data in the form of BMI timeline and simulate the 3D female body shape for non-pregnant, pregnant women and postpartum women. Z-Size Ladies application [31] is a tool that helps pregnant and postpartum women to simulate their 3D-body shapes. The aim of developing the Z-Size Ladies application is to be a tool that can create a precise online 3D model for non-pregnant, pregnant, and postpartum women in real-time with a simple set of input data of their weight, height, and gestational age. The tool was validated by several linear regression studies and users' survey. The simulation of the body shape from body measurements can be considered as a low-cost alternative to full-body 3D scanning [32]. Furthermore, this application can be applied to provide online clothing services. The users only put in their weight, height, and gestational age; then they will know their body shape proportions. More supporting information about Z-Size Ladies can be showed/downloaded in Supplementary Materials.

## 2. Literature Reviews

There are some relevant studies on the prediction of 3D body shape during pregnancy using multiple 3D body scans with a purpose of setting the standard sizing chart for maternity wear that addresses the changes throughout pregnancy [33]. Vaughan et al. [34] matched personal weight, height and age with the overall body shapes taken from Magnetic Resonance Imaging (MRI) images to create 3D adjustable parametric human body models using OpenGL with 3D mesh deformation along with Artificial Neural Networks (ANNs) trained and assessed with the clinical data of 23,088 patients, including pregnant and postpartum patients from the National Health and Nutrition Examination Survey (NHANES) data from 1999 to 2012. The ANNs used in their study managed to predict the anthropometric measurements with the following margins of error including subscapular skinfold thickness within 3.54 mm, waist circumference 3.92 cm, thigh circumference 2.00 cm, arm circumference 1.21 cm, calf circumference 1.40 cm, and triceps skinfold thickness of 3.43 mm. An alternative regression analysis method gave overall predictions slightly less accurate for subscapular skinfold thickness within 3.75 mm, waist circumference 3.84 cm, thigh circumference 2.16 cm, arm circumference 1.34 cm, calf circumference 1.46 cm, and triceps skinfold thickness 3.89 mm. The results showed a parametric model of the patient's body shape and ligament thickness using OpenGL and adjusted by 3D mesh deformation. However, the 3D image that resulted from the mesh deformation looked unrealistic, despite the accurate anthropometric measurements.

Haddox et al. [35] created a musculoskeletal model of a pregnant woman to simulate the changes in segmental mass and inertia distribution. It included a case of changing breast size during pregnancy. That caused pregnant ladies to fall due to the changes on the centers of the upper trunks, pelvis regions, and torso centers along with lumbar curvatures. They used datasets from 25 pregnant Caucasian ladies in six sessions and postpartum women obtained from US Air Forces Research Lab as models having BMI before pregnancy between 18.9 to 26 kg/m$^2$. That was substantially lower than the average BMI of American women at 26.5 kg/m$^2$.

Ponnalagu et al. [36] pointed out that waist circumference (WC) is a simpler anthropometric measurement that has strong association with an individual's metabolic risk level. BMI alone is not adequate since Asians have a high tendency to deposit fat at the viscera compared with their European counterparts. This explains why Asians have a higher fat percentage than Europeans despite having the same BMI. Furthermore, high waist circumference increased the risks of developing hypertension, type 2 diabetes mellitus, hypercholesterolemia, joint pain, low back pain, and hyperuricemia as mentioned in the paper by Darsini et al. [37].

Han et al. [38] investigated the cut-off points of body mass index (BMI) and waist circumference (WC) for gestational diabetes mellitus (GDM) and interactions between high BMI and high WC on the risk of GDM. They collected the data during 2010 to 2012 from 17,803 Chinese pregnant women from Tainjin who were at 4–12 week gestation. The results showed that higher than 22.5 kg/m$^2$ BMI and higher than 78.5 cm WC were the cut-off points for gestational diabetes mellitus (GDM).

Jacobson et al. [39] invented Electronic Monitoring of Mom's Schedule (eMOMS™) for monitoring improved postpartum weight, blood sugars, and breastfeeding among high BMI women who had BMI between 25 to 35 kg/m$^2$. It offered an interactive communication between patients and physicians via Facebook, FaceTime and Skype. However, this invention focused only on weight rather than taking other anthropometric variables into account to help postpartum mothers to keep other physical factors of the body in check.

Ha et al. [40] conducted studies on postpartum weight retention in relation to gestational weight gain and pre-pregnancy BMI due to the rising cases of maternal overweight and obesity in Vietnam. They studied 2030 pregnant women recruited from three cities in Vietnam who were 24–28 weeks of gestation for the analyses on gestational weight gain (GWG). In addition, they followed 1666 mothers for 12 months after delivery for the analyses on 12-month postpartum weight retention (PPWR). They recorded all pre-pregnant BMI. The results showed that both pre-pregnancy BMI and GWG were significantly associated with PPWR since those pregnancies with underweight before pregnancy and excessive GWG contributed to greater weight retention twelve months after giving birth. The measures to prevent postpartum maternal obesity should target at risk women who are underweight or overweight at the first antenatal visit and control their weight gain during the course of pregnancy.

Nagpal et al. [41] carried out analyses on postpartum weight retention (PPWR) on 150 participants while taking anthropometric variables other than BMI and weight into account, such as waist circumference, hip circumference, and waist hip ratio. The results showed the postpartum weight retention was associated with the anthropometric measurements including waist circumference, hip circumference and waist-hip ratio. Increasing waist circumference and hip circumference could be applied to make the risk assessment for developing non-communicable diseases (NCD), such as gestation diabetes, which have been rising during the post-partum period.

## 3. Methodology

The Faculty of Nursing at Chiang Mai University and Maharaj Nakorn Chiang Mai Hospital approved the ethical authorization document for this study for data collection of pregnant and postpartum mothers with the following objectives:

(A) To collect six-month forward tracking data on body weight, body circumference, chest, waist, hip, upper arms and thighs during the pregnancy and postpartum period;

(B) To carry out an accuracy test of the Z-Size Ladies program in terms of 3D shape simulation compared with women during pregnancy and postpartum period.

This study was a 'prospective study' on volunteering pregnant women and postpartum women. The samples were women from antenatal care, who used postnatal services, and took their babies for vaccination at secondary and tertiary hospitals. A sample size of 98 pregnant and 83 postpartum women was analyzed for calculating model shapes of the pregnant and postpartum women. For the 3D simulation testing, a new group of 75 pregnant women and 74 postpartum women were compared in body shape proportion. The 3D body shape simulation of pregnant and postpartum women is based on the simulation of the non-pregnant female shape simulation studied in Sinthanayothin et al. [42].

The data collection started at 12–16 weeks of pregnancy and zero weeks of postpartum. The research assistant measured the body circumferences of chest, waist, hip, thighs (left/right), and upper arms (left/right) and explained to the volunteers how to measure their body size by themselves. The measurement was delicate, therefore the interrater agreement was essential that the participants had to hold the measuring tape in the correct position and not too tight. The measurement values could vary approximately ±2 cm. The measurements were taken at home by pregnant/postpartum women subjects with the assistance of someone at home. The measurement was taken every four weeks. There would be a reminder notice from the research assistant when the schedule was approaching. Three measurements were taken at each position and the median value was recorded for each position. The volunteers sent their measured data to the research team via mobile LINE application each time they measured their body shape.

### 3.1. The Simulation of the Female Shape in Three Dimensions (Z-Size Ladies)

The 3D simulation results for non-pregnant female bodies in various weights and heights are shown in Figure 1 [42].

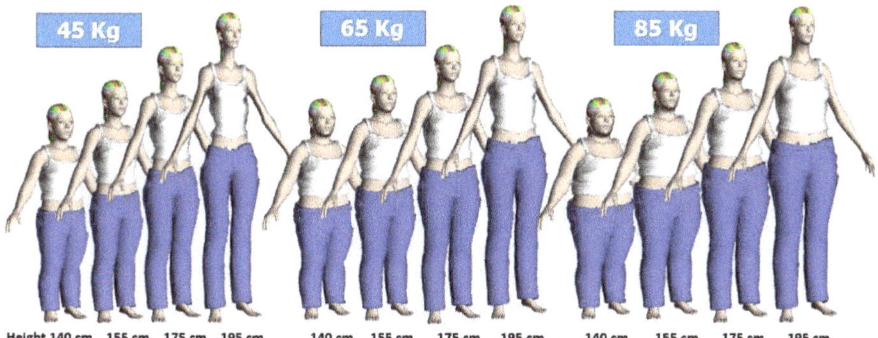


**Figure 1.** 3D body simulation for non-pregnant females using Morphing Technique.

### 3.2. The Correlation Analysis of Pregnant Women's Body Proportion Using Multiple Linear Regression of the 587 Data Collected from 98 Pregnant Women

The data from 98 pregnant women volunteers were collected and analyzed. Information of all pregnant volunteers is shown in Table 1. The data were 587 sets in total. Each data contained woman's age, pre-pregnancy weight, height, gestational age, weight gain during pregnancy, inseam (measure once at 12–16 week pregnancy), and body circumference measurements: chest, waist, hip, upper arm (left/right), and thigh (left/right). The data used for this study were from 94 women who were 12-week gestation; 98 of 16-week gestation; 91 of 20-week gestation; 82 of 24-week gestation; 79 of 28-week gestation; 78 of 32-week gestation and 65 of 36-week gestation, a total of 587 sets.

**Table 1.** Information the pregnant participants.

| Information | Range | Average | SD |
| --- | --- | --- | --- |
| Age ($Y_w$—Years) | 18–43.5 | 29.64 | 5.24 |
| Pre-Pregnancy Weight ($W_{pp}$—Kg) | 38–102 | 54.07 | 11.30 |
| Height ($H_w$—cm) | 104–174 | 157.58 | 5.97 |
| Gravida ($Gr$—child) | 1–2 | 1.39 | 0.49 |
| Pregnancy Week ($Wk_p$—Weeks) | 12–36 | 23.05 | 7.88 |
| Weight Gain ($W_g$—Kg) | −5–26 | 6.33 | 5.01 |

Wendland et al. [43] investigated the relationship between waist circumference and obesity-related pregnancy. The variables used in the correlation analysis were age, height, gravida, gestational age, uterine height, gestational BMI, and pre-pregnancy BMI. Similar work by Ricalde et al. [44] reported that some postpartum women's anthropometric was related to birth weight. Therefore, in this pregnancy study, the relationships between the body shape proportion and variables of pregnancy such as woman's age, weight, height, gestational age, weight gain during pregnancy were analyzed using multiple linear regression, which could be calculated in Excel [45].

The correlation indicates the relationship between the body shape proportion and variables of pregnancy such as woman's age, weight, height, gestational age, weight gain during pregnancy, and so on as shown in Equation (1), where Value is the proportion of pregnancy woman's body: Chest, Waist, Hip, Upper Arm, Thigh, respectively. $Y_w$ = Woman's age (Default is set to 30 in case age is unknown), $W_{pp}$ = Pre pregnancy weight (Kg), $H_w$ = Women's Height (cm), Gravida = Number of pregnancies (The default is set to 1, when pregnant for the first time), $Wk_p$ = Gestational age or Pregnant week (Weeks) and $W_g$ = Weight gain during pregnancy (Kg), respectively.

$$Value = (A \times Y_w) + (B \times W_{pp}) + (C \times H_w) + (D \times Gr) + (E \times Wk_p) + (F \times W_g) + G \tag{1}$$

Although there is no direct factor of BMI categories in our correlation analysis, the body proportions, chest, waist, hip, thigh, and upper arm circumferences were calculated using a multiple linear regression method based on 587 data collected from 98 pregnant women. However, when a user wanted to predict her 3D pregnancy shape at other gestation ages using the web app (Z-Size Ladies), weight gain during pregnancy was unknown. Therefore, weight gain during pregnancy ($W_g$) would be predicted from pre-pregnancy BMI as shown in Table 2 based on the Institute of Medicine (IOM), 2009 [46].

**Table 2.** Weight gained during pregnancy (kg) at each pregnancy stage based on pre-pregnancy BMI.

| Pre-Pregnancy BMI Type | Pre-Pregnancy BMI (Kg/m$^2$) (WHO) | Weight Gain (Kg) | Weight Gain per Week during the Quarter 2–3 (Kg/Wk) |
|---|---|---|---|
| Under weight | <18.5 | 12.73–18.18 | 0.45 (0.45–0.59) |
| Normal weight | 18.5–24.9 | 11.36–15.91 | 0.45 (0.36–0.45) |
| Over weight | 25.0–29.9 | 6.82–11.36 | 0.27 (0.23–0.32) |
| Obese | ≥30.0 | 5.00–9.09 | 0.23 (0.18–0.27) |

### 3.3. Simulation of Pregnant Women in 3D

The real-time visualization of 3D morphing of pregnant and postpartum female body shapes on the online Z-Size Ladies web application was implemented using the three.js library [47] incorporated with HTML5, JavaScript, and CSS for client-side development. Python, flask, and MySQL were employed for the server-side. Three.js was used as it was a cross-browser JavaScript library to ease the process of creating and displaying real-time 3D computer graphics and animation in the web browser.

A 3D model of a pregnant woman was created from a pregnant thin avatar shown in Figure 2 using a female model from Turbosquid [48,49] and combined with other avatars shown in Figure 3 for simulating a pregnancy figure by morphing technique.

The pregnant avatar refers to the pregnant women's simulation based on our previous studies [42]. The pregnant woman simulation is a combination of a woman shape and a pregnant shape. As the non-pregnant female simulation is a combination of a thin avatar (Figure 3A) and other avatar shapes (including Figure 3B big breast, (C) big waist, (D) big hip, (E) tall avatar, (F) long legs), so to simulate a pregnant woman, a pregnant thin avatar (Figure 2 or Figure 3G) is added.

TurboSquid is a digital media company that sells 3D models used in 3D graphics to a variety of industries, including computer games, architecture, and interactive training [48,49].

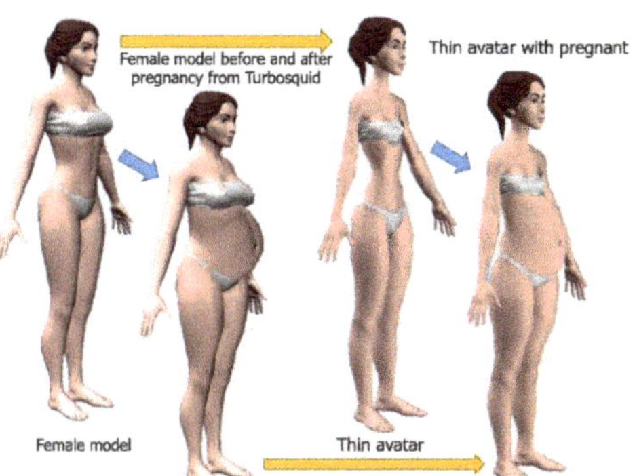



**Figure 2.** Creating a thin pregnancy avatar.

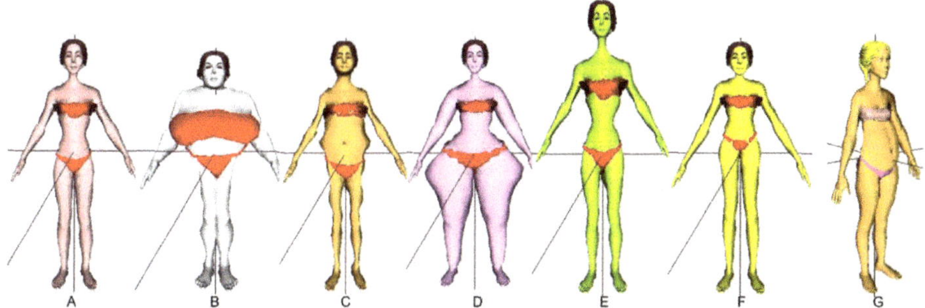



**Figure 3.** Seven avatars with accessories (eyes, eyes brows, hair, cloths): (**A**) thin, (**B**) big breast, (**C**) big waist, (**D**) big hip, (**E**) tall avatar, (**F**) long legs and (**G**) pregnant thin avatar. (Avatar (**G**)) has different view/pose as we would like to emphasize that this avatar has been added to this study while other avatars are from our previous study. Avatars (**A–F**) are shown in 'front view'. However, if avatar (**G**) is shown in only 'front view', the shape changes from pregnancy would be difficult to notice).

Details of creating a 3D non-pregnant female shape by combining thin, big breast, big waist, big hip, tall and long legs avatars can be found in the article by Sinthanayothin et al. [42]. The idea of utilizing a combination of the avatar bodies for 3D shape simulation came from the morphing technique [50]. Morphing is a geometric interpolation technique, which mixed different characteristics of the objects. The body shape simulation that adjusted only a specific part was a challenge. For example, chest or hip circumference could be set bigger or smaller with the least impact on the waist and others. Therefore, our team designed the avatars in different ways to combine the shape of the body and to be able to adjust the size of specific parts as needed. Therefore, the 3D non-pregnant female shape was created by combining thin, big breast, big waist, big hip, tall and long legs avatars using the morphing technique to make it easier to adjust only a specific part of the body.

For simulating a pregnant body shape, the morphing technique was applied as shown in Equation (2):

$$P_i = \left(1 - \sum_{i=0}^{5} K_i\right) \times A_i + (K_0 \times X_i) + (K_1 \times B_i) + (K_2 \times C_i) + (K_3 \times D_i) + (K_4 \times E_i) + (K_5 \times F_i) \tag{2}$$

where $X_i$ is pregnant thin avatar, $A_i$–$F_i$ are avatars with thin, big breast, big waist, big hip, tall and long legs, respectively.

The simulation of a non-pregnant body shape from our previous study [42] showed that the variables $K_1$–$K_5$ depended upon the BMI values. Therefore, a similar experiment was performed in this study by testing 30 pregnant female subjects whose 3D data were simulated using Equation (2) in comparison with the statical measurement from Equation (1).

In our experiment, the $K_0$–$K_5$ values are the sum between $K_{00}$–$K_{05}$ and the corresponding values between $Alp_0$–$Alp_5$ shown in Equation (3a,b). Where $K_{00}$–$K_{05}$ are depended on the values of *Chest, Waist, Hip, Height* and *Inseam* as shown in Equation (3c,d) respectively.

$$K_0 = K_{00} + Alp_0, \; K_1 = K_{01} + Alp_1, \; K_2 = K_{02} + Alp_2 \tag{3a}$$

$$K_3 = K_{03} + Alp_3, \; K_4 = K_{04} + Alp_4, \; K_5 = K_{05} + Alp_5 \tag{3b}$$

$$\text{Where } K_{00} = 0, \; K_{01} = \frac{(Chest - 57)}{(200 - 57)}, \; K_{02} = \frac{(Waist - 40)}{(160 - 40)}, \tag{3c}$$

$$K_{03} = \frac{(Hip - 68)}{(180 - 68)}, \; K_{04} = \frac{1.36 \times (Height - 165)}{(199 - 165)}, \; K_{05} = \frac{(Inseam - 78)}{(120 - 48)} \tag{3d}$$

The simulations of pregnant and postpartum women were further modified based on data from non-pregnant females from our previous study [42]. The constant values of Equation (3) were derived from the size of the designed avatars as mentioned in [42]. "The thin avatar was used as a default or initial model with the minimum values of chest, waist, and hip of approximately 57, 40, and 68 cm, respectively. The avatar with the big breast was applied to adjust the size of the chest values. The chest circumference of the avatar with the big breast was set as maximum chest values of approximately 200 cm. Similarly, for the avatar with the large waist and with the big hip, the size of the waist and hip of these avatars were set as maximum values of approximately 160 and 180 cm, respectively. For inseam, the default value for the thin avatar was approximately 73 cm. In this work, the minimum and maximum values of the inseam have been set to 48 and 120 cm, respectively. The last avatar (tall avatar) with the height of 200 cm is set to be the maximum height value."

The values of Chest, Waist, Hip, and Inseam, which are the values that defined $K00$–$K05$, were calculated from the non-pregnant female body shape according to the article by Sinthanayothin et al. [42], which can be expressed by linear equations shown in Equation (4a–d) respectively.

$$Chest = 0.872260 \times Weight - 0.437949 \times Height + 110.131573 \tag{4a}$$

$$Waist = 0.931735 \times Weight - 0.497702 \times Height + 104.780946 \tag{4b}$$

$$Hip = 0.729978 \times Weight - 0.152380 \times Height + 78.526838 \tag{4c}$$

$$Inseam = -0.059182 \times Weight + 0.547734 \times Height - 14.226815 \tag{4d}$$

From our experiments of cross-sectioning and measuring the circumference of 3D simulation figures, the $Alp_0$–$Alp_5$ were functions of the BMI, which could be calculated as the following polynomial equations. The quadratic functions derived from second-order polynomial regression and parameters from the experiment were performed to obtain a 3D pregnant woman model that was the closest to the calculated statistical value as shown in Equation (5a–f).

$$Alp_0 = (Alp_{01} \times BMI \times BMI) + (Alp_{02} \times BMI) + Alp_{03} \tag{5a}$$

$$Alp_1 = (Alp_{11} \times BMI \times BMI) + (Alp_{12} \times BMI) + Alp_{13} \tag{5b}$$

$$Alp_2 = (Alp_{21} \times BMI \times BMI) + (Alp_{22} \times BMI) + Alp_{23} \tag{5c}$$

$$Alp_3 = (Alp_{31} \times BMI \times BMI) + (Alp_{32} \times BMI) + Alp_{33} \tag{5d}$$

$$Alp_4 = (Alp_{41} \times BMI \times BMI) + (Alp_{42} \times BMI) + Alp_{43} \tag{5e}$$

$$Alp_5 = (Alp_{51} \times BMI \times BMI) + (Alp_{52} \times BMI) + Alp_{53} \tag{5f}$$

$Alp_{XY}$ is a variable that depends on the gestational age ($Wkp$), so it can be written as a quadratic function shown in Equation (6a).

$$Alp_{XY} = (Alp_{XYC} \times Wkp \times Wkp) + (Alp_{XYB} \times Wkp) + Alp_{XYA} \tag{6a}$$

where $Alp_{XYA}$, $Alp_{XYB}$ and $Alp_{XYC}$ are constants calculated by polynomial fitting shown in Table 3. These values were applied to the morphing equations to obtain 3D pregnant women model closed to the calculated statistical value, as shown in Equation (1).

**Table 3.** Correlation coefficient $Alp_{XY}$ used in morphing equations.

| $Alp_{XY}$ | $Alp_{XYA}$ | $Alp_{XYB}$ | $Alp_{XYC}$ |
|---|---|---|---|
| $Alp_{01}$ | $-6.956 \times 10^{-3}$ | $4.582 \times 10^{-4}$ | $-8.577 \times 10^{-6}$ |
| $Alp_{02}$ | $4.671 \times 10^{-1}$ | $-2.857 \times 10^{-2}$ | $5.314 \times 10^{-4}$ |
| $Alp_{03}$ | $-6.945$ | $3.811 \times 10^{-1}$ | $-6.870 \times 10^{-3}$ |
| $Alp_{11}$ | $-1.033 \times 10^{-4}$ | $-2.141 \times 10^{-5}$ | $6.106 \times 10^{-7}$ |
| $Alp_{12}$ | $5.440 \times 10^{-3}$ | $9.985 \times 10^{-4}$ | $-2.975 \times 10^{-5}$ |
| $Alp_{13}$ | $-1.019 \times 10^{-1}$ | $-9.765 \times 10^{-3}$ | $3.342 \times 10^{-4}$ |
| $Alp_{21}$ | $-3.492 \times 10^{-4}$ | $9.483 \times 10^{-5}$ | $-2.375 \times 10^{-6}$ |
| $Alp_{22}$ | $1.685 \times 10^{-2}$ | $-7.334 \times 10^{-3}$ | $1.539 \times 10^{-4}$ |
| $Alp_{23}$ | $-1.692 \times 10^{-1}$ | $1.215 \times 10^{-1}$ | $-2.524 \times 10^{-3}$ |
| $Alp_{31}$ | $-1.053 \times 10^{-5}$ | $2.944 \times 10^{-5}$ | $-3.744 \times 10^{-7}$ |
| $Alp_{32}$ | $-2.220 \times 10^{-3}$ | $-2.386 \times 10^{-3}$ | $4.237 \times 10^{-5}$ |
| $Alp_{33}$ | $1.524 \times 10^{-2}$ | $4.118 \times 10^{-2}$ | $-9.049 \times 10^{-4}$ |
| $Alp_{41}$ | $1.083 \times 10^{-4}$ | $-2.414 \times 10^{-5}$ | $6.383 \times 10^{-7}$ |
| $Alp_{42}$ | $-4.416 \times 10^{-3}$ | $1.477 \times 10^{-3}$ | $-3.730 \times 10^{-5}$ |
| $Alp_{43}$ | $8.753 \times 10^{-2}$ | $-2.232 \times 10^{-2}$ | $5.554 \times 10^{-4}$ |
| $Alp_{51}$ | $-1.808 \times 10^{-4}$ | $9.690 \times 10^{-6}$ | $6.453 \times 10^{-7}$ |
| $Alp_{52}$ | $1.370 \times 10^{-2}$ | $-1.370 \times 10^{-3}$ | $-1.147 \times 10^{-5}$ |
| $Alp_{53}$ | $-1.656 \times 10^{-1}$ | $2.241 \times 10^{-2}$ | $-1.634 \times 10^{-5}$ |

If $Wkp$ (the gestational age) is less than 12 weeks, $Alp_0$ can be calculated as shown in Equation (6b).

$$Alp_0 = Alp_0 \times Wkp/12 \tag{6b}$$

*3.4. The Correlation Analysis of Postpartum Women's Body Proportion Using Multiple Linear Regression of the 503 Data Collected from 83 Postpartum Women*

The data from 83 postpartum women volunteers were collected and analyzed. Information of all postpartum volunteers was shown in Table 4. The data were 503 sets in total. Each data contains the woman's age, pre-pregnancy weight, height, gravida, baby weight, postpartum week, postpartum weight, inseam (measure once at zero weeks of postpartum), and body circumference measurements: chest, waist, hip, upper arm (left/right), and thigh (left/right). The data used for this study were from 81 women of 0-week postpartum; 76 of 4-week postpartum; 73 of 8-week postpartum; 72 of 12-week postpartum; 72 of 16-week postpartum; 70 of 20-week postpartum and 59 of 24-week postpartum, a total of 503 sets.

**Table 4.** Information the postpartum participants.

| Information | Range | Average | SD |
|---|---|---|---|
| Age ($Y_w$—Years) | 17.1–45.25 | 29.05 | 5.17 |
| Pre-Pregnancy Weight ($W_{pp}$—Kg) | 38–95 | 56.35 | 12.10 |
| Height ($H_w$—cm) | 142–173 | 156.85 | 5.99 |
| Gravida ($Gr$—child) | 1–2 | 1.41 | 0.49 |
| Baby Weight ($W_b$—Kg) | 2.1–4.02 | 2.97 | 0.42 |
| Postpartum Week ($Wk_{ppt}$—Weeks) | 0–24 | 11.37 | 7.94 |
| Postpartum Weight ($W_{ppt}$—Kg) | 35.5–117 | 59.51 | 12.19 |

The correlation indicates the relationship between the body shape proportion and variables of postpartum such as woman's age, pre-pregnancy weight, height, gravida, baby weight, postpartum week and postpartum weight, as shown in Equation (7), which is similar to Equation (1), however, with postpartum parameters. The values are the proportion of a postpartum woman's body: Chest, Waist, Hip, Upper Arm, Thigh, respectively. $Y_w$ = Woman's age (Default is set to 30 in case age is unknown), $W_{pp}$ = Pre pregnancy weight (Kg), $H_w$ = Height (cm), $Gr$ = Number of pregnancies (The default is set to one, when pregnant for the first time), $W_b$ = Baby weight in Kg (The default is set to three in case baby weight is unknown), $Wk_{ppt}$ = Postpartum week (Weeks) and $W_{ppt}$ = Postpartum weight (Kg), respectively.

$$Value = (A \times Y_w) + (B \times W_{pp}) + (C \times H_w) + (D \times Gr) + (E \times W_b) \\ + (F \times Wk_{ppt}) + (G \times W_{ppt}) + H \tag{7}$$

The postpartum weight ($W_{pp}$) from measurement was already used as an independent variable in calculating the body circumference of postpartum women according to Equation (7). However, when a user wants to predict her 3D postpartum shape at other postpartum weeks using web app (Z-Size Ladies), postpartum weight is unknown. In the case of calculating the postpartum weight as a dependent variable, it would be complicated since it involved many factors such as BMI, pre-pregnancy weight, gestational weight gain, baby weight, and postpartum age. Moreover, data must be divided into four groups according to BMI types (underweight, normal weight, overweight and obese). Therefore, data from 83 postpartum women must also be divided into four groups, resulting in less than 30 postpartum women in each group. Data with n < 30 may not be sufficient for statistical analysis calculations [51].

Therefore, in the postpartum simulation application, the postpartum weight ($Wpp$) was predicted from the review articles. Theanansuk and Lertbunnaphong [52] concluded that the mean weight retention at the sixth week postpartum in Thai singleton pregnancy with normal pre-pregnancy BMI was 4.99 Kg. Cheng and Schmitt [53,54] discussed the postpartum weight retention in Asia and reviewed other articles showing that the postpartum weight retention at 0–24 weeks with inversion was approximately 7.4–2.5 Kg. Huang [8] studied 602 postpartum Taiwanese women and provided gestational weight gain (GWG), body weight retention and BMI at six months postpartum. Therefore, GWG from Huang [8] was compared to the values from IOM 2009 [46] to calculate GWG as shown in Table 5. The calculated weight retention results at six-month postpartum are shown in Table 6. Comparing results from six-month weight retention calculated for this study by applying the rule of three in arithmetic in comparison with the corresponding results from Huang [8] and IOM 2009 [46] have shown the weight retention. The results implied that females with extreme levels of BMI were slightly more vulnerable from higher weight retention than Taiwan females with corresponding BMI types, while the results were on the reverse for the case of normal weight and overweight.

**Table 5.** Comparison of gestational weight gain (GWG) for different types of BMI, obtained from two sources [8,46].

| Pre-Pregnancy BMI Type | Pre-Pregnancy BMI (Kg/m$^2$) (WHO) | Weight Gain (Kg) | GWG (Average Total Weight Gain (Kg) from IOM) | GWG (Kg) from Huang et al., 2010 |
| --- | --- | --- | --- | --- |
| Under weight | <18.5 | 12.73–18.18 | 15.455 | 14.36 |
| Normal weight | 18.5–24.9 | 11.36–15.91 | 13.635 | 14.37 |
| Overweight | 25.0–29.9 | 6.82–11.36 | 9.09 | 13.07 |
| Obese | ≥30.0 | 5.00–9.09 | 7.045 | 11.15 |

**Table 6.** Comparison of weight retention at six-month postpartum for different types of BMI.

| Pre-Pregnancy BMI Type | Pre-Pregnancy BMI (Kg/m$^2$) (WHO) | Weight Gain (Kg) (IOM 2009) | Weight Retention at 6-mo Postpartum (Huang et al., 2010) | Weight Retention at 6-mo Postpartum (Apply in This Study) |
| --- | --- | --- | --- | --- |
| Under weight | <18.5 | 12.73–18.18 | 3.32 | 3.573 |
| Normal weight | 18.5–24.9 | 11.36–15.91 | 2.57 | 2.430 |
| Overweight | 25.0–29.9 | 6.82–11.36 | 1.67 | 1.161 |
| Obese | ≥30.0 | 5.00–9.09 | −0.29 | −0.183 |

An article by American Pregnancy Association (APA) [55] provided an average pregnancy weight gain distribution in a total of 30 pounds (13.63 Kg) as shown in Table 7.

**Table 7.** Average pregnancy weight gain distribution in a total of 13.63 Kg (30 pounds) suggested by APA. About half belongs to Mom and the other half belongs to the baby.

| Pregnancy Weight Gain Distribution | Weight (Kg) | Belongs to Mom or Baby |
| --- | --- | --- |
| The weight of the baby by the end of pregnancy | 3.4 | Baby |
| The weight of the placenta | 0.68 | Baby |
| Attributed to increased fluid volume | 1.82 | Baby |
| Increased blood volume | 1.82 | Baby/Mom |
| The weight of the uterus | 0.91 | Mom |
| The weight of breast tissue | 0.91 | Mom |
| Maternal stores of fat, protein and other nutrients | 3.18 | Mom |
| The amniotic fluid | 0.91 | Mom |

From the above assumption, the weight retention at zero weeks (or delivery date) would be about half of the GWG (Average total weight gain from IOM), which were 7.7275, 6.8175, 4.545 and 3.5225 for pre-pregnancy BMI of underweight, normal weight, overweight and obese, respectively. Therefore, the weight retention from 0–24 weeks can be calculated by fitting graphs with different BMI types, as shown in Figure 4. The corresponding equations could be expressed as shown in Equations (8)–(11) with $X$ referring to postpartum weeks while $Y_{UW}$, $Y_{NW}$, $Y_{OW}$, $Y_{OB}$ are postpartum weight retention for the case of pre-pregnancy BMI type: underweight, normal weight, overweight and obese, respectively. In addition, postpartum weight was calculated as a summation of pre-pregnancy weight and weight retention.

$$Y_{UW} = 0.0064X^2 - 0.3268X + 7.7275 \tag{8}$$

$$Y_{NW} = 0.0068X^2 - 0.3452X + 6.8175 \tag{9}$$

$$Y_{OW} = 0.0052X^2 - 0.2662X + 4.5450 \tag{10}$$

$$Y_{OB} = 0.0057X^2 - 0.2915X + 3.5225 \tag{11}$$

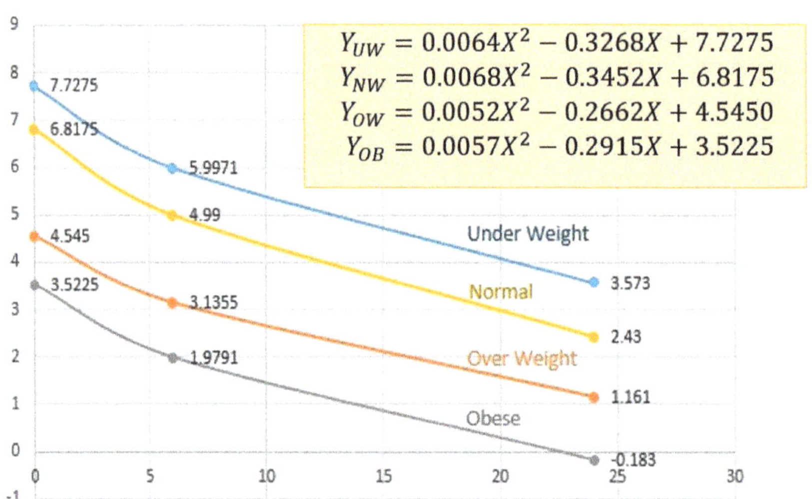


**Figure 4.** Weight retention estimation from 0 to 24 weeks postpartum, with pre-pregnancy BMI types.

### 3.5. Simulation of Postpartum Women in 3D

The 3D model of a postpartum woman was calculated in a similar way to the non-pregnant female [42], although with different parameters. For simulating a postpartum body shape, the morphing technique was applied, as in Equation (12), which was similar to Equation (2).

$$P_i = (1 - \sum_{i=1}^{5} K_i) \times A_i + (K_1 \times B_i) + (K_2 \times C_i) + (K_3 \times D_i) + (K_4 \times E_i) + (K_5 \times F_i) \quad (12)$$

where $A_i$–$F_i$ are avatars with thin, big breast, big waist, big hip, tall and long legs, respectively. From the experiment, it was found that the $K_1$–$K_5$ values were also proportional to the BMI and also depended on the postpartum weeks as well. In our experiment, a second order polynomial fitting curve and parameters from the experiment were performed as shown in Equation (13a–c), which is similar to Equation (3a–d):

$$K_1 = K_{01} + Alp_1, \quad K_2 = K_{02} + Alp_2, \quad K_3 = K_{03} + Alp_3, \quad K_4 = K_{04} + Alp_4, \quad K_5 = K_{05} + Alp_5 \quad (13a)$$

$$\text{Where } K_{01} = \frac{(Chest - 57)}{(200 - 57)}, \quad K_{02} = \frac{(Waist - 40)}{(160 - 40)}, \quad (13b)$$

$$K_{03} = \frac{(Hip - 68)}{(180 - 68)}, \quad K_{04} = \frac{1.36 \times (Height - 165)}{(199 - 165)}, \quad K_{05} = \frac{(Inseam - 78)}{(120 - 48)} \quad (13c)$$

The values of Chest, Waist, Hip and Inseam are the values calculated from non-pregnant female body shape according to the article by Sinthanayothin, et al. [42]. The values could be expressed as linear equations shown in Equation (14a–d), which are similar to the ones shown in Equation (4a–d).

$$Chest = 0.872260 \times Weight - 0.437949 \times Height + 110.131573 \quad (14a)$$

$$Waist = 0.931735 \times Weight - 0.497702 \times Height + 104.780946 \quad (14b)$$

$$Hip = 0.729978 \times Weight - 0.152380 \times Height + 78.526838 \quad (14c)$$

$$Inseam = -0.059182 \times Weight + 0.547734 \times Height - 14.226815 \quad (14d)$$

From the experiments of cross-sectioning and measuring the circumference of 3D simulation figures, $Alp_1$–$Alp_5$ were functions of the body mass index (BMI), which could be calculated as the following morph equations, which were quadratic functions shown in Equation (15a–f). They are similar to those shown in Equation (5a–e).

$$Alp_1 = (Alp_{11} \times BMI \times BMI) + (Alp_{12} \times BMI) + Alp_{13} \tag{15a}$$

$$Alp_2 = (Alp_{21} \times BMI \times BMI) + (Alp_{22} \times BMI) + Alp_{23} \tag{15b}$$

$$Alp_3 = (Alp_{31} \times BMI \times BMI) + (Alp_{32} \times BMI) + Alp_{33} \tag{15c}$$

$$Alp_4 = (Alp_{41} \times BMI \times BMI) + (Alp_{42} \times BMI) + Alp_{43} \tag{15d}$$

$$Alp_5 = (Alp_{51} \times BMI \times BMI) + (Alp_{52} \times BMI) + Alp_{53} \tag{15e}$$

$Alp_{XY}$ is a variable that depends on the postpartum weeks ($Wk_{ppt}$), so it can be written as a quadratic function shown in Equation (16) and similar to Equation (6a).

$$Alp_{XY} = (Alp_{XYC} \times Wkppt \times Wkppt) + (Alp_{XYB} \times Wkppt) + Alp_{XYA} \tag{16}$$

where $Alp_{XYA}$, $Alp_{XYB}$ and $Alp_{XYC}$ are the constants calculated by polynomial fitting shown in Table 8. These values were applied to the morphing equations to obtain 3D postpartum women model closed to the calculated statistical value, as shown in Equation (7).

**Table 8.** $Alp_{XY}$ correlation coefficients used in morphing equations for 3D postpartum woman model.

| $Alp_{XY}$ | $Alp_{XYA}$ | $Alp_{XYB}$ | $Alp_{XYC}$ |
|---|---|---|---|
| $Alp_{11}$ | $3.942 \times 10^{-4}$ | $-6.943 \times 10^{-6}$ | $1.720 \times 10^{-7}$ |
| $Alp_{12}$ | $-2.351 \times 10^{-2}$ | $5.974 \times 10^{-4}$ | $-1.915 \times 10^{-5}$ |
| $Alp_{13}$ | $2.716 \times 10^{-1}$ | $-8.912 \times 10^{-3}$ | $3.394 \times 10^{-4}$ |
| $Alp_{21}$ | $-1.201 \times 10^{-3}$ | $6.315 \times 10^{-5}$ | $-1.935 \times 10^{-6}$ |
| $Alp_{22}$ | $6.153 \times 10^{-2}$ | $-2.718 \times 10^{-3}$ | $8.334 \times 10^{-5}$ |
| $Alp_{23}$ | $-5.699 \times 10^{-1}$ | $1.173 \times 10^{-2}$ | $-5.636 \times 10^{-4}$ |
| $Alp_{31}$ | $1.668 \times 10^{-5}$ | $-1.037 \times 10^{-5}$ | $3.959 \times 10^{-7}$ |
| $Alp_{32}$ | $-4.785 \times 10^{-3}$ | $6.078 \times 10^{-4}$ | $-1.986 \times 10^{-5}$ |
| $Alp_{33}$ | $7.614 \times 10^{-2}$ | $-7.619 \times 10^{-3}$ | $1.400 \times 10^{-4}$ |
| $Alp_{41}$ | $-6.535 \times 10^{-5}$ | $2.052 \times 10^{-5}$ | $-9.603 \times 10^{-7}$ |
| $Alp_{42}$ | $3.961 \times 10^{-3}$ | $-5.569 \times 10^{-4}$ | $3.121 \times 10^{-5}$ |
| $Alp_{43}$ | $-2.846 \times 10^{-2}$ | $1.740 \times 10^{-3}$ | $-1.908 \times 10^{-4}$ |
| $Alp_{51}$ | $-3.400 \times 10^{-4}$ | $-1.750 \times 10^{-5}$ | $7.106 \times 10^{-7}$ |
| $Alp_{52}$ | $1.718 \times 10^{-2}$ | $2.525 \times 10^{-4}$ | $-1.226 \times 10^{-5}$ |
| $Alp_{53}$ | $-1.914 \times 10^{-1}$ | $1.244 \times 10^{-2}$ | $-4.478 \times 10^{-4}$ |

## 4. Results

*4.1. Statistical Correlation and 3D Modeling Simulation for Pregnant Female Body Shape*

The results of the correlation analysis of pregnant women shape with independent variables using multiple linear regression of the 587 data collected from 98 pregnant women are shown in Table 9. Coefficient values were calculated according to the woman's age, pre-pregnancy weight, height, gestational age, and weight gain during pregnancy, which were applied to the statistical calculation of pregnant female body shape in Equation (1).

**Table 9.** Coefficient values for the multiple linear regression of pregnant women.

| Value | A | B | C | D | E | F | G |
|---|---|---|---|---|---|---|---|
| Chest | −0.016 | 0.674 | −0.132 | 0.090 | 0.100 | 0.457 | 69.896 |
| Waist | 0.108 | 0.752 | −0.203 | 1.315 | 0.591 | 0.840 | 59.394 |
| Hip | −0.022 | 0.736 | −0.012 | −0.491 | −0.074 | 0.970 | 58.071 |
| Upper Arm | 0.017 | 0.306 | −0.075 | 0.004 | 0.016 | 0.272 | 21.423 |
| Thigh | 0.031 | 0.516 | −0.026 | −0.064 | −0.066 | 0.668 | 29.115 |

Figure 5 shows the results of the pregnant women 3D simulations at various gestational ages for women with four types of pre-pregnancy BMI: Underweight (BMI < 18.5); Normal weight (18.5 ≤ BMI ≤ 24.9); Overweight (25.0 ≤ BMI ≤ 29.9); and Obese (BMI ≥ 30.0).

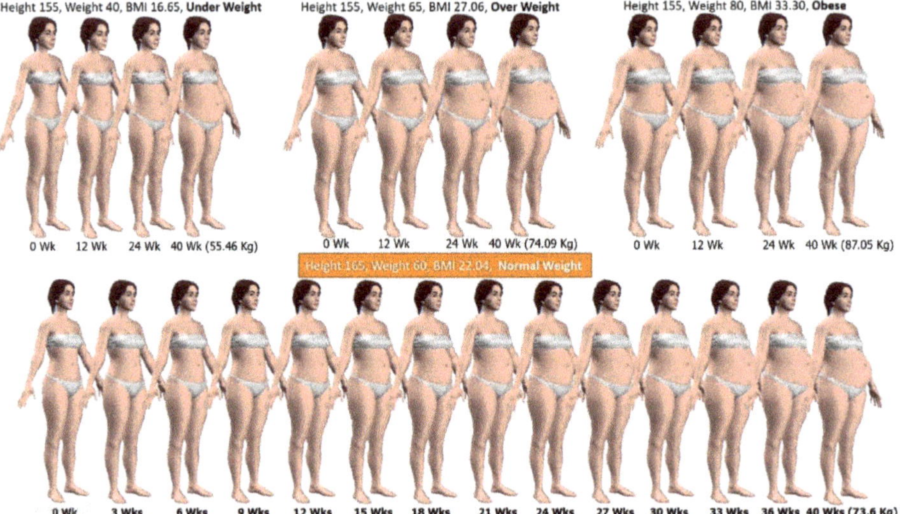


**Figure 5.** The simulated 3D models before pregnancy and during pregnancy 12, 24, and 40 weeks with four types of pre-pregnancy BMI: underweight, normal weight, overweight and obese.

*4.2. Statistical Correlation and 3D Modeling Simulation for Postpartum Female Body Shape*

The results of the correlation analysis of postpartum women's shape with independent variables using multiple linear regression of the 503 data collected from 83 postpartum females are shown in Table 10. Coefficient values were calculated according to the woman's age, pre-pregnancy weight, height, gravida, baby weight, postpartum week, and postpartum weight, which were applied to the statistical calculation of postpartum female body shape in Equation (7).

**Table 10.** Coefficient values for the multiple linear regression of postpartum women.

| Value | A | B | C | D | E | F | G | H |
|---|---|---|---|---|---|---|---|---|
| Chest | −0.036 | 0.033 | 0.013 | 0.689 | −0.066 | −0.066 | 0.550 | 53.038 |
| Waist | 0.041 | −0.064 | −0.213 | 2.880 | −2.098 | −0.396 | 0.870 | 75.729 |
| Hip | 0.058 | −0.074 | −0.180 | −0.022 | −0.387 | −0.196 | 0.780 | 83.935 |
| Upper Arm | 0.002 | −0.014 | −0.164 | 0.984 | −0.533 | −0.013 | 0.297 | 35.833 |
| Thigh | −0.046 | 0.162 | −0.124 | −1.189 | 1.112 | −0.072 | 0.390 | 41.065 |

Figure 6 shows the results of 3D simulation of postpartum women from the postpartum week of zero to 24 weeks with Height of 165 cm, Pre-pregnancy Weight of 60 Kg, Baby weight of 3 Kg.

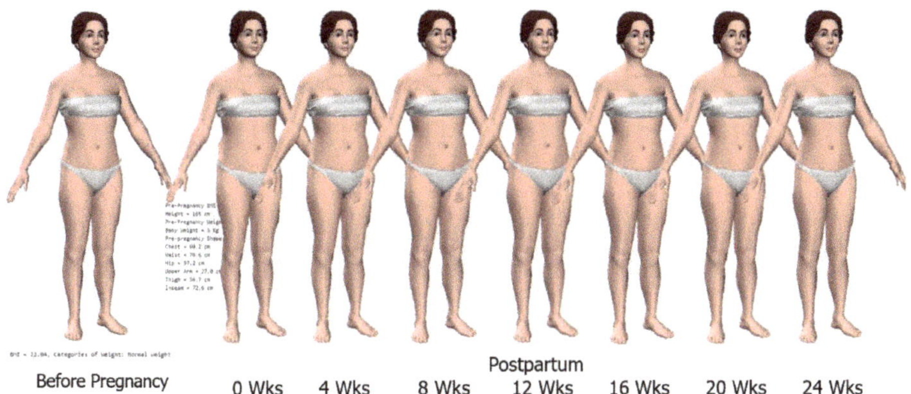



**Figure 6.** 3D simulation models of postpartum women.

### 4.3. The Accuracy Test on 3D Modeling of Pregnant Women

For accuracy test on pregnant simulation, the average and standard deviation of the measurements, Pearson coefficients and relative errors were calculated to measure the association and agreement between pairs of measurement methods. Furthermore, confidence interval plots were examined to assess and compare the results of the two methods.

The accuracy test on 3D modeling of pregnant women was divided into two parts:

A.  Comparison of body measurements in centimeters of chest, waist, hip, upper arm, thigh, and inseam between 3D models of pregnant women and the calculated statistical values for 30 datasets, with the mean age of $28.22 \pm 4.69$ years, the height of $159.13 \pm 4.78$ cm, the pre-pregnancy weight of $61.82 \pm 18.19$ Kg and gestational age of 12–36 weeks with the mean at $24 \pm 9.9$ weeks. Z = Z-Size Ladies statistical values, B = Manual cross-section on 3D model and measured body circumferences, Chest = chest, Waist = waist, Hip = hip, Thigh = thigh circumference, Upper Arm = upper arm circumference and Inseam = leg length, respectively. Where Avg is the mean, SD is the standard deviation, L CI and U CI are the lower and upper bounds of the 95% confidence interval, R Error is the relative error and Corr is the correlation, as shown in Table 11. Also 95% Confidence Interval (CI) plot for the mean measurements of chest, waist, hip, upper arm, thigh, and inseam of pregnant women between Z-Size Ladies statistical values and manual cross-section 3D values are shown in Figure 7.

B.  Comparison of chest, waist, hip, and arm thigh in centimeters between the values calculated from the statistical data and the values obtained from 75 pregnant volunteers using a tape measurement at Maharaj Nakorn Chiang Mai Hospital, with the mean age of $29.72 \pm 4.95$, the height of $156.97 \pm 5.32$ cm, the pre-pregnancy weight of $55.37 \pm 10.01$ Kg and the mean gestational age at 6–39 weeks with the mean of $25.71 \pm 9.9$ weeks. When Z = Z-Size Ladies statistical values, M = Manual, Weight Z = estimated maternal weight with App, Weight M = actual mother's weight, Chest = chest, Waist = waist, Hip = hip circumference and Upper Arm = upper arm circumference. Avg is the mean, SD is the standard deviation, L CI and U CI are the lower and upper bounds of the 95% confidence interval, R Error is the relative error and Corr is the correlation, as shown in Table 12. Also 95% Confidence Interval (CI) plot for the mean measurements of weight, chest, waist, hip, and upper arm of pregnant women between Z-Size Ladies statistical values and tape measurements are shown in Figure 8.

**Table 11.** Comparing the average results of chest, waist, hip, upper arm circumference, thigh, and inseam, with the standard deviation, the lower and upper bounds of the 95% confidence interval, the relative error, and the correlation between the Z-Size Ladies statistical value and the circumference values measured cross-sectionally on 3D modelling for the case of pregnant women.

| | Chest | | Waist | | Hip | | Upper Arm | | Thigh | | Inseam | |
|---|---|---|---|---|---|---|---|---|---|---|---|---|
| | **Z** | **B** | **Z** | **B** | **Z** | **B** | **Z** | **B** | **Z** | **B** | **Z** | **B** |
| Avg | 94.51 | 95.30 | 96.92 | 97.06 | 103.98 | 103.93 | 30.70 | 31.60 | 59.80 | 58.98 | 69.51 | 69.93 |
| SD | 10.49 | 10.86 | 14.05 | 14.36 | 11.22 | 12.42 | 4.66 | 4.63 | 7.79 | 7.06 | 2.56 | 2.91 |
| L CI | 90.76 | 91.41 | 91.89 | 91.92 | 99.96 | 99.49 | 29.03 | 29.94 | 57.01 | 56.45 | 68.59 | 68.89 |
| U CI | 98.26 | 99.19 | 101.95 | 102.20 | 108.00 | 108.37 | 32.37 | 33.26 | 62.59 | 61.51 | 70.43 | 70.97 |
| R Error | 0.836% | | 0.144% | | 0.048% | | 2.931% | | 1.371% | | 0.604% | |
| Corr | 0.989 | | 0.962 | | 0.961 | | 0.968 | | 0.902 | | 0.960 | |

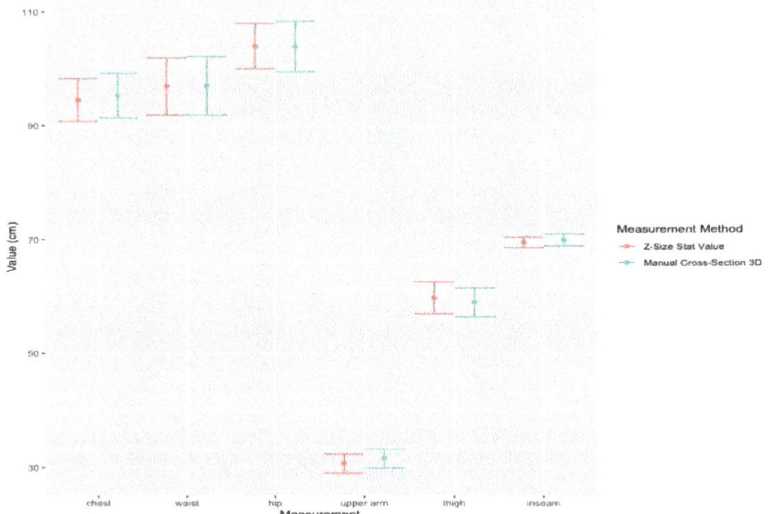


**Figure 7.** 95% Confidence Interval (CI) plot for the mean measurements (30 data from pregnant) of chest, waist, hip, upper arm, thigh, and inseam between Z-Size Ladies statistical values and manual cross-section 3D values.

**Table 12.** Comparing the average weight measurement, chest, waist, hip, and upper arm circumference with the standard deviation, the lower and upper bounds of the 95% confidence interval, the relative error and the correlation between the Z-Size Ladies statistical value and the value measured by tape measurement on 75 pregnant volunteers.

| | Weight | | Chest | | Waist | | Hip | | Upper Arm | |
|---|---|---|---|---|---|---|---|---|---|---|
| | **Z** | **M** | **Z** | **M** | **Z** | **M** | **Z** | **M** | **Z** | **M** |
| Avg | 62.94 | 63.40 | 92.36 | 92.27 | 96.10 | 96.26 | 101.65 | 101.11 | 29.83 | 29.05 |
| SD | 10.41 | 12.11 | 7.71 | 8.40 | 13.23 | 12.77 | 9.30 | 8.89 | 3.51 | 3.35 |
| L CI | 60.58 | 60.66 | 90.62 | 90.37 | 93.11 | 93.37 | 99.55 | 99.10 | 29.04 | 28.29 |
| U CI | 65.30 | 66.14 | 94.10 | 94.14 | 99.09 | 99.15 | 103.75 | 103.12 | 30.62 | 29.81 |
| R Error | 0.725% | | 0.098% | | 0.166% | | 0.534% | | 2.685% | |
| Corr | 0.960 | | 0.940 | | 0.963 | | 0.917 | | 0.893 | |

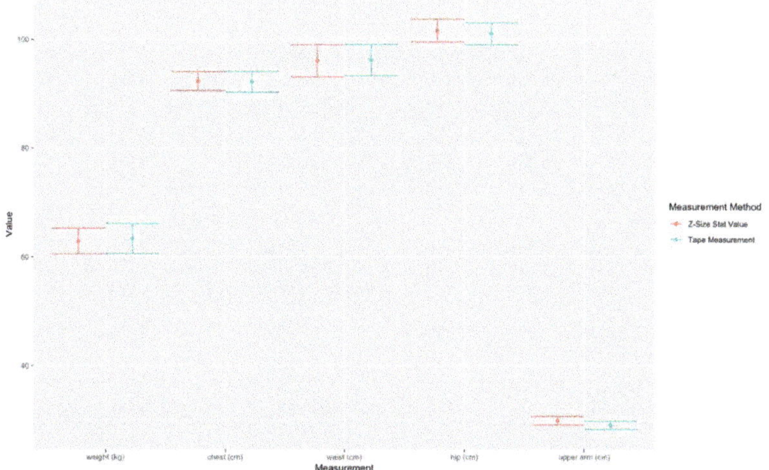

**Figure 8.** 95% Confidence Interval (CI) plot for the mean measurements (from 75 pregnant volunteers) of weight, chest, waist, hip, and upper arm between Z-Size Ladies statistical values and tape measurements.

### 4.4. The Accuracy Test on 3D Modeling of Postpartum Women

Similarly to Section 4.3, in order to calculate the accuracy test for postpartum simulation, the average and standard deviation of the measurements, Pearson coefficients and relative errors were calculated to measure the association and agreement between pairs of measurement methods. Furthermore, confidence interval plots were examined to assess and compare the results of the two methods.

The accuracy test on 3D modeling of postpartum women was divided into two parts:

A. Comparison of body proportion in centimeters of chest, waist, hip, upper arm, thigh, and inseam between 3D models of postpartum women and the calculated statistical values for 30 datasets, with the mean age of 27.4 ± 6.18 years, the height of 155.43 ± 5.67 cm, the pre-pregnancy weight of 57.13 ± 47.09 Kg and postpartum age of 0–24 weeks with the mean at 9 ± 12 weeks. Z = Z-Size Ladies statistical values, B = Manual cross-section on 3D model and measured body circumferences, Chest = chest, Waist = waist, Hip = hip, Thigh = thigh circumference, Upper Arm = upper arm circumference and Inseam = leg length. Avg is the mean, SD is the standard deviation, L CI and U CI are the lower and upper bounds of the 95% confidence interval, R Error is the relative error and Corr is the correlation, as shown in Table 13. Also 95% Confidence Interval (CI) plot for the mean measurements of chest, waist, hip, upper arm, thigh, and inseam of postpartum women between Z-Size Ladies statistical values and manual cross-section 3D values are shown in Figure 9.

B. Comparison of chest, waist, hip, and upper arm in centimeters between the values calculated from statistical data and the values obtained from 74 postpartum volunteers using a tape measurement at Maharaj Nakorn Chiang Mai Hospital, with the mean age of 29.90 ± 5.67, the height of 156.94 ± 6.07 cm, the pre-pregnancy weight of 57.26 ± 13.92 Kg and the mean postpartum age at 0–24 weeks with the mean of 7.02 ± 5.12 weeks. Z = Z-Size Ladies statistical values, M = Manual, Weight Z = estimated maternal weight with App, Weight M = actual mother's weight, Chest = chest, Waist = waist, Hip = hip circumference and Upper Arm = upper arm circumference, respectively. Avg is the mean, SD is the standard deviation, L CI and U CI are the lower and upper bounds of the 95% confidence interval, R Error is the relative error and Corr is the correlation, as shown in Table 14. Also 95% Confidence Interval

(CI) plot for the mean measurements of weight, chest, waist, hip, and upper arm of postpartum women between Z-Size Ladies statistical values and tape measurements are shown as in Figure 10.

**Table 13.** Comparing the average results of chest, waist, hip, upper arm circumference, thigh, and inseam, with standard deviation and the correlation between the Z-Size Ladies statistical value and the circumference values measured cross-sectionally on 3D modeling for the case of postpartum women.

| | Chest | | Waist | | Hip | | Upper Arm | | Thigh | | Inseam | |
|---|---|---|---|---|---|---|---|---|---|---|---|---|
| | Z | B | Z | B | Z | B | Z | B | Z | B | Z | B |
| Avg | 90.08 | 89.71 | 84.70 | 85.35 | 96.34 | 96.31 | 26.77 | 29.01 | 53.90 | 56.50 | 67.38 | 67.14 |
| SD | 7.17 | 7.85 | 10.23 | 10.75 | 8.31 | 8.98 | 2.77 | 1.89 | 6.32 | 3.09 | 2.62 | 3.24 |
| L CI | 87.51 | 86.90 | 81.04 | 81.50 | 93.37 | 93.10 | 25.78 | 28.33 | 51.64 | 55.39 | 66.44 | 65.98 |
| U CI | 92.65 | 92.52 | 88.36 | 89.20 | 99.31 | 99.52 | 27.76 | 29.69 | 56.16 | 57.61 | 68.32 | 68.30 |
| R Error | 0.411% | | 0.767% | | 0.031% | | 8.368% | | 4.824% | | 0.356% | |
| Corr | 0.963 | | 0.979 | | 0.992 | | 0.882 | | 0.969 | | 0.985 | |

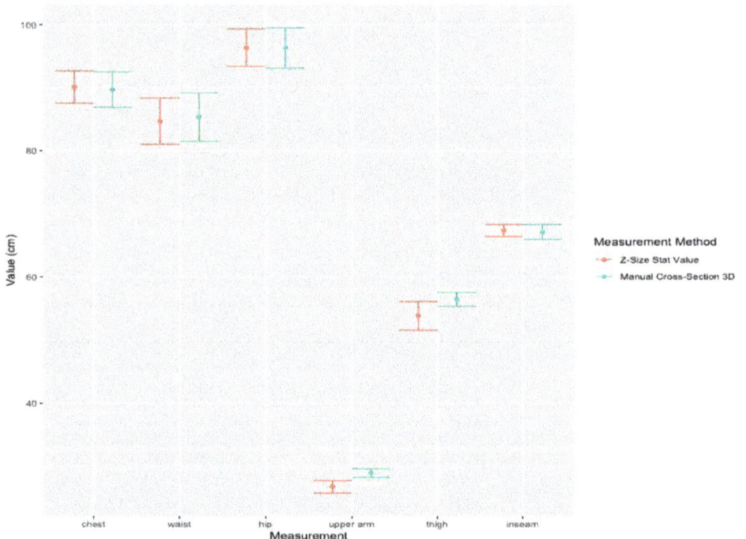


**Figure 9.** 95% Confidence Interval (CI) plot for the mean measurements (30 data from postpartum) of chest, waist, hip, upper arm, thigh, and inseam between Z-Size Ladies statistical values and manual cross-section 3D values.

**Table 14.** Comparing the average weight measurement, chest, waist, hip and upper arm circumference with the standard deviation and the correlation between the Z-Size Ladies statistical value and the value measured by tape measurement on 74 postpartum volunteers.

| | Weight | | Chest | | Waist | | Hip | | Upper Arm | |
|---|---|---|---|---|---|---|---|---|---|---|
| | Z | M | Z | M | Z | M | Z | M | Z | M |
| Avg | 61.92 | 61.29 | 91.06 | 92.18 | 88.16 | 91.77 | 98.63 | 100.56 | 26.70 | 28.46 |
| SD | 10.16 | 12.69 | 6.59 | 10.56 | 9.22 | 11.95 | 7.95 | 10.33 | 3.06 | 4.42 |
| L CI | 59.61 | 58.40 | 89.56 | 89.77 | 86.06 | 89.05 | 96.82 | 98.21 | 26.00 | 27.45 |
| U CI | 64.23 | 64.18 | 92.56 | 94.59 | 90.26 | 94.49 | 100.44 | 102.91 | 27.40 | 29.47 |
| R Error | 1.028% | | 1.215% | | 3.934% | | 1.919% | | 6.184% | |
| Corr | 0.891 | | 0.847 | | 0.828 | | 0.901 | | 0.736 | |

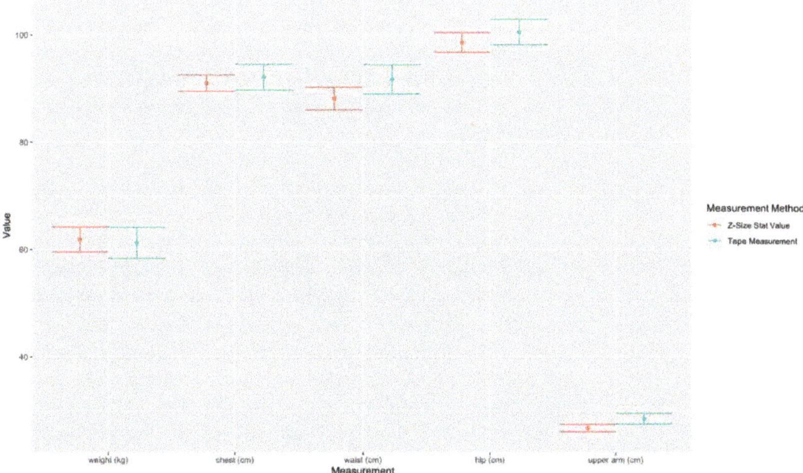

**Figure 10.** 95% Confidence Interval (CI) plot for the mean measurements (74 postpartum volunteers) of weight, chest, waist, hip, and upper arm between Z-Size Ladies statistical values and tape measurements.

*4.5. Satisfaction Test of the Developed Tool That Helps Pregnant/Postpartum Women to Simulate Their 3D-Body Shapes, Based on Height, Weight, and Gestational Age (Web App Z-Size Ladies)*

Survey results of the satisfaction test of the web app Z-Size Ladies by 149 pregnant and postpartum volunteers at Maharat Nakorn Chiang Mai Hospital are shown in Table 15, the highest score of each item was five.

**Table 15.** Satisfaction test for using Z-Size Ladies, surveyed from 149 pregnant and postpartum volunteers.

| Information | Rating Stars |
| --- | --- |
| This application is interesting. | 4.66 |
| Working efficiency such as fast response. | 4.44 |
| Ease of use. | 4.32 |
| Layout, keypad size, icon placement on screen. | 4.19 |
| Would you recommend this app to other pregnant women? | 4.51 |
| Do you think you will use the app again during pregnancy or after delivery? | 4.56 |
| How many stars would you rate the average for this app? | 4.50 |

## 5. Discussion

The results of the correlation analysis of pregnant/postpartum women's body shape with independent variables: pre-pregnancy weight; height; gestational age/ postpartum duration; and weight gain are shown in Tables 9 and 10, respectively. Errors that could occur during statistical modeling in this study could come from the multiple regression models in the independent variables [56] such as pre-pregnancy weight, height, age, gestation age, and body circumferences measurements data (chest, waist, hip). Our regression models assumed that those variables and data were obtained from measurement without errors. Moreover, errors could come from a small sample size, which might lead to insignificant results, whereas too large a sample size may increase the risk of harming volunteer subjects and might cause them discomfort [57].

Also, our developed technique can simulate 3D body shapes of women during pregnancy and postpartum in various gestational ages, BMI, and postpartum duration as shown in Figures 5 and 6, respectively. The pregnancy simulation included various gestational ages starting from 12–40 weeks with four types of pre-pregnancy BMI: underweight; normal weight; overweight and obese; and the postpartum at 0–24 weeks. For pregnancy simulation, different types of pre-pregnancy BMI indicates differences in weight gain during pregnancy. Therefore, 3D simulation of pregnant women was simulated at various gestational ages for women with four types of pre-pregnancy BMI according to IOM 2009 [46].

Note that for Figures 7 and 9, the y-axis is in cm and the x-axis represents body circumference measurements. For Figures 8 and 10, the y-axis displays in kg for weight and cm for circumferences and, the units are in the square blanket under the x-axis after each parameter value. The values on the y-axis are quite wide in range due to the different sizes of the upper arm and hip being quite significant.

Comparing results from the accuracy test on body measurements between the statistical values from this study (Z-Size Ladies) and the corresponding results taken from the manual measurement of the cross-section of the pregnant 3D model taken from Z-Size Ladies on 30 datasets are shown in Table 11. The accuracy is a measure of the degree of closeness of the measured or calculated value to its actual value. The percent relative errors are less than 3% with the maximum error being the upper arm (Relative error = 2.931%). However, the results show a strong correlation with the overlap plots of 95% confident interval between the results from Z-Size Ladies statistical values with the manual cross-section measurements of 3D models (Corr > 0.9). It implies that the results from the statistic values of Z-Size Ladies are comparable to the results from the manual cross-section measurements of the 3D model.

The accuracy test on body measurements between Z-Size Ladies statistic values and the manual tape measurements from 75 volunteers with the gestational age of 6–39 weeks is shown in Table 12. Table 12 shows all relative errors less than 3%, and the maximum error is in the upper arm (Relative error = 2.685%). It indicates a high correlation and some overlapping plots of 95% confident interval between the results from Z-Size Ladies and the results from the tape measurement (Corr > 0.89) even though the correlation is slightly less than the cross-section measurement on the 3D models. This is due to the locations in the manual tape measurements and the locations of measurements by Z-Size Ladies causing the variations. The highest correlation with the least relative error is at the chest and waist measurements due to the relative ease of locating the level for girt measurements of the chest and waist. However, the measurement of the upper arm has the lowest correlation value (Cor = 0.893) due to the difficulties in locating the places for upper arm measurements under the armpits.

Comparing results from the accuracy test on body measurements between the statistical values from Z-Size Ladies and the corresponding results taken from the manual measurement of the cross-section of postpartum 3D models taken from Z-Size Ladies for 30 datasets are shown in Table 13. The maximum relative error belongs to the position of the upper arm and thigh with 8.368% and 4.824%, respectively. Also, the lowest correlation and less overlapping plots of 95% confident interval are at the upper arm and thigh measurements due to the difficulty in locating the exact location for manual measurements of upper arms and thighs, which are near the armpits and crotch.

The accuracy test on body measurements between Z-Size Ladies statistic values and the manual tape measurements from 74 volunteers with postpartum age of 0–24 weeks is shown in Table 14. The maximum relative error belongs to the position of the upper arm (Relative error = 6.184%) and waist (Relative error = 3.934%). The 95% confident interval shows non overlapping in the position of the upper arm. The measurement of the upper arm also has the lowest correlation value (Cor = 0.736) as well.

Comparison the accuracy test between 3D modeling of pregnant and postpartum women, the 3D pregnant simulation shows a higher correlation between statistical values and 3D body measurements, less error, and the 95% confident interval plots show the intervals overlapped better than the postpartum shape simulation. Furthermore, Figures 9 and 10 show the upper arm relative error rate is relatively high and the 95%

confidence interval plots show the intervals are not overlapped. This indicates that the 3D simulation of pregnant women is more accurate than the simulation of postpartum women in this study. This may be due to postpartum women beginning to work after the delivery of their babies and, as such, the arm, thighs, and other muscles become distinctly different from the calculated values, leading to variations in the upper arm and other body measurements.

This study presents a 3D model shape simulation of pregnant and postpartum women. The data of woman's anthropometric measurements in different gestational and postpartum stages were collected. Based on the work of our previous study [42], pregnancy data was included to generate models to predict the shape of women at specific pregnancy and postpartum periods, based on pre-pregnancy measurements. The work led to the creation of a web application (Z-Size Ladies) to display 3D pregnancy and postpartum models, allowing women to input their metrics and observe the simulation. The website was validated through a survey from the users and received positive satisfaction scores from pregnant and postpartum women as illustrated in Table 15.

The limitations of this study were the resource deficiencies and the small sample size. Our research was a long-term perspective study that collected data six months before and after pregnancy. It was time-consuming and demanding to the participants. Therefore, the drop-out rate was high. The study excluded all possible aspects, such as our volunteers were pregnant and postpartum women with single pregnancy, who may not have regular exercise and may not have disabilities. In addition, this study included only a sample of Asians and did not include any foreigners.

## 6. Conclusions

Our web app (Z-size Ladies) accurately predicts the body proportions of pregnant and postpartum women based on a woman's age (years), pre-pregnancy weight (Kg), height (cm), gravida (number of pregnancies, the default is one), pregnancy/gestational week (weeks) and weight gain (program predicted automatically with adjustable personalized input from the user). The experiment results have shown that Z-Size Ladies could generate 3D models of pregnant participants, as well as postpartum participants, with high accuracy and could be considered as a lower-cost alternative method to the use of a full-body 3D scanner. However, more participants are needed to ensure continuity and high statistics for the study and to improve the accuracy of 3D models for pregnant and postpartum women. Better algorithms for 3D data reconstruction on the obscured sections, such as armpits and thighs, would be required for improving the accuracy of upper arm and thigh measurements.

**Supplementary Materials:** The following supporting information can be showed/downloaded at: Z-Size Ladies Website: https://zsize.openservice.in.th/; Video: https://vimeo.com/660604220 (accessed on 28 December 2021).

**Author Contributions:** Conceptualization, C.S. and P.X.; methodology, C.S. and W.B.; software, C.S. and N.W.; validation, C.S., W.B., P.X. and N.R.; formal analysis, C.S., W.B., P.X. and N.R.; investigation, C.S. and P.X.; resources, P.X.; data curation, P.B., S.C. and P.X.; writing—C.S. and W.B.; writing—review and editing, C.S., D.G. and W.B.; visualization, C.S.; supervision, C.S. and P.X. All authors have read and agreed to the published version of the manuscript.

**Funding:** This research was funded by National Science and Technology Development Agency (NSTDA), Thailand.

**Institutional Review Board Statement:** The study was conducted in accordance with the Declaration of Helsinki, and approved by the Institutional Review Board (or Ethics Committee) of Faculty of Nursing, Chiang Mai University (Research ID: 143-2017; Study code: FULL055-2017 and date of approval: 6 October 2017) and Maharaj Nakorn Chiang Mai Hospital, Faculty of Medicine, Chiang Mai University (Research ID: 5134; Study code: NONE-2560-05134 and date of approval: 6 October 2017). for studies involving humans.

**Informed Consent Statement:** Informed consent was obtained from all subjects involved in the study.

**Data Availability Statement:** Not applicable.

**Acknowledgments:** We would like to express our deepest gratitude to K. Nipharat, former an internship from Panyaniwat Management Institute, Thailand. Currently, he works as a freelancer for NECTEC in assistance of updating our websites and services. Most importantly, thank you C. Junlouchai for taking care of Z-Size Ladies' server. Finally, the authors are grateful to S. Seraphin at the Professional Authorship Center, Thailand National Science and Technology Development Agency (NSTDA) for fruitful discussions on the manuscript preparation.



# References


1. Poston, L.; Harthoorn, L.; Van Der Beek, E.M. Obesity in Pregnancy: Implications for the Mother and Lifelong Health of the Child. A Consensus Statement. *Pediatr. Res.* **2011**, *69*, 175–180. [CrossRef] [PubMed]
2. Chen, C.; Xu, X.; Yan, Y. Estimated global overweight and obesity burden in pregnant women based on panel data model. *PLoS ONE* **2018**, *13*, e0202183. [CrossRef] [PubMed]
3. Black, K.I.; Schneuer, F.; Gordon, A.; Ross, G.P.; Mackie, A.; Nassar, N. Estimating the impact of change in pre-pregnancy body mass index on development of Gestational Diabetes Mellitus: An Australian population-based cohort. *Women Birth* **2022**, 1–7. [CrossRef] [PubMed]
4. Persson, M.; Pasupathy, D.; Hanson, U.; Westgren, M.; Norman, M. Pre-pregnancy body mass index and the risk of adverse outcome in type 1 diabetic pregnancies: A population-based cohort study. *BMJ Open* **2012**, *2*, 1–8. [CrossRef]
5. Heude, B.; Thiébaugeorges, O.; Goua, V.; Forhan, A.; Kaminski, M.; Foliguet, B.; Schweitzer, M.; Magnin, G.; Charles, M.A. Pre-pregnancy body mass index and weight gain during pregnancy: Relations with gestational diabetes and hypertension, and birth outcomes. *Matern. Child Health J.* **2012**, *16*, 355–363. [CrossRef]
6. Li, N.; Liu, E.; Guo, J.; Pan, L.; Li, B.J.; Wang, P.; Liu, J.; Wang, Y.; Liu, G.S.; Baccarelli, A.A.; et al. Maternal Prepregnancy Body Mass Index and Gestational Weight Gain on Pregnancy Outcomes. *PLoS ONE* **2013**, *8*, 1–7. [CrossRef]
7. Phillips, J.K.; Higgins, S.T. Maternal Applying behavior change techniques to weight management during pregnancy: Impact on perinatal outcomes. *Prev. Med.* **2017**, *104*, 133–136. [CrossRef]
8. Huang, T.T.; Wang, H.S.; Dai, F.T. Effect of pre-pregnancy body size on postpartum weight retention. *Midwifery* **2010**, *26*, 222–231. [CrossRef]
9. Chang, M.Y.; Kuo, C.H.; Chiang, K.F. The effects of pre-pregnancy body mass index and gestational weight gain on neonatal birth weight in Taiwan. *Int. J. Nurs. Midwifery* **2010**, *2*, 28–34. [CrossRef]
10. Van Der Pligt, P.; Willcox, J.; Hesketh, K.D.; Ball, K.; Wilkinson, S.; Crawford, D.; Campbell, K. Systematic review of lifestyle interventions to limit postpartum weight retention: Implications for future opportunities to prevent maternal overweight and obesity following childbirth. *Obes Rev.* **2013**, *14*, 792–805. [CrossRef]
11. Arizona WIC (Women Infant and Children) Nutrition Care Guidelines: Breastfeeding and Postpartum Women. 2015. Available online: https://azdhs.gov/documents/prevention/azwic/face-to-face/2015/jan/6-Breastfeeding-and-Postpartum-Women.pdf (accessed on 28 December 2021).
12. Sui, Z.; Turnbull, D.; Dodd, J. Effect of body image on gestational weight gain in overweight and obese women. *Women Birth* **2013**, *26*, 267–272. [CrossRef] [PubMed]
13. Hill, B.; Skouteris, H.B.; McCabe, M.; Fuller-Tyszkiewicz, M. Body image and gestational weight gain: A prospective study. *J. Midwifery Women's Health* **2013**, *58*, 189–194. [CrossRef] [PubMed]
14. Simona, F.P.; Elisabeta, R.L.; Cristian, R.M. Relation Between Body Shape and Body Mass Index. *Procedia Soc. Behav. Sci.* **2015**, *197*, 1458–1463. [CrossRef]
15. Kang, N.E.; Kim, S.J.; Oh, Y.S.; Jang, S.E. The effects of body mass index and body shape perceptions of South Korean adults on weight control behaviors; Correlation with quality of sleep and residence of place. *Nutr Res. Pract.* **2020**, *14*, 160–166. [CrossRef]
16. Wells, J.C.; Treleaven, P.; Cole, T.J. BMI compared with 3-dimensional body shape: The UK National Sizing Survey. *Am. J. Clin. Nutr.* **2007**, *85*, 419–425. [CrossRef]
17. Pregnant Women, Fetuses and Neonates as Vulnerable Population. Human Research Protection Program. 2009. Available online: https://cphs.berkeley.edu/policies_procedures/sc501.pdf (accessed on 28 December 2021).
18. Pregnancy and Security Screening. Frequently Asked Questions in HPS Specialists in Radiation Protection. 2016. Available online: https://hps.org/publicinformation/ate/faqs/pregnancyandsecurityscreening.html (accessed on 28 December 2021).
19. Wells, J.; Treleaven, P.; Charoensiriwath, S. Body shape by 3-D photonic scanning in Thai and UK adults: Comparison of national sizing surveys. *Int. J. Obes.* **2012**, *36*, 148–154. [CrossRef]
20. Wells, J.C.; Charoensiriwath, S.; Treleaven, P. Reproduction, aging, and body shape by three-dimensional photonic scanning in Thai men and women. *Am. J. Hum. Biol.* **2011**, *23*, 291–298. [CrossRef]
21. Body Visualizer. Available online: https://bodyvisualizer.com (accessed on 28 December 2021).
22. Body Visualizer. Available online: https://ps.is.mpg.de/code/bmi-visualizer (accessed on 28 December 2021).

23. Black, M.J.; Balan, A.O.; Weiss, A.W.; Sigal, L.; Loper, M.M.; St. Clair, T. Method and apparatus for estimating body shape. U.S. Patent Application No. 20100111370, 6 May 2010.
24. Black, M.J.; Balan, A.O.; Weiss, A.W.; Sigal, L.; Loper, M.M.; St. Clair, T. Method and apparatus for estimating body shape. U.S. Patent No. 9189886, 17 November 2015.
25. Piryankova, I.V.; Stefanucci, J.K.; Romero, J.; De La Rosa, S.; Black, M.J.; Mohler, B.J. Can I Recognize My Body's Weight? The Influence of Shape and Texture on the Perception of Self. *ACM Trans. Appl. Percept.* **2014**, *11*, 1–18. [CrossRef]
26. Thaler, A.; Geuss, M.N.; Mölbert, S.M.; Giel, K.E.; Streuber, S.; Romero, J.; Black, M.J.; Mohler, B.J. Body size estimation of self and others in females varying in BMI. *PLoS ONE* **2018**, *13*, 1–24. [CrossRef]
27. Pujades, S.; Mohler, B.; Thaler, A.; Tesch, J.; Mahmood, N.; Hesse, N.; Bülthoff, H.H.; Black, M.J. The Virtual Caliper: Rapid Creation of Metrically Accurate Avatars from 3D Measurements. *IEEE Trans. Vis. Comput. Graph.* **2019**, *25*, 1887–1897. [CrossRef]
28. Loper, M.; Mahmood, N.; Romero, J.; Pons-Moll, G.; Black, M.J. SMPL: A skinned multi-person linear model. *ACM Trans. Graph.* **2015**, *34*, 1–16. [CrossRef]
29. Osman, A.A.; Bolkart, T.; Black, M.J. Star: Sparse trained articulated human body regressor. In Proceedings of the Computer Vision–ECCV 2020: 16th European Conference, Proceedings, Part VI 16, Glasgow, UK, 23–28 August 2020; pp. 598–613.
30. SizeThailand (In Thai). Available online: http://www.sizethailand.org (accessed on 28 December 2021).
31. Z-Size Ladies: Web Application for BMI Timeline & Program Assesses Ladies' Sizes and Facial Images. 2021. Available online: https://zsize.openservice.in.th (accessed on 28 December 2021).
32. Gallucci, A.; Znamenskiy, D.; Petkovic, M. Prediction of 3D Body Parts from Face Shape and Anthropometric Measurements. *J. Image Graph.* **2020**, *8*, 67–77. [CrossRef]
33. Balasubramanian, M.; Robinette, K. Longitudinal anthropometric changes of pregnant women: Dynamics and prediction. *Int. J. Fash. Des. Technol. Educ.* **2020**, *13*, 231–237. [CrossRef]
34. Vaughan, N.; Dubey, V.N.; Wee, M.Y.K.; Isaacs, R. Parametric model of human body shape and ligaments for patient-specific epidural simulation. *Artif. Intell. Med.* **2014**, *62*, 129–140. [CrossRef]
35. Haddox, A.G.; Hausselle, J.; Azoug, A. Changes in segmental mass and inertia during pregnancy: A musculoskeletal model of the pregnant woman. *Gait Posture* **2020**, *62*, 389–395. [CrossRef]
36. Ponnalagu, S.D.; Bi, X.Y.; Henry, C.J. Is waist circumference more strongly associated with metabolic risk factors than waist-to-height ratio in Asians? *J. Nut.* **2019**, *60*, 30–34. [CrossRef]
37. Darsini, D.; Hamidah, H.; Notobroto, H.B.; Cahyono, E.A. Health risks associated with high waist circumference: A systematic review. *J. Public Health Res.* **2020**, *9*, 94–100. [CrossRef]
38. Han, Q.; Shao, P.; Leng, J.; Zhang, C.; Li, W.; Liu, G.; Zhang, Y.; Li, Y.; Li, Z.; Ren, Y.; et al. L Interactions between general and central obesity in predicting gestational diabetes mellitus in Chinese pregnant women: A prospective population-based study in Tianjin, China. *J. Diabetes* **2018**, *10*, 59–67. [CrossRef]
39. Jacobson, L.T.; Collins, T.C.; Lucas, M.; Zackula, R. Electronic Monitoring of Mom's Schedule (eMOMS™): Protocol for a feasibility randomized controlled trial to improve postpartum weight, blood sugars, and breastfeeding among high BMI women. *Contemp. Clin. Trials Commun.* **2020**, *18*, 1–13. [CrossRef]
40. Ha, A.V.V.; Zhao, Y.; Pham, N.M.; Nguyen, C.L.; Nguyen, P.T.H.; Chu, T.K.; Tang, H.K.; Binns, C.W.; Lee, A.H. Postpartum weight retention in relation to gestational weight gain and pre-pregnancy body mass index: A prospective cohort study in Vietnam. *Obes. Res. Clin. Pract.* **2019**, *13*, 143–149. [CrossRef]
41. Nagpal, S.; Chandrashekarappa, S.; Chakrashali, S.; Rakshitha, J.; Murth, N.; Ramaiah, M. Exploring the hidden part of the iceberg: Post-partum weight retention among mothers and its association with sociodemographic, cultural and behavioural factors. *Clin. Epidemiol. Glob. Health* **2021**, *9*, 62–68. [CrossRef]
42. Sinthanayothin, C.; Bholsithi, W.; Gansawat, D.; Wongwaen, N.; Xuto, P.; Ratisoontorn, N.; Bunporn, P.; Charoensiriwath, S. Simulation of Three-dimensional Female Body Shapes with Proportional Representation for Various Weights and Heights. *Simulation* **2020**, *96*, 851–866. [CrossRef]
43. Wendland, E.M.; Duncan, B.B.; Mengue, S.S.; Nucci, L.B.; Schmidt, M.I. Waist circumference in the prediction of obesity-related adverse pregnancy outcomes. *Cad. Saude Publica* **2007**, *23*, 391–398. [CrossRef] [PubMed]
44. Ricalde, A.E.; Velásquez-Melendez, G.; Tanaka, A.C.; de Siqueira, A.F. Mid-upper arm circumference in pregnant women and its relation to birth weight. *Rev. Saude Publica* **1998**, *32*, 112–117. [CrossRef]
45. Cameron, A.C. EXCEL 2007: Multiple Regression. 2007. Available online: http://cameron.econ.ucdavis.edu/excel/ex6 1multipleregression.html (accessed on 28 December 2021).
46. Institute of Medicine, US; National Research Council, US. *Summary in Weight Gain during Pregnancy: Reexamining the Guidelines*, 1st ed.; Rasmussen, K.M., Yaktine, A.L., Eds.; National Academies Press: Washington, DC, USA, 2009. [CrossRef]
47. Three JS. 2018. Available online: https://threejs.org (accessed on 28 December 2021).
48. Turbosquid: Pregnant Woman by Motion Cow. Available online: https://www.turbosquid.com/3d-models/pregnant-woman-pregnancy-3d-max/693351 (accessed on 28 December 2021).
49. Turbosquid. 3D Models for Professionals. Available online: https://www.turbosquid.com/ (accessed on 28 December 2021).
50. Kang, J.Y.; Lee, B.S. Application of morphing technique with mesh-merging in rapid hull form generation. *Int. J. Nav. Archit. Ocean. Eng.* **2012**, *4*, 228–240. [CrossRef]

51. Sharma, A. Is n = 30 Really Enough? A Popular Inductive Fallacy Among Data Analysts. 2020. Available online: https://towardsdatascience.com/is-n-30-really-enough-a-popular-inductive-fallacy-among-data-analysts-95661669dd98 (accessed on 28 December 2021).
52. Theanansuk, M.; Lertbunnaphong, T. Postpartum Weight Retention in Thai Singleton Pregnant Women with Normal Pre-pregnancy Body Mass Index. *J. Obstet. Gynaecol.* **2008**, *16*, 221–226.
53. Cheng, H.R.; Walker, L.O.; Tseng, Y.F.; Lin, P.C. Post-partum weight retention in women in Asia: A systematic review. *Obes. Rev.* **2011**, *12*, 770–780. [CrossRef]
54. Schmitt, N.M.; Nicholson, W.K.; Schmitt, J. The association of pregnancy and the development of obesity-results of a systematic review and meta-analysis on the natural history of postpartum weight retention. *Int. J. Obes.* **2007**, *31*, 1642–1651. [CrossRef]
55. Pregnancy Weight Gain. Available online: https://americanpregnancy.org/healthy-pregnancy/pregnancy-health-wellness/pregnancy-weight-gain/ (accessed on 28 December 2021).
56. Errors-in-Variables Models. *Wikipedia*. 2022. Available online: https://en.wikipedia.org/wiki/Errors-in-variables_models (accessed on 28 December 2021).
57. Biau, D.J.; Kernéis, S.; Porcher, R. Statistics in Brief: The Importance of Sample Size in the Planning and Interpretation of Medical Research. *Clin. Orthop. Relat. Res.* **2008**, *466*, 2282–2288. [CrossRef]

 *sensors*

MDPI

*Article*

# Automatic Object Detection Algorithm-Based Braille Image Generation System for the Recognition of Real-Life Obstacles for Visually Impaired People

Dayeon Lee and Jinsoo Cho *

IT Convergence Engineering and Computer Convergence Major, Gachon University, Seongnam 13120, Korea; lidy030@gachon.ac.kr
* Correspondence: jscho@gachon.ac.kr

**Abstract:** The global prevalence of visual impairment due to diseases and accidents continues to increase. Visually impaired individuals rely on their auditory and tactile senses to recognize surrounding objects. However, accessible public facilities such as tactile pavements and tactile signs are installed only in limited areas globally, and visually impaired individuals use assistive devices such as canes or guide dogs, which have limitations. In particular, the visually impaired are not equipped to face unexpected situations by themselves while walking. Therefore, these situations are becoming a great threat to the safety of the visually impaired. To solve this problem, this study proposes a living assistance system, which integrates object recognition, object extraction, outline generation, and braille conversion algorithms, that is applicable both indoors and outdoors. The smart glasses guide objects in real photos, and the user can detect the shape of the object through a braille pad. Moreover, we built a database containing 100 objects on the basis of a survey to select objects frequently used by visually impaired people in real life to construct the system. A performance evaluation, consisting of accuracy and usefulness evaluations, was conducted to assess the system. The former involved comparing the tactile image generated on the basis of braille data with the expected tactile image, while the latter confirmed the object extraction accuracy and conversion rate on the basis of the images of real-life situations. As a result, the living assistance system proposed in this study was found to be efficient and useful with an average accuracy of 85% a detection accuracy of 90% and higher, and an average braille conversion time of 6.6 s. Ten visually impaired individuals used the assistance system and were satisfied with its performance. Participants preferred tactile graphics that contained only the outline of the objects, over tactile graphics containing the full texture details.

**Keywords:** image processing; object detection; artificial intelligence; blind; braille system

**Citation:** Lee, D.; Cho, J. Automatic Object Detection Algorithm-Based Braille Image Generation System for the Recognition of Real-Life Obstacles for Visually Impaired People. *Sensors* **2022**, *22*, 1601. https://doi.org/10.3390/s22041601

Academic Editors: Hossein Anisi, Vahid Abolghasemi and Saideh Ferdowsi

Received: 29 December 2021
Accepted: 25 January 2022
Published: 18 February 2022



## 1. Introduction

The leading cause of visual impairment can be congenital or a result of accidents, aging, or diseases. In addition, the number of people with acquired vision loss is increasing because of urban environmental factors resulting from the development of electronic devices [1,2]. A survey made by the World Health Organization (WHO) in 2020 indicated that approximately 2.2 billion people, which accounts for 28.22% of the global population, are visually impaired (i.e., near or distance visual impairment) [3,4].

Visually impaired people rely on their auditory perception and somatosensation—primarily sound and braille—to obtain information from the environment; they use assistive devices such as canes to recognize obstacles. However, although 28.22% of the global population accounts for visually impaired individuals [5,6], accessible facilities are not universally installed, leading to issues of social discrimination due to the limitations of their activities. Particularly, they cannot face unexpected situations outdoors independently, thereby restricting their activities to indoors or in their neighborhood. Accessible facilities such as tactile pavements and tactile signs are not appropriately installed in all institutions. Moreover, some countries do not provide support for assistive devices. In addition, most artworks, such as paintings and sculptures, cannot be touched to preserve them, making it difficult for visually impaired people to enjoy cultural activities through their imagination alone with tactile brochures. Therefore, researchers conducted numerous studies to help them become self-sufficient in their daily lives. In particular, studies on providing information via braille have recently gained attention. However, most of these studies focused on tactile maps or graphic image braille conversion. A system is needed worldwide to ease their daily lives because it is difficult to assist the visually impaired individuals in real life.

This study proposes a living assistance system based on images of the surroundings and objects that visually impaired people want to experience in real life that are captured by smart glasses. The system stores object information using an object detection algorithm to provide voice guidance when the user goes outdoors. Moreover, the system provides an object image braille conversion service using an object extraction algorithm when indoors and carrying a braille pad. The braille data are generated as binary data to enable use in various braille pads, and the images are generated at three degrees of expression to enable users to recognize the shapes at different types. The accuracy of the proposed system is calculated by comparing the example tactile image with the expected tactile image on the basis of the braille data, and the usefulness of the system is evaluated by comparing the object detection results in real-life images and the execution time.

## 2. Related Research

Researchers conducted various studies regarding the living assistance for visually impaired people. Previous studies were focused on the generation of tactile signs and maps as navigation aids for the visually impaired, image conversion, and the development of tactile image output devices for braille pads. However, there is a lack of studies on the generation of tactile images based on real-life images or systems that assist with real-life outdoor activities, such as the automatic object detection voice guidance system proposed in this study.

### *2.1. Similar Research*

### 2.1.1. Tactile Graphics

Tactile maps and images are generated through image processing based on general maps to create tactile maps. Tactile maps are the most provided navigation aid for the visually impaired people by public institutions. However, tactile maps are gradually being provided by various institutions, fueling further research on their development.

Kostopoulos et al. [7] proposed a method for generating tactile maps based on a map image created by reading the road names written on a map via OCR and converting it into a road image, as shown in Figure 1. Although the proposed system for creating tactile maps can quickly recognize roads on the basis of the road names, it cannot detect alleys without a name. Moreover, OCR is slow and limited although it is faster than the existing algorithms.

Zeng et al. [8] developed an interactive map in which the user can zoom in and out, as shown in Figure 2. They allowed users to explore the tactile map by dividing it into zoom levels. However, a post-experiment survey found that visually impaired people preferred maps with only two zoom levels, and the usage time increased due to various factors such as the production of the interactive map, the guidance of the selections, and the selection.

Moreover, Krufkaf et al. [9] proposed an advanced braille conversion algorithm for vector graphics on the basis of previous studies. The algorithm extracted object boundaries using the outline information of the graphic based on the vector graphics hierarchical characteristics. The levels are classified on the basis of the extracted boundaries, and the multi-level braille is converted to a braille tablet using the tiger advantage braille printer program [10]. Although the proposed multi-level braille conversion system can provide meaningful results, it is difficult to apply to real-life objects using vector graphics, as shown in Figure 3.

(a) Primary map image     (b) Obtained road network structure     (c) Obtained 3D map model

**Figure 1.** Map image-based tactile map production method [7].

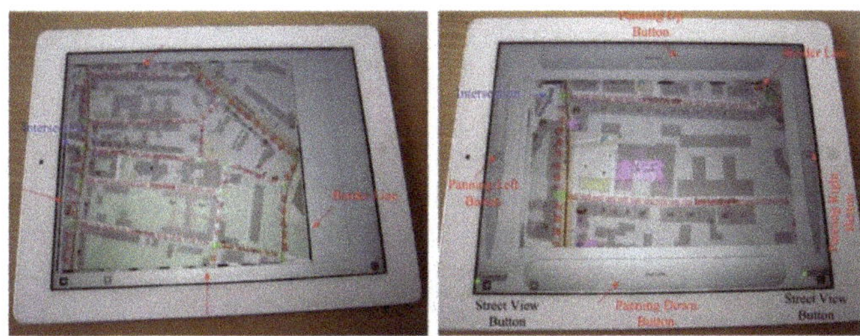

(a) The screenshots of Map2 on iPad

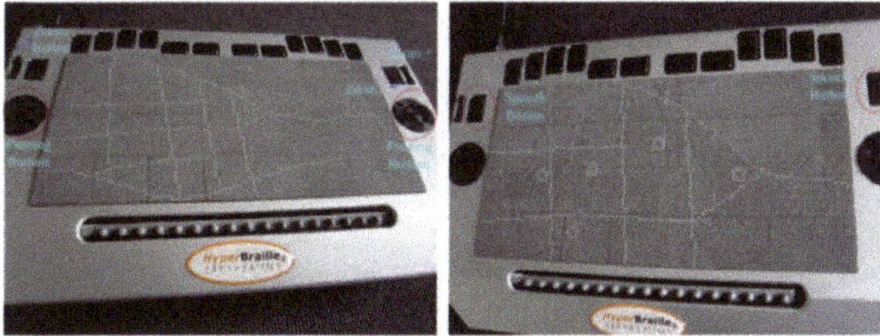

(b) Screenshots of Map3 on the HBMap system

**Figure 2.** iPad and HBMap system-based interactive maps [8].

**Figure 3.** Outputs of proposed method for the vector graphic [9].

In Korea, Kim et al. [11] investigated braille conversion on the basis of images captured via a webcam. The locations with and without data are compared to identify characters in the image by analyzing the images using MATLAB. Figure 4 shows the evaluation of the recognition level according to the font size, font type, and camera performance. In addition, an algorithm was developed by configuring an optimal environment based on the evaluation results. Although their research showed significant results, the system can only convert numbers and uppercase English letters, and it cannot identify objects other than letters or recognize Korean letters.

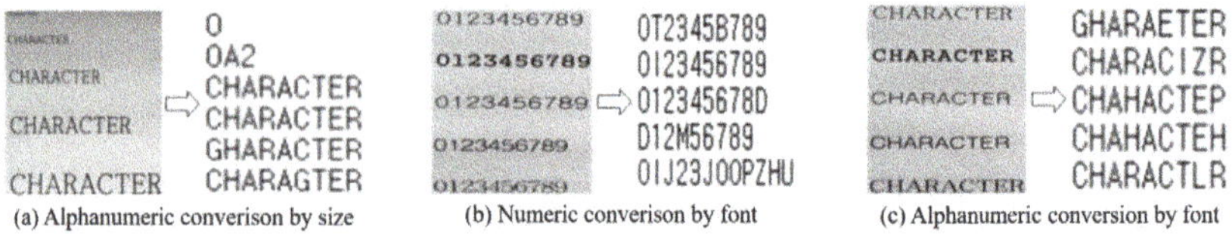

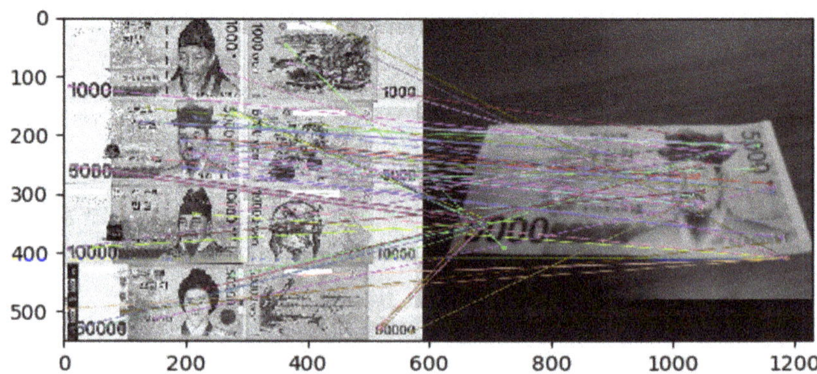

(a) Alphanumeric converison by size     (b) Numeric converison by font     (c) Alphanumeric conversion by font

**Figure 4.** Image conversion according to font size and font [11].

Lee et al. [12] developed a banknote recognition system using Raspberry Pi as a camera. The process consisted of two steps (i.e., extraction and matching). The researchers compared the extraction algorithms SIFT, SURF, and ORB; they adopted SIFT because it yielded the highest recognition rate. The system achieved high accuracy even when changing the shooting method or in unsuitable environments (e.g., low light or rotated banknote) by generating vector images using extreme values as features. Nevertheless, the brute-force algorithm requires extensive time for recognition, as shown in Figure 5, making it unsuitable for this study, which uses many objects.

**Figure 5.** Keypoints Matching Using the Brute-Force Algorithm [12].

### 2.1.2. Braille Pad

Researchers have made several attempts to output tactile images by combining a haptic device with a braille display [13].

Kim, S. et al. [14,15] proposed a 2D braille display to output data in the digital accessible information system (DAISY) and the electronic publication (EPUB) formats. They developed the braille pad for outputting braille information and the technology for tactile image conversion, as shown in Figure 6. Tactile image tests were conducted using simulators, and the tactile image conversion technology quantizes and binarizes data to convert graphs, graphic images, and even photos, enabling them to obtain significant results.

Prescher et al. [16] proposed a PDF-editor-based braille pad and braille conversion system. The user interface (UI) for displaying and editing PDF content was designed to show on one screen using a horizontally long touch-enabled braille pad. As both the content and editing UI are displayed on one screen, excessive information is provided at once, making it difficult for first-time users. Moreover, it can only translate the diagrams and text input, which are in PDF files rather than images, although it can display diagrams as shown in Figure 7.

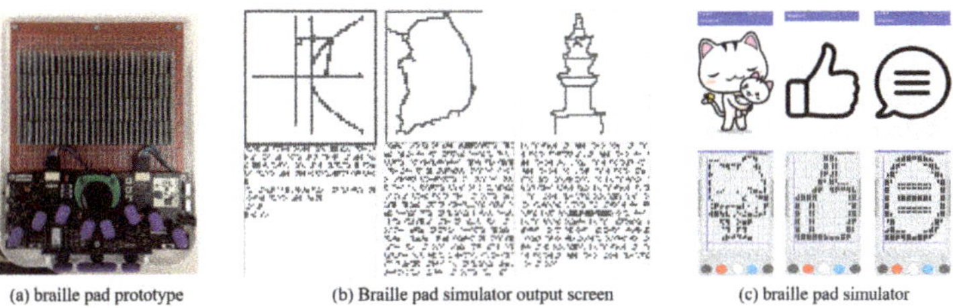

(a) braille pad prototype      (b) Braille pad simulator output screen      (c) braille pad simulator

**Figure 6.** Braille pad prototype and Output screen [14,15].

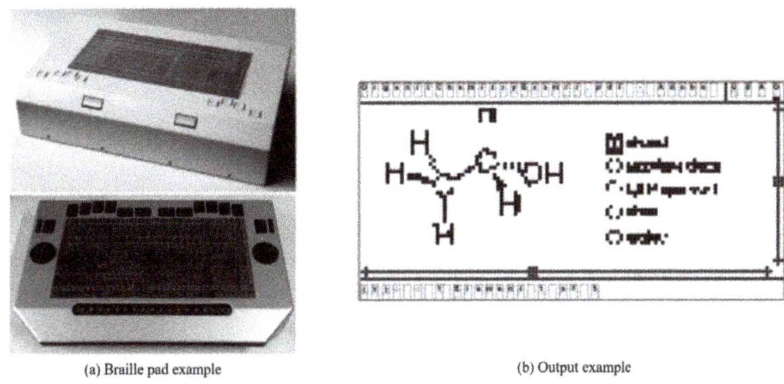

(a) Braille pad example      (b) Output example

**Figure 7.** Braille pad and Output example [16].

2.1.3. Supplementation and Service

In addition, various products and services are being researched to assist the visually impaired.

Kłopotowska et al. [17] studied architectural typhlographics and developed them through multi-criteria analysis by integrating the characteristics of braille maps and architectures (Figure 8). The study results show the future growth potential of typhlographics on the basis of its social values of enabling tourism for the visually impaired in addition to its broad utility in the development of tactile architectural drawings such as diversification of architectural education and interior design.

Morad [18] studied the assistive devices that receive location coordinates via the global positioning system (GPS) and process data through a PIC controller to output specific voice messages stored in the device for visually impaired people. The study aimed to develop an affordable and easy-to-use assistive device that helps the visually impaired people find their way on their own as they listen to the voice messages through the headset. It received a positive response from them when the device was used by people with visual impairments.

On the other hand, Fernandes et al. [19] proposed a radiofrequency identification (RFID)-based cane navigation system to guide people with visual impairments by using the RFID device installed under the road. The navigation system provides audio navigation assistance to reach the desired destination through the route calculation and location tracking using the RFID tags once the user inputs a specific destination in the cane. It is considered to have a significant growth potential owing to its higher accuracy than GPS and the easy-to-update feature of the navigation system.

Liao et al. [20] proposed the integration of the GPS and RFID technologies to develop a system for indoor use in order to address the shortcoming of the GPS system used. This hybrid system receives location data based on GPS and fine tunes the specific location data with RFID, which was developed to provide walking assistance to users. The study results are expected to facilitate the development of the walking assistance system for the visually impaired individuals and the enhancement of GPS accuracy.

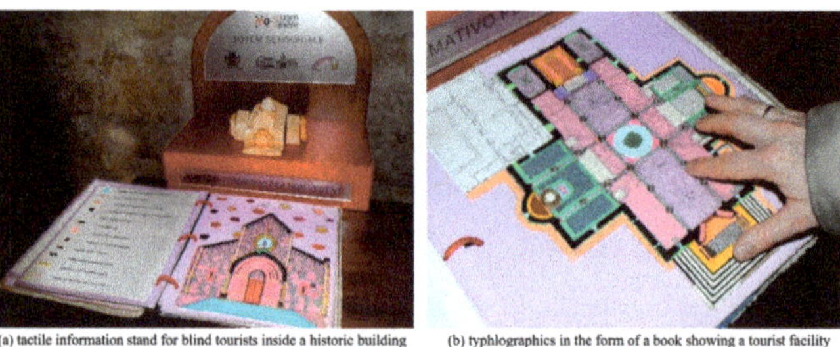

(a) tactile information stand for blind tourists inside a historic building          (b) typhlographics in the form of a book showing a tourist facility

**Figure 8.** Typhlogics in the form of a book that shows tactile information tables and tourist facilities for blind tourists [17].

*2.2. Algorithms*

2.2.1. YOLO

You Only Look Once v3(YOLOv3), a Darknet-53 network-based object detection algorithm, passes through layers of various sizes and compares them with object characteristics analyzed in the dataset to detect objects [21–23]. This study used YOLOv3 for object detection to identify objects within the line of sight of users. YOLOv3 has undergone several versions of development, making it more accurate than other algorithms [24–27]. In addition, it is fast and specialized for real-time detection as it searches only once, enabling an object detection from images in real time. According to the study of Redmon et al. [23]. YOLOv3 yielded an mAP of 57.9% in a COCO dataset test, demonstrating the high speed and accuracy of the algorithm. Figures 9 and 10 shows the YOLOv3 operating structure and the network structure, respectively. The method detected through the network is shown in Figure 11 and is expressed by Equation (1).

$$
\begin{aligned}
b_x &= \sigma(t_x) + c_x \\
b_y &= \sigma(t_y) + c_y \\
b_w &= p_w e^{t_w} \\
b_h &= p_h e^{t_h}
\end{aligned}
\tag{1}
$$

2.2.2. Grabcut

The GrabCut algorithm allows more effective object feature classification and ease of use than previous algorithms [28,29], such as Magic Wand, Intelligent Scissors, Bayes Matte, Knockout2, and GraphCut. This algorithm is used to separate the detected objects from the background, exploiting its advantages of high speed and extraction accuracy with only user-specified regions. Through GraphCut-based segmentation, the color values between pixels are calculated. A color model is generated on the basis of the color values of the model, and the foreground and background are separated via segmentation, as shown in Figure 12. After adding a mask to distinguish the foreground and background on the basis of the selection of the user, the separated foreground can be re-extracted, as shown in Figure 13.

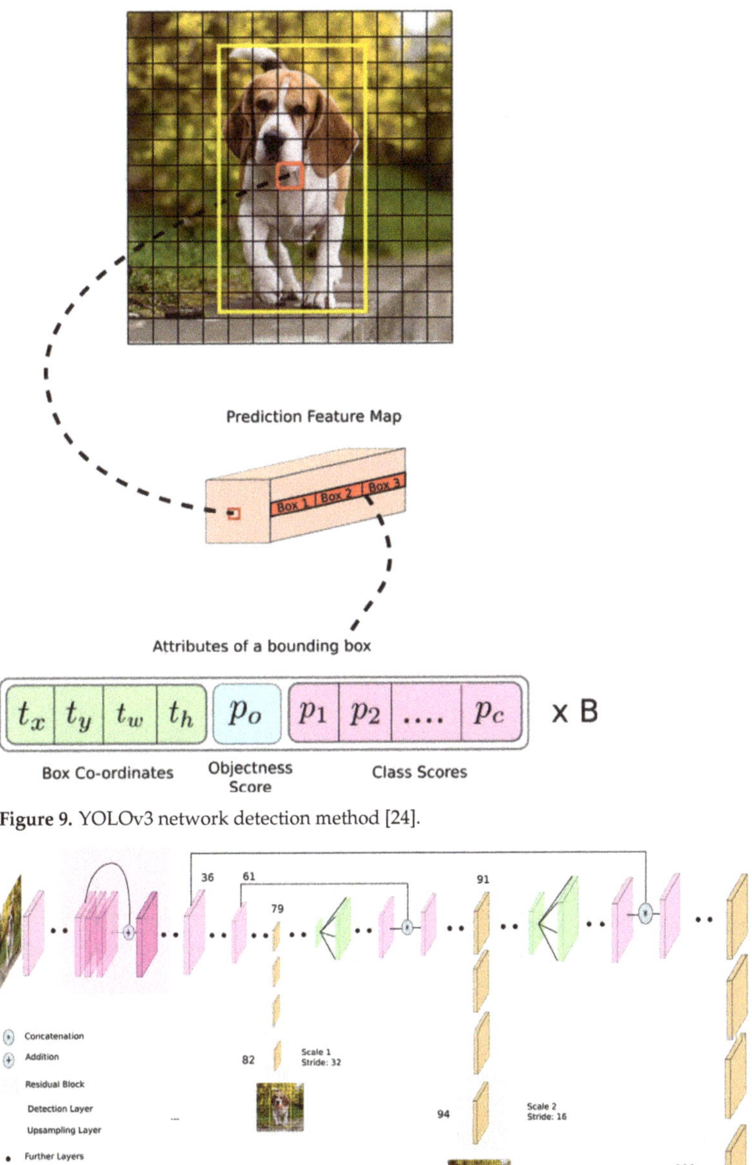



**Figure 9.** YOLOv3 network detection method [24].



**Figure 10.** YOLOv3 network architecture [24].

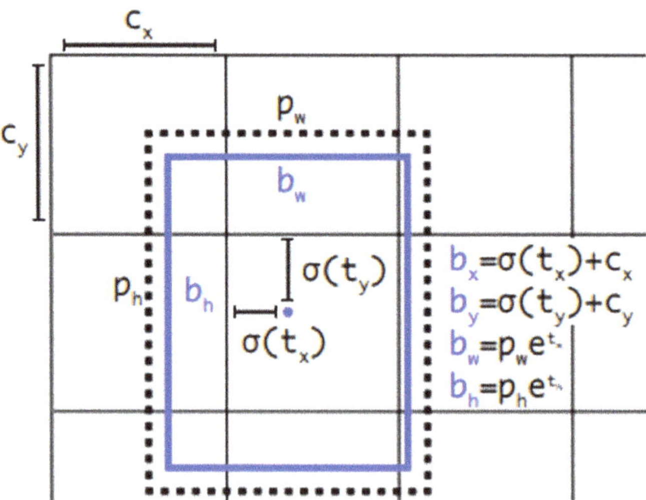

**Figure 11.** Numerical expression of YOLOv3 object detection [23].

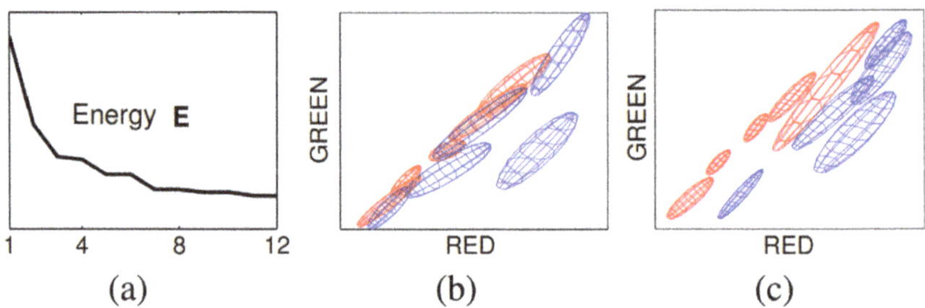


**Figure 12.** GrabCut principle image-Convergence of iterative minimization. (**a**) The energy E for the llama example converges over 12 iterations. The GMM in RGB colour space (side-view showing R,G) at initialization (**b**) and after convergence (**c**) [28].

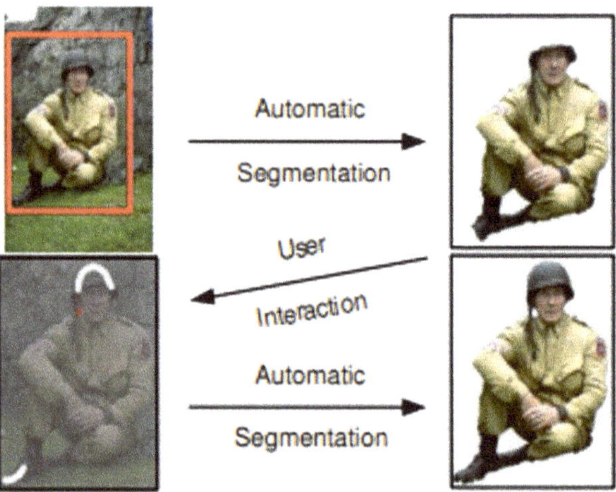


**Figure 13.** Grabcut example image [28].

### 2.2.3. Canny

In contrast to Contour, which is a contour line detection algorithm that generates boundary lines based on the height of the boundary detection target [30], Canny identifies the boundary values of the object to generate an outline [31]. In comparison to previous algorithms for generating outlines, Canny is fast and applicable to color images. Therefore, it was used to generate outlines for converting the extracted object to braille. In addition, new criteria were added to prevent it from generating abnormal outlines to achieve a low error rate and stable and improved system performance. Additionally, the criteria of existing algorithms are strengthened, and a parametric closed outline generation technique is provided through numerical optimization. Accordingly, additional criteria were hypothesized, and various equations and operators were used to satisfy the hypotheses. Figure 14 shows the results of this application, indicating its suitability as an outline generation algorithm.

**Figure 14.** Canny example image [31].

## 3. System Design and Configuration of Use Environments

The automatic-object-detection-algorithm-based braille conversion system for the living assistance of the visually impaired mainly targets visually impaired people including those with limited sight who typically use braille since the system is fully operated by smartphones. The images of surrounding environment and objects are captured with smart glasses, and the braille images are generated on the braille pads. The relevant objects are captured through smart glasses, and the tactile image is the output on a braille pad. Figure 15 shows the structure of the system, which is operated through a smartphone. To detect objects, it is connected to smart glasses via Bluetooth using the smartphone. The camera screen of the smart glasses and the screen of the desired field of view are confirmed through the smartphone and a shooting request is sent when the smart glasses are connected. When the shooting request reaches the smart glasses, it takes a photo with the built-in camera and sends it to the smartphone. The location and name of the objects in the photo are transmitted through the smart glasses and confirmed via TTS when the system performs object detection at the request of the user. The image is converted to braille, and the braille data are transmitted to the braille pad to allow the user to confirm the shape of the object. Once the transmission is completed, the user can recognize the shape of the object with the tactile image generated through the braille pad.

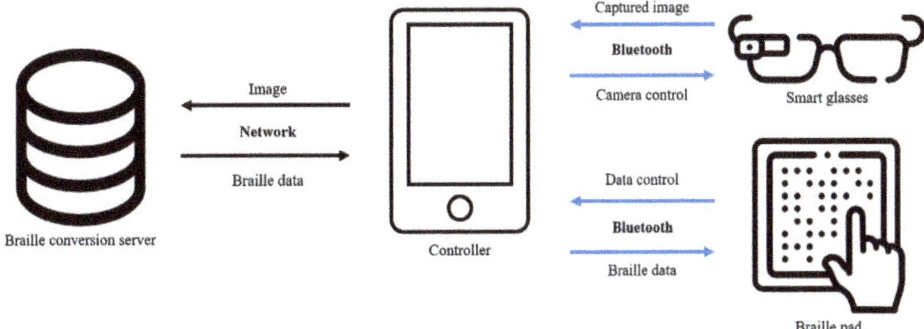


**Figure 15.** System schematic.

## 4. System Configuration

Table 1 shows the configuration of the proposed system in five steps: shooting, object detection, object extraction, outline generation, and braille conversion. The algorithms for all steps except shooting are constructed on an integrated server to increase the processing speed and store and use various image data. Each step can be separately executed through a smartphone on the basis of the scope of use and selections of the user. Moreover, only the result data are stored on the smartphone. The data from each step are maintained until the step is executed again. Figure 16 presents the overall process of the system.

**Table 1.** Requirements of proposed system.

| Function | Description |
| --- | --- |
| Image shooting | Capture photo of user-specified field of view and generate image |
| | Transfer to image controller and store |
| | Receive voice guidance data at user request |
| Object detection | Learn object images in database defined by system administrator |
| | Generate object recognition model |
| | Recognize objects based on image and store result image |
| | Store analysis result data |
| | Transmit voice guidance data at user request |
| Object extraction | Extract objects from image based on data |
| | Resize and store extracted object images |
| Outline generation | Preprocess image |
| | Calculate average color values based on extracted object images |
| | Generate object outline based on color values |
| Braille conversion | Analyze generated outline and create braille data |
| | Analyze resolution of linked braille pad |
| | Convert data size to braille pad resolution |

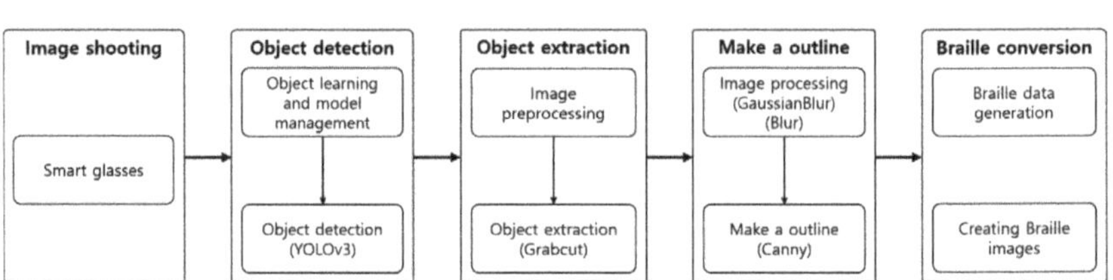


**Figure 16.** process.

### 4.1. Object Detection

In the object detection step, the YOLOv3 algorithm was used to detect a variety of objects in real time. Figure 17 shows the results from the application of the system to a real object.

**Figure 17.** Object detection example [32].

*4.2. Object Extraction*

The extraction step was configured using Python, and image processing algorithms used were from OpenCV. The objects were extracted using GrabCut after preprocessing the image. Figure 18 shows the structure of the object extraction step.

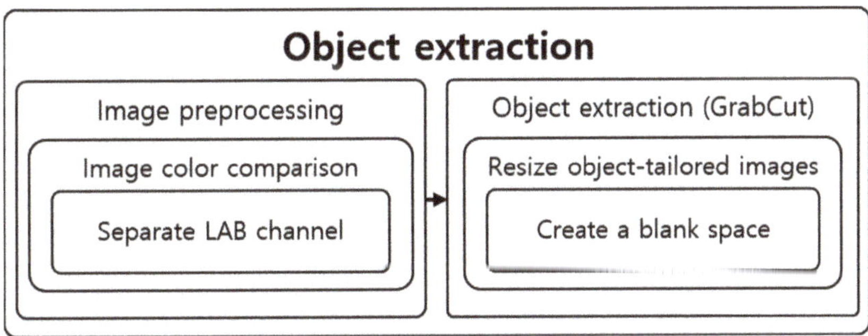


**Figure 18.** Structural diagram of object extraction steps.

4.2.1. Image Preprocessing

The contrast of the entire image, which refers to the difference in brightness between bright and dark areas in an image, is enhanced to clearly distinguish the colors of the detected image. An image with a small difference in brightness between bright and dark areas has a low contrast value, while an image with a large difference in brightness between bright and dark areas has a high contrast value. The contrast value refers to the contrast ratio. To increase the contrast value, dark areas must be darkened by increasing the color value of the pixels, and bright areas must be brightened by lowering the color values of the pixels.

Although there are various algorithms for increasing contrast value, the most basic technique is to multiply each pixel by a value based on the desired brightness of 1.0 [33,34]. Multiplication techniques are categorized into two methods: multiplying a MAT and using the saturate equation through the clip algorithm. However, they are not suitable for this study because these methods are mainly used on grayscale images to adjust only the brightness values. Instead, we examined algorithms used for colored images. The contrast of colored images is adjusted using a histogram equalization algorithm [35]. In addition, histogram smoothing converts a colored image composed of RGB channels into YCrCb channels and separates them into individual Y, Cr, and Cb channels, respectively, as shown in Figure 19. Y represents the luminance component, while Cr and Cb represent the chrominance components. Histogram equalization is applied to the separated luminance channels to increase the contrast value of the image.

(a) Before image            (b) Histogram equalized image

**Figure 19.** Histogram equalization Example.

Histogram equalization can be applied to an image composed of RGB channels to increase the contrast of the image. It increases the contrast by converting a colored image composed of RGB channels to YCrCb channels and separating them into individual Y, Cr, and Cb channels, respectively. Y represents the luminance component, while Cr and Cb represent the chrominance components. The contrast of color images is increased by applying the histogram equalization in the separated luminance component.

However, histogram equalization adjusts the contrast value of the entire image at once, making the bright areas very bright and dark areas very dark. This results in an unbalanced image overall. The CLAHE algorithm, which separately adjusts the brightness of specific areas in the image, was used to adjust the average brightness while increasing the contrast value [36]. To apply CLAHE, the image is converted to the LAB format and separated into individual channels to separate it into colored and grayscale [37] images. Channel L represents the brightness of the light and is expressed as a black and white image, while channels A and B represent the degree of color. Channel A represents magenta and green, and channel B represents blue and yellow. Moreover, the images are sequentially searched on the basis of the specified grid size, and the contrast value is adjusted to increase the contrast value in channel L and the black and white images. The channels are combined and converted back to the RGB format for other image processing after searching all images and adjusting the contrast value. Using the image from Figure 17, the contrast of an image was increased (Figure 20) through image channel separation, as shown in Figure 21.

(a) Channel L

(b) Channel A

(c) Channel B

**Figure 20.** LAB Image by channel.

**Figure 21.** Contrasted image.

### 4.2.2. Object Extraction

The stored object location information is imported to extract objects from the image whose contrast was increased in the preprocessing step. Approximately 10 is added to or subtracted from each x and y value in the stored object location information to distinguish the surrounding pixels easily, as shown in Figure 22. GrabCut is used for object extraction. A black background is generated around it when an object is extracted, leaving only the object. In addition, the image size is reduced on the basis of the location information to fit the image size to the object and save it. Figure 23 shows the result of using the GrabCut algorithm.

### 4.3. Outline Generation

It is hard for users who have difficulty distinguishing objects to recognize objects with large amounts of information at once. Therefore, the tactile image generation was divided into three types depending on the desired type of expression of the user. These three types were "Out," which displays only the outermost part such that the user can recognize the overall shape of the object; "Feature," to ensure that the user can recognize the inner boundaries and form of the object; and "Detail," which displays all information even the text in the object. Figure 24 shows the structure of the outline generation step.

(a) Indicating location information

(b) Redefining scope

**Figure 22.** Object measurement range.

**Figure 23.** Object extracted image.

4.3.1. Image Processing

In the image processing step for generating the outline, the noise was removed and the colors were averaged. GaussianBlur was performed to remove the noise created by increasing the contrast and other noise [38]. GaussianBlur is used to remove large noise, while averaging [39,40] removes small components and detailed features, such as letters and shapes. Each algorithm was performed with varying degrees of frequency and intensity depending on the outline generation type selected by the user.

In the Out mode, starting from the $7 \times 7$ kernel and sigma 0, the algorithm was run as it gradually reduces the search size to ensure an iterative and powerful preprocessing and to completely remove noise, features, and information. In the Feature mode, starting from the $5 \times 5$ kernel and sigma 0, the algorithm was run as it gradually reduces the search size to moderately remove noise and information. On the other hand, in the Detail mode, it searched with a 3 × 3 kernel and sigma 0 to remove noise while maintaining features and information. Figure 25 shows the image processing results.

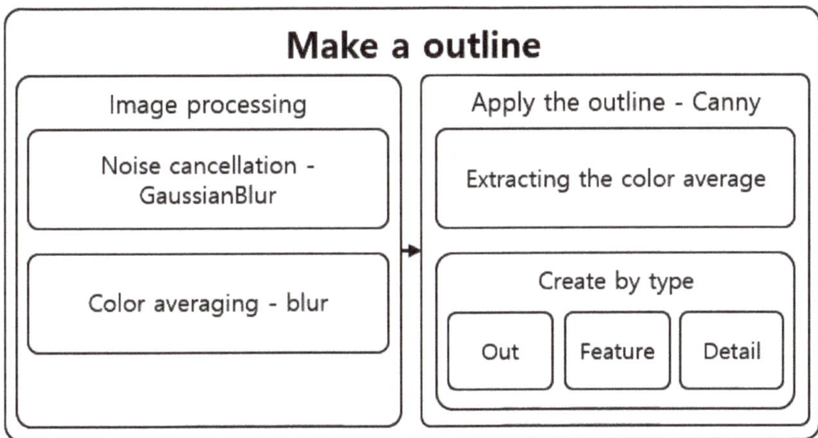

**Figure 24.** Outline creation step structure.

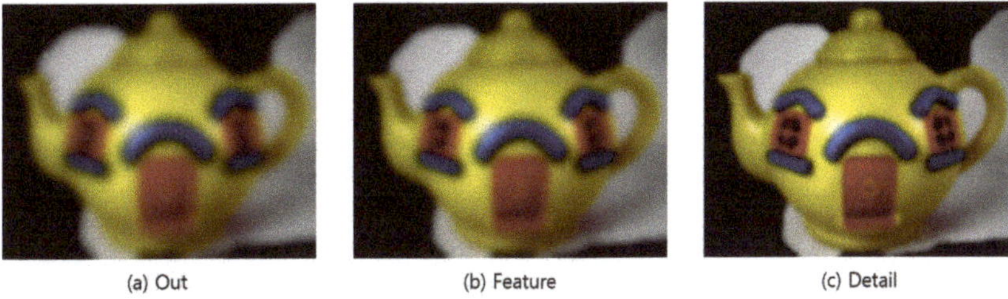



**Figure 25.** Image after performing.

4.3.2. Outline Generation

Canny [31] was used because it had a higher speed than Contour [30] although both Contour and Canny yielded similar accuracies for the outline generation algorithm. To generate the outline for each mode, the arrangement average and standard deviation of the images are calculated based on the noise-removed image, and the sum is set to a maximum value, so that the outline generation degree varies depending on the value range and mode. The morphology operations erosion and dilation were used to remove noise and small outlines remaining in the generated outline image. For braille conversion, the thickness was increased three times to confirm the line region, and the generated outlines were stored as individual images according to the mode. The thickness was increased three-fold by repeating the morphology dilation operation [41] three times, and the generated outlines were saved as an individual image on the basis of the mode to clearly define the lines for braille conversion. Figure 26 shows the result of outline generation.

*4.4. Braille Conversion*

Finally, the braille data were generated in the braille conversion step. For the data size, an image with a horizontal or vertical size of 416 was used as an input in the detection network in YOLOv3. The transformed image was resized on the basis of the detected location information of the object in the object detection process. Moreover, the data size was converted through braille data resizing on the basis of the received braille pad resolution when the braille pad was connected, ensuring that the output braille fitted the braille pad.

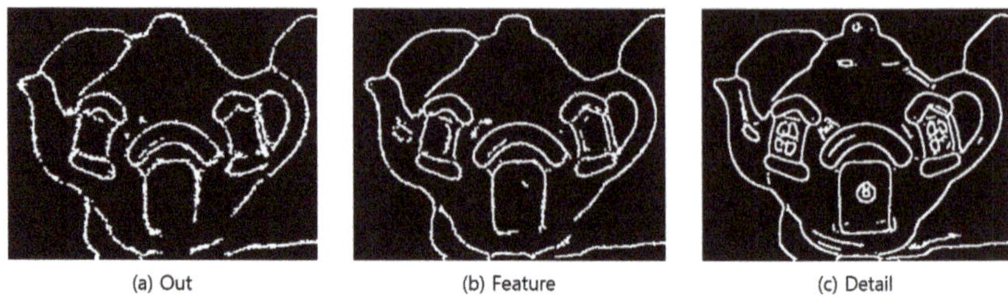

(a) Out       (b) Feature       (c) Detail

**Figure 26.** Outline image—Step-by-step image completed up to thickness increase.

n × n Comparison Conversion

At this stage, the outline images generated through Python were imported and converted to braille data to create braille images. Colored values in which the color and brightness can be identified were searched via array comparison because images in Python are expressed as an array. A two-dimensional array of the same size as the image was generated to perform the search. The image was searched in a 5 × 5 pixel neighborhood, and it was checked whether there are color data in the center pixel (>0), as shown in Figure 27. A value of 1 was stored in the same location as the generated two-dimensional array if there are color data. An array was finally generated by searching the entire image, which is stored for transmission to the braille pad. The tactile image was generated by the same technique; the image was searched, and a circle was created in areas with a value. Figure 28 shows the results of braille transformation through comparative transformation.

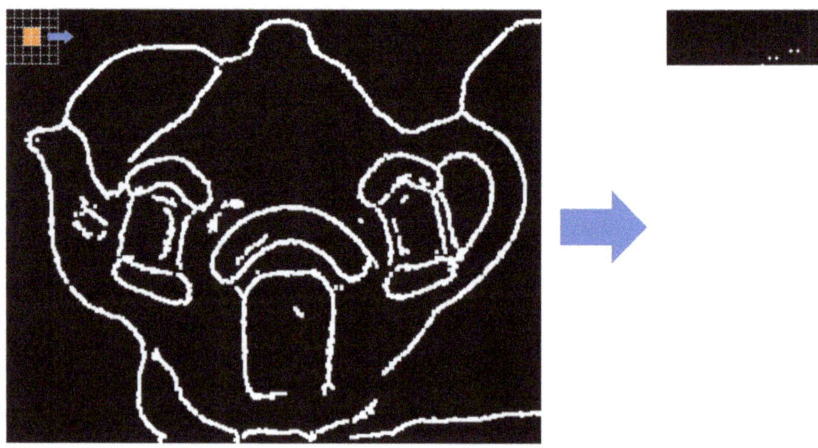

**Figure 27.** Example of braille conversion process.

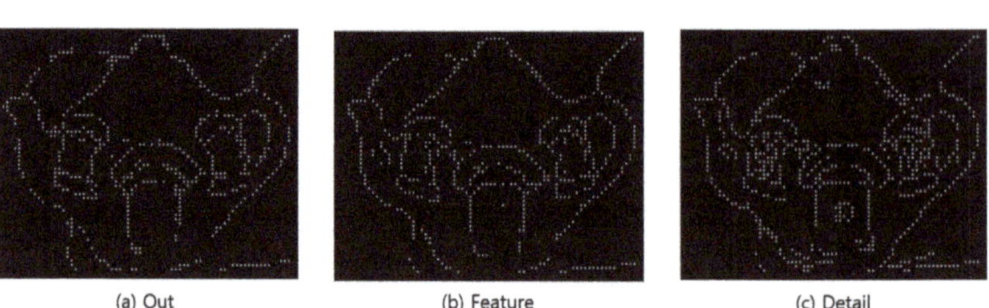

(a) Out       (b) Feature       (c) Detail

**Figure 28.** Generated braille image.

## 5. Experiment and Evaluation

We evaluated the accuracy and usefulness of the tactile image generated by the proposed system. To evaluate the accuracy, the expected result images and the system result images for a variety of objects were compared. On the other hand, to evaluate the usefulness, the execution time of the system was calculated using photos in diverse situations that can be confirmed in real life, which verified the applicability of the system in real life.

### 5.1. Experiment

#### 5.1.1. Object Data Generation

The highly well-known and stable Microsoft COCO dataset [42] was used as the basic dataset because the dataset was required for object detection through YOLOv3. Table 2 lists the selected objects. Additionally, based on the COCO dataset object list, objects that give visually impaired people discomfort were added according to survey results, thus forming a dataset with 100 types of objects. The survey was conducted among visually impaired people in Korea at a welfare center. Table 3 summarizes the results.

#### 5.1.2. Accuracy Evaluation

To select the objects for the evaluation criteria, the objects that the visually impaired frequently use or encounter in real life were categorized into the following: (1) "indoors" and "outdoors" and (2) based on their sizes (i.e., large, medium, and small), resulting in a total of six objects. The following size criteria were applied: objects difficult to hold in the hands were classified as large, objects that can be held with two hands as medium, and objects that can be held with one hand as small. For the objects that are most frequently encountered outdoors, "car" was selected for large, "fire hydrant" for medium, and "traffic cone" for small. On the other hand, "closet" was selected for large, "chair" for medium, and "comb" for small for the objects that are most frequently used indoors. Figures 29 and 30 show the comparison between the expected data and actual object results. The actual results were compared with [43,44] the braille for the "Detail" mode to verify the expression of details in the images.

#### 5.1.3. Usefulness Evaluation

To evaluate the usefulness, based on the three photos with the themes of "walking," "eating," and "washing face," the conversion time in each step was measured and averaged, and the identified objects were compared with [43,44] the detected object list. Only the name and location value of the object closest to the user were used when there were duplicate objects in the braille conversion step, thus performing braille conversion without any duplicate objects. Figure 31 shows the photos used for the evaluation, converted photos, and detected objects list, with conversion times of 5.8, 4.5, and 7.4 s, respectively.

### 5.2. Overall Evaluation

The main object was compared with the expected generated data to evaluate the accuracy of the tactile image. We verified the amount of time needed for conversion to evaluate the usefulness of the system.

In the accuracy evaluation, the expected result image was visually compared with the resulting image of the system, and the accuracy of the generated tactile image was measured. The results showed that the final image has an average accuracy of 85% which is similar to that of the expected image.

In the usefulness evaluation, the list of detected objects was compared and the conversion time was measured on the basis of the photos of three situations that users can encounter in real life. For the objects detected in photos of real-life situations, the results indicated an accuracy of approximately >90%. By excluding duplicate objects, the average time needed to convert the objects was less than 6.6 s, exhibiting that it can be quickly used in real life.

**Table 2.** COCO dataset object list [42].

| | | | | | | | | | |
|---|---|---|---|---|---|---|---|---|---|
| Person | Backpack | Umbrella | Handbag | Tie | Suitcase | Bicycle | Car | Motorcycle | Airplane |
| Bus | Train | Truck | Boat | Traffic light | Fire hydrant | Stop sign | Parking meter | Bench | Bird |
| Cat | Dog | Goose | Sheep | Cow | Elephant | Bear | Zebra | Giraffe | Frisbee |
| Skis | Snowboard | Sports ball | Kite | Baseball bat | Baseball glove | Skateboard | Surfboard | tennis racket | Bottle |
| Wine glass | Cup | Fork | Knife | Spoon | Bowl | Banana | Apple | Sandwich | Orange |
| Broccoli | Carrot | Hot dog | Pizza | Donut | Cake | Chair | Couch | Potted plant | Bed |
| Dining table | Toilet | TV | Laptop | Mouse | Remote | Keyboard | Cell phone | Microwave | Oven |
| Toaster | Sink | Refrigerator | Book | Clock | Vase | Scissors | Teddy bear | Hair drier | Toothbrush |

**Table 3.** List of selected objects.

| | | | | | | | | | |
|---|---|---|---|---|---|---|---|---|---|
| Person | Backpack | Umbrella | Handbag | Tie | Suitcase | Bicycle | Car | Motorcycle | Airplane |
| Bus | Train | Truck | Traffic light | Fire hydrant | Subway | Bench | Bird | Cat | Dog |
| Sports ball | Skateboard | Bottle | Wind glass | Cup | Fork | Knife | Spoon | Bowl | Chair |
| Tissu | Potted plant | Bed | Dining table | Toilet | TV | Laptop | Mouse | Remote | Keyboard |
| Cell phone | Microwave | Sink | Refrigerator | Book | Clock | Pillow | Scissors | Toothbrush | Toothpaste |
| Hair drier | Braille pad | Tree | Street lamp | Utility pole | Manhole | Vending machine | Elevator | Standing board | Escalator |
| Shampoo | Conditioner | Lottion | Stair | Traffic cone | Bollard | Radio | Desk | Whellchair | Eletric rice cooker |
| Gas cooker | Closet | Washing machine | Teapot | Electric fan | Comb | Bookmark | Soap | Glasses | Key |
| Shoes | Shower | Tumbler | Walking stick | Plate | Pencil | Electric kettle | Pen | Eraser | Earphones |
| Towel | Chopsticks | Meat | Fish | Hat | Rice | Kimchi | Bread | Cushin | Mattress |

This system can output tactile images generated on the basis of braille data of objects with a shape similar to those of real-life objects, yielding significant results.

Ten visually impaired individuals were satisfied with the performance of the assistance system. Moreover, they preferred the Out type, which simplifies the tactile information in a straightforward manner, over the Detail type, which converts the real objects of complex composition.

**(a) Actual image**

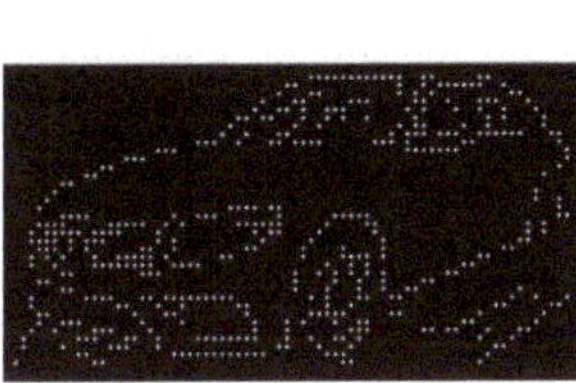

**(b) Expected result**

**Figure 29.** Comparison of expected and actual data(Outdoors).

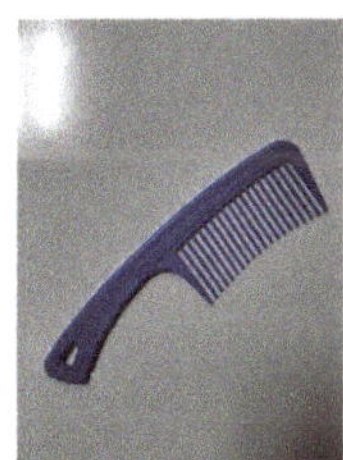

**(a) Actual Image**

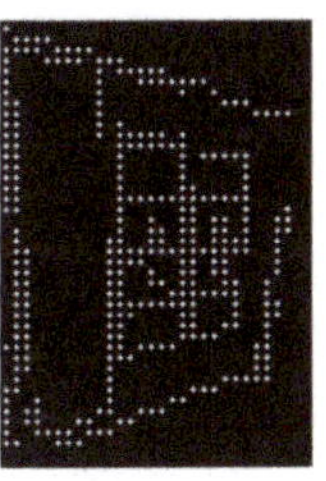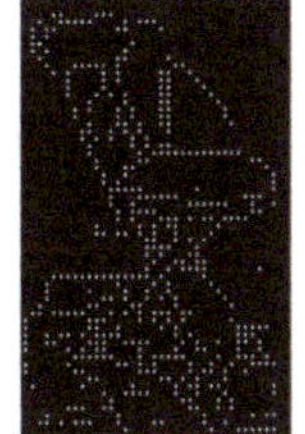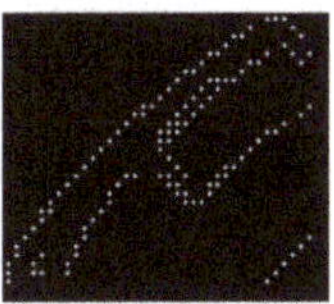

**(b) Expected result**

**Figure 30.** Comparison of expected and actual data(Indoors).

**(a) Panoramic photos in action**

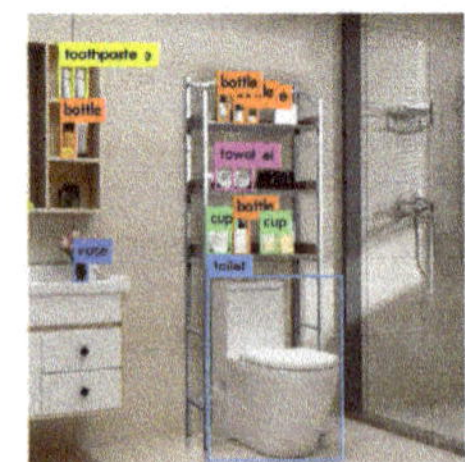

traffic cone, car, motorcycle, plant, person     towel, cup, plate, flowerpot, chair, dining table, orange, cake     toilet bowl, towel, potted plant, cup, bottle, toothpaste

**(b) Detection result**

**Figure 31.** As a result of applying it to real life photos.

## 6. Conclusions and Discussion

### 6.1. Conclusions

The proposed system was designed to inform visually impaired people about the types of obstacles in their field of view and to help them recognize their shapes. The system used an AI algorithm with high processing speed to quickly guide the user and integrated a simple image processing algorithm to provide tactile images in a short time. This study proposes a new and simple type of assistive device for visually impaired people who usually use braille, including people with limited sight. However, new algorithms or the latest technologies were not applied in the proposed system. The proposed braille conversion algorithm yielded an accuracy of 85% in relation to the expected result, demonstrating its usefulness. By excluding duplicate objects, approximately 12 out of 13 objects that can be confirmed in real life were detected on average. In addition, the conversion took an average of 6.6 s, indicating that the system is sufficient for use in real life.

### 6.2. Discussion

This study proposes a living assistance system that is applicable both indoors and outdoors by integrating object recognition, object extraction, outline generation, and braille conversion algorithms. According to the experiments and evaluations, we found that the system developed on the basis of the database tailor-made to the needs of visually impaired people (includes people with limited sight), who usually use braille, was useful.

However, some limitations of this study include the object extraction results obtained through GrabCut using the coordinates of the detected objects with YOLOv3 did not match with the real object. Moreover, some images other than the object image are left, indicating an inaccuracy in the braille conversion.

Therefore, we plan to perform primary development research to further improve the accuracy of the system and to generate and apply YOLOv3-based object masks although a more advanced system may require additional conversion time. Furthermore, we plan to conduct secondary development research to convert detected objects to icons and reflect the areas of improvement found from tests.

**Author Contributions:** D.L. designed, contributed to system model and implemented testbed. J.C. contributed to paper review and formatting. All authors have read and agreed to the published version of the manuscript.

**Funding:** This research was supported by the Basic Science Research Program through the National Research Foundation of Korea (NRF) funded by the Ministry of Education (2019R1F1A105775713) and This work was supported by the Gachon University research fund of 2020 (GCU-202008460007).

**Institutional Review Board Statement:** Not applicable.

**Informed Consent Statement:** Informed consent was obtained from all subjects involved in the study.

**Data Availability Statement:** Publicly available datasets were analyzed in this study. COCO dataset can be found here: https://cocodataset.org/ (accessed on 30 September 2021).

**Acknowledgments:** This research was supported by the Basic Science Research Program through the National Research Foundation of Korea (NRF) funded by the Ministry of Education (2019R1F1A105775713) and This work was supported by the Gachon University research fund of 2020 (GCU-202008460007).

**Conflicts of Interest:** The authors declare no conflict of interest.

# References


1. Korea Institute for Health and Social Affairs. Cause and Timing of Disability. 2019. Available online: https://kosis.kr/statHtml/statHtml.do?orgId=331&tblId=DT_33109_F37&conn_path=I2 (accessed on 30 September 2021).
2. The Online Database of Health Reporting (GBE). The Information System of the Federal Health Report and Statement of the Federal Republic of Germany. 2015. Available online: https://www.gbe-bund.de/gbe/ (accessed on 1 October 2021).
3. World Health Organization (WHO); World Bank. *World Report on Disability 2011*. 2011. Available online: https://www.who.int/disabilities/world_report/2011/report.pdf (accessed on 2 October 2021).
4. World Health Organization (WHO). Blindness and Vision Impairment. 2021. Available online: https://www.who.int/en/news-room/fact-sheets/detail/blindness-and-visual-impairment (accessed on 14 October 2021).
5. Ministry of Health and Welfare. The Number of Registered Disabled Persons by Type of Disability and Gender Nationwide. 2020. Available online: https://kosis.kr/statHtml/statHtml.do?orgId=117&tblId=DT_11761_N001&conn_path=I2 (accessed on 14 October 2021).
6. German Federal Statistical Office. People with Severe Disabilities with ID (Absolute and 100 per Person). (Population of 1000). Features: Years, Region, Type of Disability, Degree of Disability. 2019. Available online: https://www.gbe-bund.de/gbe/pkg_isgbe5.prc_menu_olap?p_uid=gast&p_aid=21134557&p_sprache=D&p_help=0&p_indnr=218&p_indsp=&p_ityp=H&p_fid= (accessed on 15 October 2021).
7. Kostopoulos, K.; Moustakas, K.; Tzovaras, D.; Nikolakis, G. Haptic Access to Conventional 2D Maps for the Visually Impaired. In Proceedings of the 2007 3DTV Conference, Kos, Greece, 7–9 May 2007; pp. 1–4. [CrossRef]
8. Zeng, L.; Miao, M.; Weber, G. Interactive audio-haptic map explorer on a tactile display. *Interact. Comput.* **2015**, *27*, 413–429. [CrossRef]
9. Krufka, S.E.; Barner, K.E.; Aysal, T.C. Visual to tactile conversion of vector graphics. *IEEE Trans. Neural Syst. Rehabil. Eng.* **2007**, *15*, 310–321. [CrossRef] [PubMed]
10. Krufka, S.E.; Barner, K.E. A user study on tactile graphic generation methods. *Behav. Inf. Technol.* **2006**, *25*, 297–311. [CrossRef]
11. Kim, H.J.; Kim, Y.C.; Park, C.J.; Oh, S.J.; Lee, B.J. Auto Braille Translator using Matlab. *J. Korea Inst. Electron. Commun. Sci.* **2017**, *12*, 691–700.
12. Jiwan Lee, J.A.; Lee, K.Y. Development of a raspberry Pi-based banknote recognition system for the visually impaired. *J. Soc.-Bus. Stud.* **2018**, *23*, 21–31. [CrossRef]
13. Hahn, M.E.; Mueller, C.M.; Gorlewicz, J.L. The Comprehension of STEM Graphics via a Multisensory Tablet Electronic Device by Students with Visual Impairments. *J. Vis. Impair. Blind.* **2019**, *113*, 404–418. [CrossRef]
14. Kim, S.; Park, E.S.; Ryu, E.S. Multimedia vision for the visually impaired through 2d multiarray braille display. *Appl. Sci.* **2019**, *9*, 878. [CrossRef]
15. Kim, S.; Yeongil Ryu, J.C.; Ryu, E.S. Towards Tangible Vision for the Visually Impaired through 2D Multiarray Braille Display. *Sensors* **2019**, *19*, 5319. [CrossRef]
16. Prescher, D.; Bornschein, J.; Köhlmann, W.; Weber, G. Touching graphical applications: Bimanual tactile interaction on the HyperBraille pin-matrix display. *Univers. Access Inf. Soc.* **2018**, *17*, 391–409. [CrossRef]
17. Kłopotowska, A.; Magdziak, M. Tactile Architectural Drawings—Practical Application and Potential of Architectural Typhlographics. *Sustainability* **2021**, *13*, 6216. [CrossRef]
18. H.Morad, A. GPS Talking For Blind People. *J. Emerg. Technol. Web Intell.* **2010**, *2*, 239–243. [CrossRef]
19. Fernandes, H.; Filipe, V.; Costa, P.; Barroso, J. Location based Services for the Blind Supported by RFID Technology. *Procedia Comput. Sci.* **2014**, *27*, 2–8. [CrossRef]

20. Liao, C.; Choe, P.; Wu, T.; Tong, Y.; Dai, C.; Liu, Y. RFID-Based Road Guiding Cane System for the Visually Impaired. In *Cross-Cultural Design. Methods, Practice, and Case Studies*; Springer: Berlin/Heidelberg, Germany, 2013; pp. 86–93.
21. Redmon, J.; Divvala, S.; Girshick, R.; Farhadi, A. You only look once: Unified, real-time object detection. In Proceedings of the IEEE Conference on Computer Vision and Pattern Recognition, Las Vegas, NV, USA, 26 June–1 July 2016; pp. 779–788.
22. Redmon, J.; Farhadi, A. YOLO9000: Better, faster, stronger. In Proceedings of the IEEE Conference on Computer Vision and Pattern Recognition, Honolulu, HI, USA, 21–26 July 2017; pp. 7263–7271.
23. Redmon, J.; Farhadi, A. YOLOv3: An Incremental Improvement. *arXiv* **2018**, arXiv:1804.02767.
24. Kathuria, A. What Is New in YOLO v3? 2018. Available online: https://towardsdatascience.com/yolo-v3-object-detection-53fb7d3bfe6b (accessed on 6 September 2021).
25. Lee, S.; Lee, G.; Ko, J.; Lee, S.; Yoo, W. Recent Trends of Object and Scene Recognition Technologies for Mobile/Embedded Devices. *Electron. Telecommun. Trends* **2019**, *34*, 133–144.
26. Poudel, S.; Kim, Y.J.; Vo, D.M.; Lee, S.W. Colorectal disease classification using efficiently scaled dilation in convolutional neural network. *IEEE Access* **2020**, *8*, 99227–99238. [CrossRef]
27. Siddiqui, Z.A.; Park, U.; Lee, S.W.; Jung, N.J.; Choi, M.; Lim, C.; Seo, J.H. Robust powerline equipment inspection system based on a convolutional neural network. *Sensors* **2018**, *18*, 3837. [CrossRef]
28. Rother, C.; Kolmogorov, V.; Blake, A. "GrabCut" interactive foreground extraction using iterated graph cuts. *ACM Trans. Graph. TOG* **2004**, *23*, 309–314. [CrossRef]
29. OpenCV. Interactive Foreground Extraction Using GrabCut Algorithm. Available online: https://docs.opencv.org/master/d8/d83/tutorial_py_grabcut.html (accessed on 7 September 2021).
30. Ghuneim, A.G. Contour Tracing. 2000. Available online: http://www.imageprocessingplace.com/downloads_V3/root_downloads/tutorials/contour_tracing_Abeer_George_Ghuneim/author.html (accessed on 7 September 2021).
31. Canny, J. A computational approach to edge detection. *IEEE Trans. Pattern Anal. Mach. Intell.* **1986**, *PAMI-8*, 679–698. [CrossRef]
32. Berry, S. Big Yellow Teapot. Available online: https://www.flickr.com/photos/unloveable/2388661262 (accessed on 7 September 2021).
33. Gonzalez, R.C.; Woods, R.E. *Digital Image Processing*; Pearson/Prentice Hall: Hoboken, NJ, USA, 2008.
34. Laughlin, S. A Simple Coding Procedure Enhances a Neuron's Information Capacity. *Z. Für Naturforschung C* **1981**, *36*, 910–912. [CrossRef]
35. Hummel, R.A. Image Enhancement by Histogram transformation. *Comput. Graph. Image Process.* **1975**, *6*, 184–195. [CrossRef]
36. Pizer, S.M.; Amburn, E.P.; Austin, J.D.; Cromartie, R.; Geselowitz, A.; Greer, T.; ter Haar Romeny, B.; Zimmerman, J.B.; Zuiderveld, K. Adaptive histogram equalization and its variations. *Comput. Vis. Graph. Image Process.* **1987**, *39*, 355–368. [CrossRef]
37. International Commission on Illumination. *Colorimetry*; CIE Technical Report; Commission Internationale de l'Eclairage: Vienna, Austria, 2004.
38. Haddad, R.; Akansu, A. A class of fast Gaussian binomial filters for speech and image processing. *IEEE Trans. Signal Process.* **1991**, *39*, 723–727. [CrossRef]
39. Kalman, R.E. A New Approach to Linear Filtering and Prediction Problems. *Trans. ASME–J. Basic Eng.* **1960**, *82*, 35–45. [CrossRef]
40. Gonzalez, R.; Wintz, P. *Digital Image Processing*, 2nd ed.; Addison-Wesley: Boston, MA, USA, 1987.
41. OpenCV. Eroding and Dilating. Available online: https://docs.opencv.org/4.x/db/df6/tutorial_erosion_dilatation.html (accessed on 8 September 2021).
42. Lin, T.Y.; Maire, M.; Belongie, S.; Bourdev, L.; Girshick, R.; Hays, J.; Perona, P.; Ramanan, D.; Zitnick, C.L.; Dollár, P. Microsoft COCO: Common Objects in Context. *arXiv* **2014**, arXiv:1405.0312.
43. Park, J.H.; Whangbo, T.K.; Kim, K.J. A novel image identifier generation method using luminance and location. *Wirel. Pers. Commun.* **2017**, *94*, 99–115. [CrossRef]
44. Wang, H.; Li, Z.; Li, Y.; Gupta, B.; Choi, C. Visual saliency guided complex image retrieval. *Pattern Recognit. Lett.* **2020**, *130*, 64–72. [CrossRef]

*Article*

# Acceleration of Magnetic Resonance Fingerprinting Reconstruction Using Denoising and Self-Attention Pyramidal Convolutional Neural Network

Jia-Sheng Hong [1], Ingo Hermann [2], Frank Gerrit Zöllner [2], Lothar R. Schad [2], Shuu-Jiun Wang [3,4,5], Wei-Kai Lee [1], Yung-Lin Chen [6], Yu Chang [6] and Yu-Te Wu [5,6,*]

1   Department of Biomedical Imaging and Radiological Sciences, National Yang Ming Chiao Tung University, Taipei 112, Taiwan; eternityjh.be06@nycu.edu.tw (J.-S.H.); l850818.be07@nycu.edu.tw (W.-K.L.)
2   Computer Assisted Clinical Medicine, Mannheim Institute for Intelligent Systems in Medicine, Medical Faculty Mannheim, Heidelberg University, 68167 Mannheim, Germany; Ingo.Hermann@medma.uni-heidelberg.de (I.H.); frank.zoellner@medma.uni-heidelberg.de (F.G.Z.); Lothar.Schad@medma.uni-heidelberg.de (L.R.S.)
3   Department of Neurology, Neurological Institute, Taipei Veterans General Hospital, Taipei 112, Taiwan; sjwang@vghtpe.gov.tw
4   College of Medicine, National Yang Ming Chiao Tung University, Taipei 112, Taiwan
5   Brain Research Center, National Yang Ming Chiao Tung University, Taipei 112, Taiwan
6   Institute of Biophotonics, National Yang Ming Chiao Tung University, Taipei 112, Taiwan; thomaschen83.be08@nycu.edu.tw (Y.-L.C.); changyu97@gm.ym.edu.tw (Y.C.)
*   Correspondence: ytwu@ym.edu.tw

**Citation:** Hong, J.-S.; Hermann, I.; Zöllner, F.G.; Schad, L.R.; Wang, S.-J.; Lee, W.-K.; Chen, Y.-L.; Chang, Y.; Wu, Y.-T. Acceleration of Magnetic Resonance Fingerprinting Reconstruction Using Denoising and Self-Attention Pyramidal Convolutional Neural Network. *Sensors* **2022**, *22*, 1260. https://doi.org/10.3390/s22031260

Academic Editors: Vahid Abolghasemi, Hossein Anisi and Saideh Ferdowsi

Received: 1 December 2021
Accepted: 5 February 2022
Published: 7 February 2022

**Publisher's Note:** MDPI stays neutral with regard to jurisdictional claims in published maps and institutional affiliations.

**Abstract:** Magnetic resonance fingerprinting (MRF) based on echo-planar imaging (EPI) enables whole-brain imaging to rapidly obtain T1 and T2* relaxation time maps. Reconstructing parametric maps from the MRF scanned baselines by the inner-product method is computationally expensive. We aimed to accelerate the reconstruction of parametric maps for MRF-EPI by using a deep learning model. The proposed approach uses a two-stage model that first eliminates noise and then regresses the parametric maps. Parametric maps obtained by dictionary matching were used as a reference and compared with the prediction results of the two-stage model. MRF-EPI scans were collected from 32 subjects. The signal-to-noise ratio increased significantly after the noise removal by the denoising model. For prediction with scans in the testing dataset, the mean absolute percentage errors between the standard and the final two-stage model were 3.1%, 3.2%, and 1.9% for T1, and 2.6%, 2.3%, and 2.8% for T2* in gray matter, white matter, and lesion locations, respectively. Our proposed two-stage deep learning model can effectively remove noise and accurately reconstruct MRF-EPI parametric maps, increasing the speed of reconstruction and reducing the storage space required by dictionaries.

**Keywords:** magnetic resonance fingerprinting; echo-planar imaging; T1 and T2* relaxation times; denoising convolutional neural network; self-attention; feature pyramid network

## 1. Introduction

Quantitative magnetic resonance (MR) relaxometry can quantify the relaxation time (e.g., T1, T2, T2* relaxation time) to clarify the physical and pathological properties of human tissues [1]. Quantitative MR relaxometry was reported to increase accuracy and precision compared with conventional weighted magnetic resonance imaging (MRI) in detecting lesions, and it can even synthesize traditional weighted images [2,3]. However, clinical applications of quantitative MR relaxometry are limited by the length of the imaging procedure required to estimate the tissue relaxation time; moreover, motion artifacts can interfere with the results, and the procedure does not meet the needs for clinical scheduling efficiency. Magnetic resonance fingerprinting (MRF) is an approach for designing the rapid quantitative sequence [4]. MRF has the advantage of providing quantitative images of multiple types of relaxation times simultaneously in a relatively short imaging time (several minutes). However, because MRF image reconstruction requires comparison with a vast computer simulation database (dictionary matching), the extended image reconstruction time has become a considerable challenge in the development of MRF [5].

The dictionary matching process is computationally expensive and requires storage space for the simulation database, which hinders clinical applications of MRF. Thus, optimizing the MRF signal matching process is crucial. Toward this aim, dimension reduction algorithms, such as singular value decomposition, were the first to be used. Studies have used singular value decomposition to project the database into low-dimensional space, speeding up the MRF signal matching process by 3.4–4.8 times that of using only the inner-product method [6,7]. Compared with approaches reducing the dimensionality of the database, a model trained by deep learning can eliminate the storage usage of the MRF simulation database and achieve near real-time reconstructions. Recent studies on the use of deep learning to accelerate the MRF reconstruction process have included the use of a one-dimensional (1D) neural network, a convolutional neural network (CNN), and a recurrent neural network (RNN) to train models for learning the simulated information [8–10]. Studies have also modeled the reconstructed images in a two-dimensional (2D) fashion by using the data after matching the dictionary with the scanned images [11–13]. Moreover, deep learning models can combine multiple tasks, including the reconstruction of MRF parametric maps, preprocessing, and tissue segmentation, thus reducing computation times from hours to seconds [13]. Deep learning is therefore an efficient method for MRF image reconstruction. In addition to deep learning studies of MRF reconstruction, one study used generative adversarial networks to speed up the generation of simulation data [14]. As the graphics hardware and deep learning algorithms mature, MRI imaging techniques can be optimized with deep learning to improve computational performance and thus increase the feasibility of clinical applications [2].

Most deep learning studies for the MRF image reconstruction have developed their models based on the original MRF protocol by Ma et al., which has a signal with long time steps (a thousand-time points) [4]. Therefore, most models are designed to reduce the time dimension. For instance, Fang et al. used a two-stage deep learning strategy that entailed first extracting features through a fully connected neural network and then training the U-Net-based model to learn the spatial distribution of the brain tissue [12]. The feature extraction step is a process of reducing high-dimensional data to low-dimensional data. Longer time steps can compensate for the effects of noise, but for MRFs with shorter time steps, such as those used in this study (35-time points), the effects of noise cannot be underestimated. Cohen et al. demonstrated the extent to which noise affected the accuracy of their model, but they did not specifically design the model for noise reduction [8]. In addition, the selection of training and testing data is another critical point for training MRF models. Cohen et al. trained their model by simulation dictionary and tested using a digital brain phantom [8]. Hoppe et al. developed their CNN-based model by simulation dictionary and tested using a quantitative phantom. Chen et al. also devised a CNN-based model and tested their model by using the human scan data from another quantitative MRI method [15]. For the study using the same MRF protocol as this study, they only used scan data and did not include the simulation dictionary for training [13]. Their model performance had a between 5% and 10% error. Ideally, the deep learning model should be trained with the simulation dictionary, and the performance of the dictionary learning model is tested with human scan data. A dictionary learning model ensures that the model has learned all the possible situations, and models tested with human scan data are more convincing. Therefore, we designed and trained our model for noise reduction and used the dictionary learning model to predict human scan data to verify the performance of the proposed model.

MRF image reconstruction is a regression task for deep learning models, and the presence of noise affects the model performance [16]. A denoising CNN model (DnCNN) was proposed for image denoising; it is highly effective in general image denoising tasks [17].

Furthermore, the model can complete denoising tasks with an unknown noise level. Because dictionary matching is performed using the 1D approach, we modified the DnCNN for 1D signal denoising for the first stage of the proposed model. For the second stage of the model, which was aimed at learning the Bloch equation simulation [18], we designed a pyramidal model to extract features of the MRF signal evolution. A pyramid CNN exhibited promising performance in object detection tasks [19], and the advantage of the pyramid architecture is that it can extract and combine features from various scales. In addition, the self-attention mechanism has been used in natural language processing and can achieve state-of-the-art performance [20]. A CNN with self-attention can associate each pixel in a 2D image to generate a global reference between pixels [21]. We thus added the self-attention layer to the model for focusing on the connection between features extracted by the CNN. The weight of important features can be enhanced through the self-attention mechanism.

This study aimed to develop a deep learning model to replace the computationally expensive inner-product method for MRF reconstruction. We investigated how precisely the proposed model learned the Bloch equation simulation [18] and the relationship between the noise and model performance with scanned data. In the present study, MRF-echo-planar imaging (MRF-EPI) was used to scan the whole brains of 32 subjects to obtain T1 and T2* parametric maps [22–24]. Herein, we propose a two-stage model that first reduces MRF signal noise and then reconstructs parametric maps of MRF by a dictionary-learning model.

## 2. Materials and Methods

### 2.1. Population

The relevant institutional review board (2019-711N) approved this study, and the subjects provided informed consent before undergoing scanning. The MRF scan was implemented using a 3T scanner (Magnetom Skyra, Siemens Healthineers, Erlangen, Germany) with 14 healthy subjects and 18 subjects with multiple sclerosis (MS). The healthy group comprised eleven men and three women (aged 22–33 years; mean: 26 years). The MS group contained seven men and eleven women (aged 23–73 years; mean: 39 years). The scans of 32 subjects were used to evaluate the proposed model and are referred to as the "scanned data".

### 2.2. Magnetic Resonance Fingerprinting Imaging and Dictionary Generation

The acquisition method used was a previously proposed and validated MRF-EPI imaging sequence [13,22–24]. The imaging parameters of the MRF sequence were as follows: in-plane spatial resolution = 1 × 1 mm$^2$; slice thickness = 2 mm; bandwidth = 998 Hz/px; GRAPPA factor = 3; partial Fourier = 5/8, variable flip angle (34°–86°), echo time (21–81.5 milliseconds [ms]), repetition time (3530–6570 ms), and fat suppression. The acquisition time was 4 min 23 s for 60 slices of the whole brain. In addition, using the same spatial resolution, fluid-attenuated inversion recovery (FLAIR) was obtained for lesion segmentation. The MRF dictionaries were generated for each slice, with 598,842 entries based on the design of MRF-EPI using the Bloch equation simulation [18]. The ranges of T1 and T2* values were 100–4000 ms and 10–3000 ms (excluding those T1 smaller than T2*), respectively, with a 2% spacing. The range of flip angle efficiency B1+ was 0.6–1.4 with a 0.05 spacing.

The T1 and T2* maps of the scanned data were reconstructed by the inner-product method based on the 2%-increment dictionary. Figure 1 displays the schematic process of the MRF imaging. There were four steps in the MRF imaging process. The first was the MRF-EPI scan, which had a total of 35 images for each slice in which each pixel can be considered as a signal with 35 values (Figure 1a). Every pixel has its specific signal evolution that depends on the T1 and T2* relaxation times for the tissue of that pixel. The second was the dictionary generation, and the simulated dictionary was generated using the Bloch equation [18], given a certain range of T1 and T2* values (Figure 1b). The third was dictionary matching, where the MRF scanned signals were matched to the simulated dictionary signals one by one using the inner product (Figure 1c). When each pixel was matched, the parametric images were obtained, as in step 4 (Figure 1d). The time required for dictionary matching in the third step depends on the size of the dictionary in the second step. The denser the dictionary is, the more signal entries there are, and the longer the matching time is. This is where the challenge of MRF image reconstruction lies.

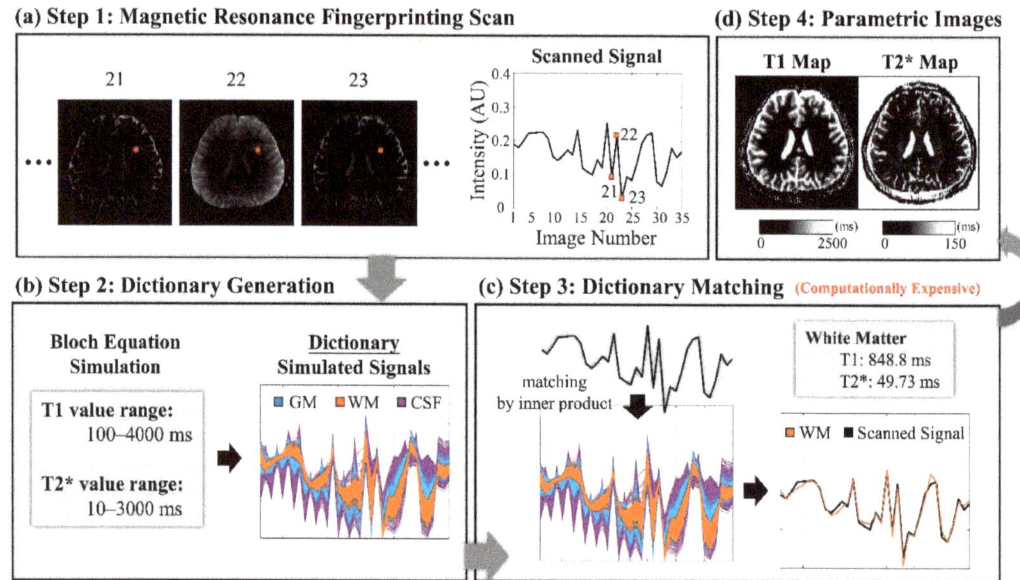


**Figure 1.** Schematic of the reconstruction for T1 and T2* maps of the magnetic resonance fingerprinting. (**a**) MRF baseline scan. (**b**) Dictionary generation process. (**c**) Dictionary matching by the inner product. (**d**) Parametric maps after matching pixel by pixel. MRF = magnetic resonance fingerprinting; AU = arbitrary unit; GM = gray matter; WM = white matter; CSF = cerebrospinal fluid.

*2.3. Dictionaries and Image Preprocessing*

We separated the dictionaries with the 2% increment in the simulation into training and validation datasets using two divisions. The first division split the training and validation datasets by the T1 and T2* value range. T1 and T2* values were 500–2500 ms and 50–1500 ms, respectively, for the training, and the other entries were used for the validation. In this division, we aimed to test whether the deep learning model was able to learn Bloch equation simulation [18] to predict relaxation times that were not in the training range.

In the second division, the training and validation datasets were divided according to the incremental spacing of the T1 and T2* values (i.e., 4%, 6%, 8%, ... , 20%). We sampled the entries by different intervals in the 2%-increment dictionary (i.e., 2, 3, 4, ... , 10) to obtain dictionaries with the mentioned increment as a training dataset and the remaining unsampled ones as a validation dataset. For instance, the 2%-increment dictionary had T1 values of 100 ms, 102 ms, 104.04 ms, ... , to the end, and T2* values of 10 ms, 10.2 ms, 10.404 ms, ... , to the end. We sampled the T1 values of 100 ms, 104.04 ms, ... , to the end, and then sampled the T2* values 10 ms, 10.404 ms, ... , to the end, obtaining a 4%-increment dictionary for training. Other unsampled entries, T1 values 102 ms, ... , to the end, and T2* values 10.2 ms, ... , to the end, were used as validation data. In this division, we aimed to test how accurate the deep learning model was in predicting the relaxation times in the training range.

To compare the reconstructed result between the standard dictionary matching and the proposed model for different tissues, manual and automatic segmentation of different brain tissues was performed. Lesion locations for the MS group were manually segmented on

FLAIR images by an expert radiologist. We used the SPM12 [25] to automatically segment the white matter (WM), gray matter (GM), and cerebrospinal fluid (CSF) from the T1 map obtained through MRF. A threshold of 80% of the maximum value was applied to the probability maps generated by SPM12 to create binary masks.

### 2.4. Noise Analysis and Denoising CNN

According to the inner product, the MRF scanned signals obtained from the subjects were first matched to the 2%-increment dictionary, which was the densest in our experiment. The matched signal from the simulated dictionary was considered as the noise-free signal. The signal without noise was subtracted from the scanned signal to obtain the residual for calculating the signal-to-noise ratio (SNR) as follows:

$$\text{SNR} = 10 \times \log_{10} \frac{\sum_{i=1}^{k} s_i^2}{\sum_{i=1}^{k} n_i^2}, \tag{1}$$

where $s$ and $n$ are the matched signal from the simulated dictionary and the residual gathered by the difference between the scanned and matched signal, respectively; $k$ is the length of the signal, which was 35 in our case. The SNR is in decibels (dB). We collected the amplitudes of residuals from 21 subjects (3 healthy subjects and 18 patients), slice by slice, and this collection was referred to as the "noise dataset" for training the denoising model. The temporal order of each residual was useless and thus discarded. The scans of the other 11 subjects were used as the testing dataset for evaluating the denoising model. Figure 2a displays a schematic of how the noise was obtained and collected.

Figure 2b displays the feedforward denoising CNN proposed for image denoising [17]. The denoising CNN was modified for noise reduction of 1D signals in this study. The proposed model began with a convolution layer followed by a rectified linear unit (ReLU) activation function and ended with a convolution layer. The model had 32 units of layers in the middle, and each unit included a convolutional layer followed by batch normalization and a ReLU. Each convolution layer had a kernel size of 3, padding of 1, and 64 channels (one channel for the final output).

The simulated dictionary signals plus randomly sampled noise from the noise dataset served as the input to train the model, and the output was the residuals (i.e., noise). The noise-free signals were obtained by subtracting the output of the model from the noisy scanned data. Independent-samples $t$ test was used to measure the difference between the SNR of the training and testing datasets. Paired-samples $t$ test was used to measure the difference in the SNR before and after denoising.

### 2.5. Pyramid CNN with Self-Attention for MRF Parametric Image Reconstruction

Figure 2c displays the deep learning model, which was based on a 1D CNN with a pyramidal structure. The dashed line extending from the green box indicates the detailed structure inside each green box. The input for the pyramid model was a 1D signal, and the outputs were T1 and T2* values. The backbone consisted of three convolutional layers with kernel sizes of 17, 11, and 7, and the number of channels was 128, 256, and 512, respectively. Each convolutional layer was followed by a ReLU activation function and then a dropout layer with 0.2 probability as a convolution block.

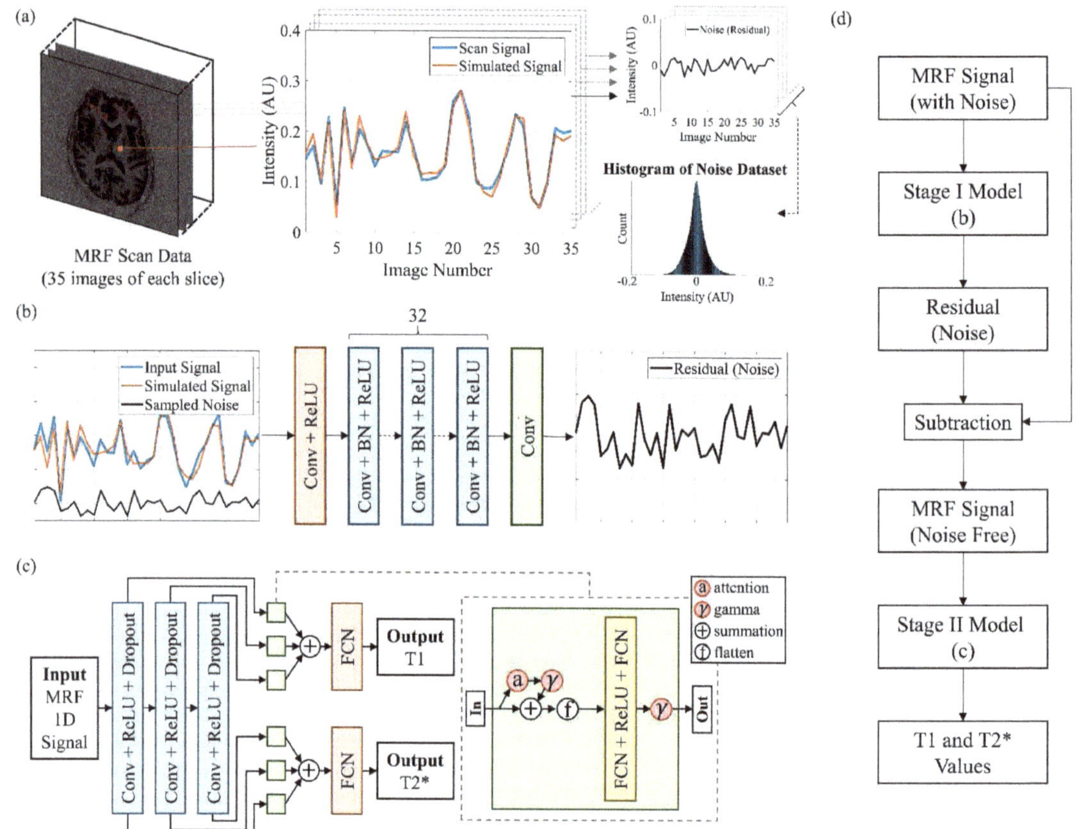

**Figure 2.** Schematic of the noise collection, denoising CNN, pyramid model, and flowchart of the two-stage model. (**a**) Collection of the noise dataset. AU = arbitrary unit. (**b**) Denoising CNN. (**c**) Weighted pyramid dual-path CNN with attention. (**d**) Flowchart of the successive process of the proposed model.

The output of T1 and T2* relaxation times had two paths. A multihead self-attention layer [20] with eight heads was first connected after each convolution block of each pathway. The expressions of the multihead self-attention are as follows:

$$\text{MultiHead}(Q, K, V) = \text{Concat}(\text{head}_1, \ldots, \text{head}_h)W^O, \tag{2}$$

$$\text{head}_i = \text{Attention}(Q_i, K_i, V_i) = \text{softmax}\left(\frac{Q_i K_i^{\;T}}{\sqrt{d_s}}\right) V_i,. \tag{3}$$

$$\text{and} \begin{cases} Q_i = X_i W_i^Q \\ K_i = X_i W_i^K \\ V_i = X_i W_i^V \end{cases}, \text{ and } X = X_1, \ldots, X_h \tag{4}$$

Equations (2)–(4) comprise the scaled dot-product self-attention with multihead. $Q$, $K$, and $V$ are the query, key, and value matrices. The corresponding matrices are, $X \in \mathbb{R}^{l \times d_{ch}(l \times h \times d_s)}$, $X_i \in \mathbb{R}^{l \times d_s}$, $W_i^Q \in \mathbb{R}^{d_s \times d_s}$, $W_i^K \in \mathbb{R}^{d_s \times d_s}$, $W_i^V \in \mathbb{R}^{d_s \times d_s}$, and $W^O \in \mathbb{R}^{h d_s \times d_{ch}}$, where $l$ is the length of the signal after each convolution block; $d_{ch}$ and $h$

are the input channels and number of heads, respectively; $d_s$ is $d_{ch}$ divided by $h$; $d_{ch}$ is 128, 256, and 512 for each convolution block; and $h$ is 8 in our implementation.

The output of the attention layer was weighted by a learnable parameter gamma and added back to its input as the input to the next layer [26]. The next layer was a flatten layer for connecting a fully connected layer with 128 output features, followed by a ReLU, and then a fully connected layer with three output features. The final output layer was a fully connected layer with one output feature, and its input was the sum of the outputs from the different scales after being weighted by the learnable parameter gamma. The output after the learnable parameter is given by:

$$Y = \gamma X^a + X \tag{5}$$

$$\text{and } Y = \sum_{i=1}^{m} \gamma_i X_i. \tag{6}$$

$Y$ in Equation (5) is the input to the flatten layer, whereas $X^a$ is the output after the attention layer. $Y$ in Equation (6) is the input for the final fully connected layer. Because three convolutional layers created separate scales, $m$ was equal to three.

The proposed model was named the weighted pyramid dual-path CNN with attention (WPDaCNN). Three other models were employed as comparisons for the proposed model. The first was a model without the weighted parameter gamma and the self-attention layer, denoted by PDCNN. The second was a model based on PDCNN but without the pyramid structure, denoted by DCNN (only the output of the third convolutional layer was considered). The final one was a model based on DCNN but with only a single path, denoted by SCNN (the output feature for the final layer of the single path became two).

### 2.6. Experimental Setup and Two-Stage CNN Framework

Figure 2d was the flowchart of the successive process of our model. The MRF signals with noise were first inputted to the stage I model to predict the noise. The denoised MRF signals were obtained by subtracting the predictive noise from the MRF signals with noise. Then, the denoised signals were inputted to the stage II model for outputting the T1 and T2* values.

The experiment was performed on a computer with an Intel Xeon W-2102 CPU and an NVIDIA Quadro P6000 24 gigabyte GPU. The deep learning models were built based on the PyTorch package (version 1.7.1+cu110) using Python 3.8.5, and the data preprocessing for dictionary generation and matching was performed by programming platform MATLAB R2020a (MathWorks; Natick, MA, USA). Statistical analysis was performed using SPSS Statistics 24 (IBM; Armonk, NY, USA).

The $L_2$ loss multiplied by 10,000 was applied to train the first stage DnCNN models. For the second-stage pyramid models, the $L_1$ loss and mean absolute percentage error (MAPE) were employed and added for training. The loss functions are as follows:

$$\text{Loss}_{\text{stageI}} = 10000 \times \frac{\sum_{i=1}^{N} \left( y_{i\,residual} - y_i^p{}_{residual} \right)^2}{N}, \tag{7}$$

$$\text{Loss}_{\text{stageII}} = \frac{\sum_{i=1}^{N} \left| y_i - y_i^p \right|}{N} + \frac{100 \times \sum_{i=1}^{N} \left| y_i - y_i^p \right| / y_i}{N}. \tag{8}$$

Equation (7) is the loss function for the first stage model, and Equation (8) is that for the second stage. We referred to the denoising study using the $L_2$ loss for training the first stage [17], and the constant 10,000 was set empirically. The $L_1$ loss for training the second stage was referenced to the literature that used the same MRF protocol as this study [13], and the MAPE term was used to balance the T1 and T2* for model learning. $N$ is the total number of values, $y_{i\,residual}$ is the true residual, $y_i^p{}_{residual}$ is the predicted residual, $y_i$ is the T1 and T2* values within the simulated dictionary, and $y_i^p$ is their predicted values. The value of the loss function corresponding to each stage was used as the error to identify the model with the lowest error.

Figure 3 presents a flowchart of our experiments. Figure 3a represents the workflow for training the DnCNN. There were 60 slices with their own unique simulated dictionaries, and thus a total of 60 models need to be trained. Because of the lengthy training time, the scanned data were split into single training and testing datasets for the experiment rather than split into multiple folds. The training and testing datasets consisted of 21 and 11 subjects from the scanned data, respectively. The noise dataset was obtained from the training dataset, as described in the Section 2.4. The DnCNN was trained for 100 epochs by inputting simulated signals plus randomly sampled values in the noise dataset. After 100 epochs, the trained model that corresponded to the lowest training error was selected as optimal. The testing dataset was then inputted to the optimal model for prediction.

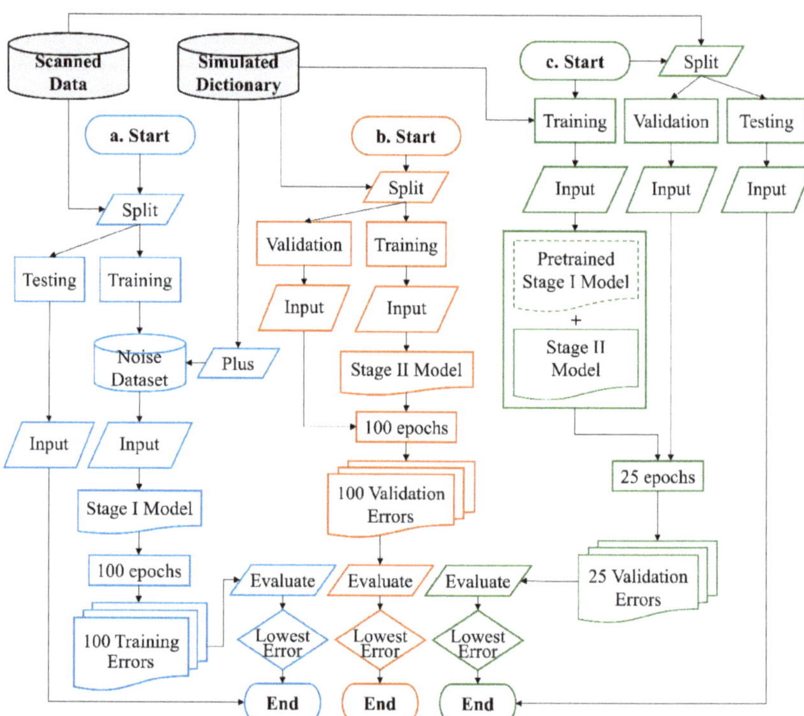

**Figure 3.** Schematic flowchart for our experiments. (**a**) Workflow for training the denoising models. (**b**) Workflow for training the pyramid models for comparison. (**c**) Workflow for training the final two-stage models. The dashed line for the pretrained stage I model indicates that the model weights were frozen and did not change during the training process.

Figure 3b indicates the workflow for training the models with different structures, namely, WPDaCNN, PDCNN, DCNN, and SCNN. The simulated dictionary was split into training and validation datasets according to the division described in the Section 2.3. Subsequently, each model was trained for 100 epochs, and the one with the lowest error for the validation dataset was selected as the optimal model.

Figure 3c displays the workflow for training the final two-stage model. We first connected the pretrained stage I model with the untrained stage II model and then froze the weights of the stage I model. The input for training was the simulated signal from the dictionary and did not pass through the denoising model while training. The scanned data was split into half for the validation dataset and another half for the testing dataset. After each epoch, the validation dataset was fed into the entire two-stage model for evaluating the error (loss) of the two-stage model. We observed that the second-stage model converged rapidly, and excessive training epochs led to overfitting; thus, only 25 epochs were set, and the optimal model was the one with the lowest validation error within 25 epochs. The testing dataset was then inputted to the optimal model for prediction.

Each whole-brain scan included 60 slices, with each slice corresponding to a distinct simulated dictionary. Hence, we trained a two-stage model for each slice with 598,842 entries as the input, and 60 models were eventually produced (Figure 3a,c). Because of the identical design concept of the pulse sequence for each slice, when we trained different models for comparison (i.e., WPDaCNN, PDCNN, DCNN, and SCNN), only one model of each type was trained by using a dictionary of the first slice (Figure 3b). During model training, the batch size was 500, and the optimizer employed was Adam with a learning rate of 0.01 and a scheduler with a 5% learning rate reduction per epoch. The intraclass correlation coefficient (ICC) was used to assess the consistency between the dictionary matching and prediction of the final two-stage models. The correlation coefficient was applied to test the mean and difference between the standards and predictions.

## 3. Results

### 3.1. SNR of Scan Data and after Denoising by DnCNN

Figure 4a displays the corresponding SNR before and after denoising. The SNR varied from slice to slice, with lower SNRs in the cranial and caudal portions and higher SNRs in the middle. Figure 4b contains two examples of the scanned signal after denoising. The noise was effectively removed after denoising, and the SNR increased (25 dB vs. 47 dB and 16 dB vs. 31 dB). Table 1 presents the SNR and statistics before and after the denoising by the DnCNN for the training and testing datasets. The results for various tissue types were obtained after applying the tissue masks that were created by the automatic and manual segmentation mentioned in the Section 2.3. The SNRs in both the training and testing datasets increased, and the increases after denoising were statistically significant ($p < 0.001$). The SNRs of GM and WM were similar, whereas the SNR of CSF was lower than that of GM and WM. Regarding the differences in mean SNR between the training and testing datasets, the $p$ values were 0.40 and 0.32 for the original and denoised SNRs, respectively. This result suggested that the mean SNRs of the training and testing datasets were not significantly different, either before or after denoising. Therefore, the model performed well in the testing dataset.

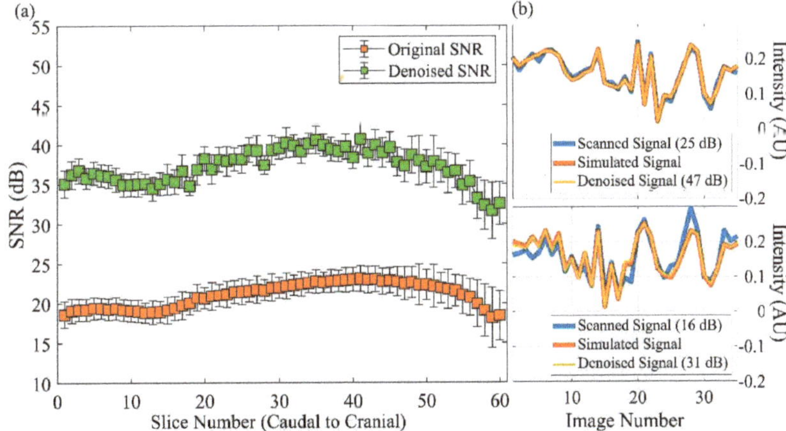


**Figure 4.** Signal-to-noise ratio (SNR) before and after denoising. (**a**) Plot of the mean (solid color box) and standard deviation (thin line bar) of the SNR in slices of the whole brain. (**b**) Two examples of the signal before and after denoising were gathered from one pixel in slice 35 (top) and slice 13 (bottom).

**Table 1.** SNR before and after the denoising model.

| | Training | | | Testing | | |
| --- | --- | --- | --- | --- | --- | --- |
| **Tissue Type** | **Original SNR** | **Denoised SNR** | ***p* Value** | **Original SNR** | **Denoised SNR** | ***p* Value** |
| Whole Brain | 21.33 ± 1.45 | 37.75 ± 1.55 | <0.001 * | 21.78 ± 1.41 | 38.31 ± 1.34 | <0.001 * |
| GM | 22.22 ± 1.51 | 39.10 ± 1.56 | <0.001 * | 23.01 ± 1.52 | 39.86 ± 1.31 | <0.001 * |
| WM | 22.44 ± 1.49 | 40.03 ± 1.68 | <0.001 * | 22.79 ± 1.69 | 40.48 ± 1.60 | <0.001 * |
| CSF | 18.97 ± 1.60 | 33.29 ± 1.59 | <0.001 * | 18.75 ± 1.30 | 33.31 ± 1.34 | <0.001 * |

GM = gray matter; WM = white matter; CSF = cerebrospinal fluid; MS = multiple sclerosis; SNR = signal-to-noise ratio. The unit of SNR is in dB. "*" indicates that the *p* value is less than 0.05.

### 3.2. Performance of the Pyramid CNN Models

For the first division, models learned well on the training dataset but poorly on the validation dataset. The mean MAPE of all models on the training dataset was 1.4%, and that on the validation dataset was 54.8%. For the second division, Figure 5 presents the pyramid model performance for different dictionary increments. As the increment increased, the losses of WPDaCNN and PDCNN increased smoothly, but the losses of DCNN and SCNN increased ruggedly. For the $L_1$ loss, WPDaCNN was the model with the optimal performance under all dictionary increments. The lowest $L_1$ loss was 10 ms and 4.5 ms for the training and validation, respectively, at the dictionary with the densest increment. Compared with SCNN, DCNN had lower losses, except for training losses at the increments of six and eight and validation losses at six.

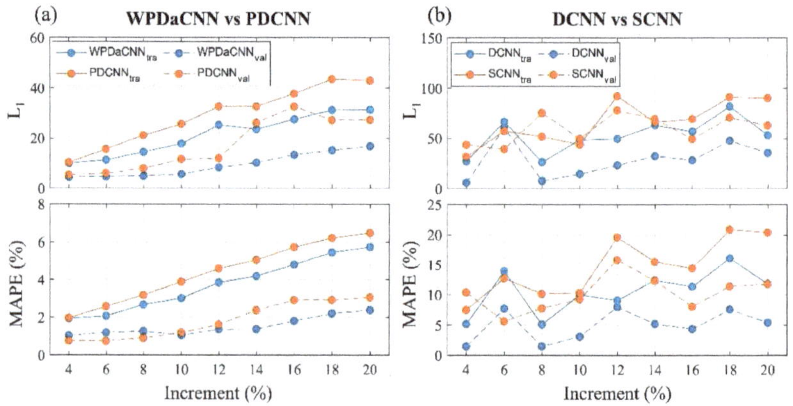

**Figure 5.** Performance of pyramid models trained under different dictionary increments. The solid line is the model prediction by training data, and the dashed line is that of the testing data. (**a**) $L_1$ and MAPE of the WPDaCNN and PDCNN models. (**b**) $L_1$ and MAPE of the DCNN and SCNN models.

### 3.3. MRF Parametric Maps Reconstruction by the Two-Stage Model

The dictionary matching using the inner product by the CPU required 1.5 min to reconstruct a slice, and the previous model with the same MRF protocol as this study by the CPU required 0.08 s [13]. The time required for the GPU with a two-stage model to reconstruct a slice was 0.02 s. Tables 2 and 3 present the statistical analysis of T1 and T2* values from 32 subjects by standard dictionary matching and that by the proposed two-stage model. Results for various tissue types were obtained by applying the corresponding tissue mask derived from the automatic and manual segmentation. In both validation and testing datasets, all ICCs were higher than 0.94 in T1 and T2* relaxation times for all tissues. The MAPE decreased by approximately a factor of two after denoising for all tissue types. In GM, WM, and MS lesions, the MAPE was less than 3.2% for T1 and 2.8% for T2* with denoising. CSF had a much higher MAPE compared with other tissue types. The overall MAPE with the denoising for the whole brain was approximately 6% and 4% for the T1 and

T2* values, respectively. Most of the overall increase in error was contributed by CSF. The previous model based on U-Net and scan data for training was MAPE of five to ten [13].

**Table 2.** Statistical analysis of T1 relaxation times between dictionary matching and the proposed two-stage model for each tissue of scanned data.

| | Standard (ms) | Predicted (ms) | ICC | MAPE (%) | Mean (ms) | Difference (ms) | R | *p* Value |
|---|---|---|---|---|---|---|---|---|
| **Validation (7 Healthy and 9 with MS)** | | | | | | | | |
| WB | 1483 ± 147 | 1461 ± 146 | 1.00 | 5.9 (12.6) | 1472 ± 145 | 22.9 ± 5.5 | 0.53 | <0.05 * |
| GM | 1278 ± 42 | 1269 ± 42 | 1.00 | 3.0 (7.1) | 1273 ± 42 | 9.2 ± 2.1 | −0.22 | 0.41 |
| WM | 803 ± 35 | 794 ± 36 | 1.00 | 3.0 (6.8) | 799 ± 36 | 8.8 ± 2.1 | −0.45 | 0.08 |
| CSF | 2993 ± 233 | 2894 ± 229 | 1.00 | 7.4 (15.1) | 2943 ± 231 | 99.5 ± 21.5 | 0.17 | 0.54 |
| MSL | 1194 ± 208 | 1192 ± 207 | 1.00 | 1.7 (3.9) | 1193 ± 207 | 1.8 ± 3.4 | 0.28 | 0.46 |
| **Testing (7 Healthy and 9 with MS)** | | | | | | | | |
| WB | 1521 ± 153 | 1496 ± 151 | 1.00 | 6.2 (12.9) | 1509 ± 152 | 25.9 ± 8.4 | 0.27 | 0.31 |
| GM | 1286 ± 42 | 1276 ± 43 | 1.00 | 3.1 (7.2) | 1281 ± 42 | 10.6 ± 4.2 | −0.16 | 0.57 |
| WM | 825 ± 51 | 816 ± 51 | 1.00 | 3.2 (7.0) | 820 ± 51 | 9.1 ± 2.5 | −0.25 | 0.35 |
| CSF | 3003 ± 224 | 2897 ± 230 | 1.00 | 7.4 (14.5) | 2950 ± 226 | 106.3 ± 29.8 | −0.22 | 0.41 |
| MSL | 1284 ± 152 | 1279 ± 151 | 1.00 | 1.9 (4.2) | 1282 ± 152 | 5.2 ± 4.4 | 0.10 | 0.81 |

WB = whole brain; GM = gray matter; WM = white matter; CSF = cerebrospinal fluid; MSL = lesion of multiple sclerosis; ICC = intraclass correlation coefficient; MAPE = mean absolute percentage error. "*" indicates that the *p* value is less than 0.05. The values in parentheses in the MAPE column are the results without noise removal. The difference is from the pairwise pixel-value difference.

**Table 3.** Statistical analysis of T2* relaxation times between dictionary matching and the proposed two-stage model for each tissue of scanned data.

| | Standard (ms) | Predicted (ms) | ICC | MAPE (%) | Mean (ms) | Difference (ms) | R | *p* Value |
|---|---|---|---|---|---|---|---|---|
| **Validation (7 Healthy and 9 with MS)** | | | | | | | | |
| WB | 95 ± 26 | 83 ± 18 | 0.97 | 4.2 (11.3) | 89 ± 22 | 11.9 ± 7.8 | 0.96 | <0.001 * |
| GM | 53 ± 3 | 53 ± 2 | 0.94 | 2.6 (6.0) | 53 ± 2 | 0.3 ± 1.2 | 0.49 | 0.05 |
| WM | 53 ± 2 | 53 ± 2 | 0.98 | 2.2 (5.8) | 53 ± 2 | 0.1 ± 0.5 | 0.35 | 0.18 |
| CSF | 245 ± 90 | 189 ± 61 | 0.96 | 9.3 (20.1) | 217 ± 75 | 56.3 ± 29.5 | 0.96 | <0.001 * |
| MSL | 78 ± 7 | 79 ± 7 | 1.00 | 2.0 (4.8) | 79 ± 7 | −0.3 ± 0.4 | −0.43 | 0.25 |
| **Testing (7 Healthy and 9 with MS)** | | | | | | | | |
| WB | 104 ± 35 | 89 ± 24 | 0.97 | 4.6 (10.5) | 96 ± 30 | 14.5 ± 11.0 | 0.96 | <0.001 * |
| GM | 53 ± 2 | 53 ± 2 | 1.00 | 2.6 (5.6) | 53 ± 2 | −0.0 ± 0.1 | 0.40 | 0.12 |
| WM | 54 ± 2 | 54 ± 2 | 1.00 | 2.3 (5.2) | 54 ± 2 | −0.1 ± 0.1 | −0.21 | 0.43 |
| CSF | 268 ± 101 | 204 ± 70 | 0.96 | 10.2 (18.9) | 236 ± 85 | 63.8 ± 34.0 | 0.93 | <0.001 * |
| MSL | 86 ± 14 | 87 ± 15 | 1.00 | 2.8 (6.5) | 87 ± 14 | −0.7 ± 0.9 | −0.71 | <0.05 * |

WB = whole brain; GM = gray matter; WM = white matter; CSF = cerebrospinal fluid; MSL = lesion of multiple sclerosis; ICC = intraclass correlation coefficient; MAPE = mean absolute percentage error. "*" indicates that the *p* value is less than 0.05. The values in parentheses in the MAPE column are the results without noise removal. The difference is from the pairwise pixel-value difference.

Figure 6 displays a Bland–Altman plot for all subjects of dictionary matching and the two-stage model. The fifth and sixth columns of Tables 2 and 3 lists the mean and difference (with standard deviations) between them. A significant positive correlation was observed for the whole brain for T1 and T2* in the validation dataset and T2* in the testing dataset. The significant positive correlation also appeared in the CSF for T2* for both validation and testing datasets. A significant negative correlation was observed for the MS lesion for T2* in the testing dataset. In both validation and testing datasets, the mean difference was less than or equal to 10 ms for GM and WM, and 5 ms for the MS lesion, for T1. The mean difference was less than or equal to 0.7 ms for GM, WM, and the MS lesion for T2* in both validation and testing datasets. Figure 7 depicts a single slice from an MS patient for the tissue masks, FLAIR, standard and predicted maps for T1 and T2*, and their corresponding difference maps. The standard maps were obtained by dictionary matching using the inner-product method, and the predicted maps were gathered by the proposed two-stage model. The MAPE for GM, WM, and MS lesions was low. The MAPE was higher for the CSF region compared with other tissue types in the difference map, especially for T2*.

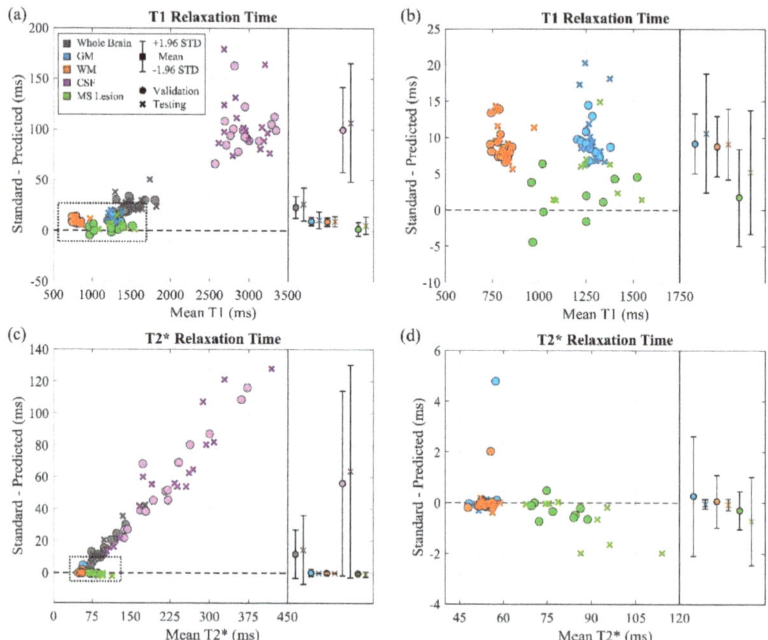

**Figure 6.** Bland-Altman plot for mean T1 and T2* relaxation times of 32 subjects (18 with MS lesions) for the whole brain and different tissues. (**a,c**) Plot for all tissues. (**b,d**) Cropped views of GM, WM, and MS lesion shown in (**a**) and (**c**).

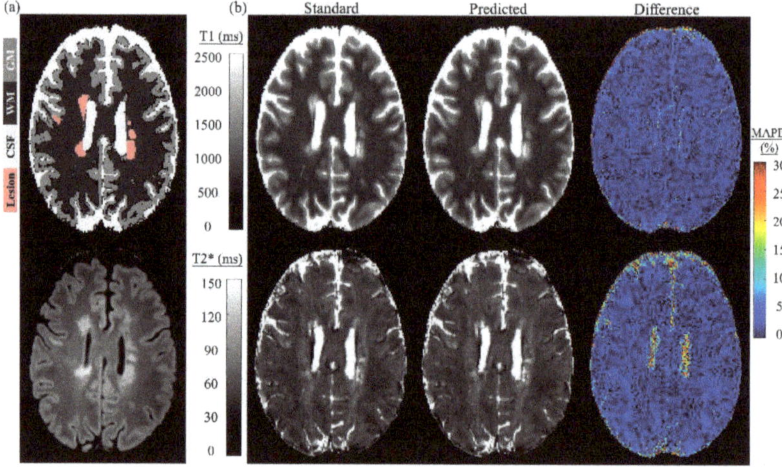

**Figure 7.** Magnetic resonance fingerprinting parametric maps of a single slice in an MS patient matched by the simulated dictionary (standard) and predicted by the proposed model. (**a**) Top is the tissue masks; bottom is the FLAIR. (**b**) Standard maps by dictionary matching, predicted maps by the proposed two-stage model, and difference maps between them.

## 4. Discussion

Herein, we propose a two-stage model for predicting parametric maps of MRF-EPI. The prediction results achieved a MAPE of equal to or less than 3% from the standard dictionary matching for GM, WM, and MS lesions. In our approach, the first stage used MRF signal denoising, and the second stage used regression of the simulated signal by the Bloch equation [18]. The model's prediction error with denoising was approximately one-half that without denoising, and in this, we demonstrated the importance of removing the noise. Furthermore, the pyramid model with self-attention learned well on the simulated signal and achieved MAPE of approximately 2% and 1% for the training and validation datasets, respectively, for the dictionary with the densest increment. Our proposed model accurately reconstructed parametric maps of MRF-EPI and can therefore replace the computationally expensive inner-product dictionary matching method.

Noise is an unavoidable problem when MRI is conducted using fast imaging techniques, and acquisition speed and SNR are perennial tradeoffs. Several approaches have been proposed for MRI denoising [27]. In general, denoising techniques are based on specific assumptions to model prior properties, such as inherent pattern redundancy and sparsity. The disadvantages of such modeling are that obtaining high performance is computationally expensive and that several manual parameters must be selected [17]. Unlike prior-based approaches, deep learning–based DnCNN is both effective and time efficient. A previous study demonstrated that by filtering the MRF baseline images, the image quality improved for parametric maps [24]. We also performed noise reduction on the MRF baseline image, but we did so on the signal evolution of each pixel instead of on 2D images. SNRs for both the training and testing datasets increased by nearly twice the original SNRs after denoising. No significant difference was observed between the training and testing datasets before and after denoising in our experiments. This result demonstrated that the DnCNN performed well in handling MRF signals with noise for both training and testing datasets. In addition, we observed a decrease in the SNR on the cranial and caudal sides, which conforms with observations in previous studies [28,29].

In learning simulated signals with different increments, the error in model prediction on both training and validation datasets increased as the increment increased. We observed that the PDCNN and WPDaCNN had fewer errors and a smoother error trend than did the DCNN and SCNN. From this result, we observed that the model with the pyramid structure was more stable than the model without the pyramid structure. In addition, in the first division type of our experiments, the model was made to learn certain T1 and T2* ranges of simulated signals and to predict the data outside the simulated scope as validation. This approach resulted in poor prediction for the validation dataset. This result demonstrated that the model did not learn the Bloch equation simulation [18] well. The model must be made to learn all the expected ranges for the simulated signals to ensure accurate predictions. Moreover, regarding the performance of the PDCNN and WPDaCNN, the validation loss was lower than the training loss at any increment, indicating that the model accurately regressed the learned data within the range of T1 and T2* contained in the training dictionary. That is, once the model learned the dictionary with a 4% increment, it was able to regress the T1 and T2* of the dictionary with a 2% increment well. Deep learning models perform better with a single output compared with multiple outputs [13]. In our experimental results, the overall performance of the DCNN was superior to that of the SCNN, which demonstrated that the dual-path for outputting T1 and T2* was beneficial in improving the model performance.

To address the problem of noise, previous studies have used Gaussian noise to test the performance of their model [8,24], but actual MRI noise distributions are non-Gaussian [30,31]. Thus, we created a noise dataset on the basis of the difference between simulated and scanned data and randomly sampled the data from this dataset to train our deep learning model. Furthermore, models trained by $L_1$ loss were reported to perform more favorably in MRF image reconstruction compared with other loss functions [13]. Hence, we concurrently used $L_1$ loss and MAPE loss to avoid the model's overfitting to either T1 or T2* values

(Equation (8)). In addition, studies have demonstrated the power of CNNs and the ability of RNNs to outperform CNNs in MRF image reconstruction [9,10]. For natural language processing, Transformer, which relies entirely on the self-attention mechanism, has been proposed as having a lower computational cost and more advanced performance than RNNs [20]. We combined a CNN with the self-attention mechanism to make the model learn the correlations among features captured by the CNN. Furthermore, the performance of a two-stage model is superior to that of a single-stage one for object detection but at the expense of computational speed [32]. Our results also indicated that the two-stage model with noise reduction outperformed the one-stage model without noise reduction. Moreover, the computation time of the model in the GPU (0.02 s) for predicting a single slice was 4500-fold faster than that of the commonly used inner-product matching in the CPU (90 s). Finally, although correlations were observed in the Bland–Altman analysis, the MAPE for clinically interesting tissues (GM, WM, and MS lesions) was less than or equal to 3%, and the mean T1 and T2* values of these tissues are consistent with those in previous studies [1,33–36].

Clinical MRI relies on qualitative imaging, which can require one hour to obtain multiple contrast weightings. Prolonged scanning is a burden for patients who cannot recline for long periods and may record motion artifacts because of patient movement. Additionally, qualitative imaging can be affected by the scanner and imaging parameters used, which hinders disease follow-up. By contrast, MRF quantitative imaging can generate multiple relaxation time maps in only a few minutes of scanning time. MRF has been demonstrated to have high repeatability and reproducibility [37,38]. MRF is a favorable approach to obtaining quantitative MR relaxation measurements. In addition, quantitative MR relaxometry can synthesize conventional contrast weightings [2,3], which can be useful for adherence to current clinical diagnostic standards. Furthermore, quantitative MRI relaxometry–based tissue segmentation was reported to have favorable repeatability [39] and can be beneficial in clinical settings for tracking the time course of a disease. With improvements addressing the drawback of the long reconstruction time of MRF, this approach is expected to replace the conventional weighted imaging currently used in clinical practice. In this study, we propose a two-stage model that is able to learn the simulated dictionary with dense increment and more quickly than dictionary matching. Our model can accelerate MRF reconstruction and thus increase the feasibility of MRF for clinical applications.

This study has some limitations. First, the gold standard we applied to evaluate the accuracy of our models was the use of the parametric maps by dictionary matching, and no other reference quantitative method was used. However, the quantification accuracy of MRF-EPI by dictionary matching was validated with a phantom and had good agreement [22,23]. Second, the prediction time we reported in the Results section was for only one slice, and approximately 30 s were required to compute 60 slices consecutively. This result was due to the continuous GPU computing also involving memory usage and data transfer time. Finally, because of the design of MRF-EPI, the simulated dictionary differed by slice. Therefore, we trained a total of 60 models corresponding to each slice, and this required training time and space to store the trained weights for the model. Approximately nine days were required to train the denoising model for stage I, and 18 h to train different pyramid models for comparison. Regarding the final two-stage model, an excessive epoch number led to poor model prediction for the scanned data because of overfitting. Therefore, we used a relatively small number of training sessions (25 epochs), and approximately two days were required for model training. Regarding storage space, space requirements were smaller compared with those for the simulated dictionary (15 megabytes for the model weights and 203 megabytes for the dictionary of each slice).

In this study, we proposed a two-stage model. The MRF signal noise reduction was for the first stage, and the T1 and T2* value prediction was for the second stage. The results showed that noise removal was very beneficial for predicting the T1 and T2* values. Compared with other studies, we used real noise and the simulation dictionary to train the model to ensure generalizability. Our proposed model was designed using a 1D architecture, which required model training for each slice. If the model is designed in 3D, a single model will be able to cover the whole brain. However, compared with the multi-model approach, the single model has fewer parameters for learning, and it is conceivable that the noise reduction performance may be worse. We used the denoising CNN proposed by Zhang et al. in 2017 [17]. Other advanced denoising deep learning models, such as a denoising autoencoder [40], are available and can be used in MRF studies in the future to improve the model performance in noise reduction. Besides, MRF using EPI fast imaging is sensitive to magnetic field inhomogeneity and can have distortion artifacts at the air-tissue interface. A common approach for distortion correction is image registration [41]. In addition, MRI image analysis often requires the segmentation of tissues such as GM, WM, CSF, and lesion to observe the correlated volumetric changes [42]. Deep learning is well established in image registration and segmentation, such as VoxelMorph [43], which used the spatial transformer function, and U-Net [44], a well-known architecture commonly used for medical image segmentation. In the future, a multi-task deep learning model for MRF can be added to specifically handle the image denoising, registration, and segmentation tasks to achieve a one-stop efficient MRF image reconstruction and enhance the value of MRF in clinical applications.

## 5. Conclusions

In conclusion, we effectively removed the noise from MRF-EPI in a 1D manner and thus improved the performance of a deep learning model in the regression task for MRF parametric map reconstruction. The proposed model achieved a prediction error equal to or less than 3% in the T1 and T2* map for tissues of clinical interest, such as GM, WM, and MS lesions. Compared with the 1.5 min required for the CPU computation using the inner-product method, the proposed model can achieve a computation speed of 0.02 s for a slice in the GPU. Our proposed two-stage model, trained with dense-increment simulated dictionaries, can accelerate image reconstruction and reduce the space required by dictionaries, thus improving imaging efficiency. Future research can target deep learning models that incorporate image processing, such as image registration and segmentation, to overcome the distortion and measure the brain volumetry for facilitating MRF in clinical applications.

**Author Contributions:** Conceptualization, J.-S.H., I.H. and F.G.Z.; methodology, J.-S.H. and I.H.; software, J.-S.H., I.H. and W.-K.L.; validation, W.-K.L., Y.-L.C. and Y.C.; formal analysis, J.-S.H.; investigation, J.-S.H. and I.H.; resources, I.H., F.G.Z. and L.R.S.; data curation, I.H., F.G.Z. and L.R.S.; writing—original draft preparation, J.-S.H.; writing—review and editing, J.-S.H., I.H., F.G.Z. and Y.-T.W.; visualization, J.-S.H.; supervision, S.-J.W. and Y.-T.W.; project administration, S.-J.W. and Y.-T.W.; funding acquisition, S.-J.W. and Y.-T.W. All authors have read and agreed to the published version of the manuscript.

**Funding:** This research and APC was funded by Taiwan's Ministry of Science and Technology, grant number MOST-110-2321-B-010-005 and MOST-110-2221-E-A49A-504-MY3.

**Institutional Review Board Statement:** The study was conducted according to the guidelines of the Declaration of Helsinki, and approved by the Institutional Review Board of Medical Faculty Mannheim, Heidelberg University (2019-711N).

**Informed Consent Statement:** Informed consent was obtained from all subjects involved in the study.

**Data Availability Statement:** The data presented in this study is not publicly available due to patient privacy concerns; publication would not be covered by the IRB statement.

**Acknowledgments:** The authors acknowledge Wallace Academic Editors for editing this manuscript.

**Conflicts of Interest:** The authors declare no conflict of interest. The funders had no role in the design of the study; in the collection, analyses, or interpretation of data; in the writing of the manuscript, or in the decision to publish the results.

## References


1. Cheng, H.L.; Stikov, N.; Ghugre, N.R.; Wright, G.A. Practical Medical Applications of Quantitative MR Relaxometry. *J. Magn. Reson. Imaging* **2012**, *36*, 805–824. [CrossRef]
2. Feng, L.; Ma, D.; Liu, F. Rapid MR Relaxometry Using Deep Learning: An Overview of Current Techniques and Emerging Trends. *NMR Biomed.* **2020**, e4416. [CrossRef]
3. Ji, S.; Yang, D.; Lee, J.; Choi, S.H.; Kim, H.; Kang, K.M. Synthetic MRI: Technologies and Applications in Neuroradiology. *J. Magn. Reson. Imaging* **2020**. [CrossRef]
4. Ma, D.; Gulani, V.; Seiberlich, N.; Liu, K.; Sunshine, J.L.; Duerk, J.L.; Griswold, M.A. Magnetic Resonance Fingerprinting. *Nature* **2013**, *495*, 187–192. [CrossRef]
5. McGivney, D.F.; Boyacıoğlu, R.; Jiang, Y.; Poorman, M.E.; Seiberlich, N.; Gulani, V.; Keenan, K.E.; Griswold, M.A.; Ma, D. Magnetic Resonance Fingerprinting Review Part 2: Technique and Directions. *J. Magn. Reson. Imaging* **2020**, *51*, 993–1007. [CrossRef]
6. McGivney, D.F.; Pierre, E.; Ma, D.; Jiang, Y.; Saybasili, H.; Gulani, V.; Griswold, M.A. SVD Compression for Magnetic Resonance Fingerprinting in the Time Domain. *IEEE Trans Med. Imaging* **2014**, *33*, 2311–2322. [CrossRef]
7. Yang, M.; Ma, D.; Jiang, Y.; Hamilton, J.; Seiberlich, N.; Griswold, M.A.; McGivney, D. Low Rank Approximation Methods for MR Fingerprinting with Large Scale Dictionaries. *Magn. Reson. Med.* **2018**, *79*, 2392–2400. [CrossRef]
8. Cohen, O.; Zhu, B.; Rosen, M.S. MR Fingerprinting Deep RecOnstruction NEtwork (DRONE). *Magn. Reson. Med.* **2018**, *80*, 885–894. [CrossRef]
9. Hoppe, E.; Thamm, F.; Korzdorfer, G.; Syben, C.; Schirrmacher, F.; Nittka, M.; Pfeuffer, J.; Meyer, H.; Maier, A. Magnetic Resonance Fingerprinting Reconstruction Using Recurrent Neural Networks. *Stud. Health Technol.* **2019**, *267*, 126–133. [CrossRef]
10. Hoppe, E.; Körzdörfer, G.; Würfl, T.; Wetzl, J.; Lugauer, F.; Pfeuffer, J.; Maier, A. Deep Learning for Magnetic Resonance Fingerprinting: A New Approach for Predicting Quantitative Parameter Values from Time Series. *Ger. Med. Data Sci. Vis. Bridges* **2017**, *243*, 202–206. [CrossRef]
11. Balsiger, F.; Jungo, A.; Scheidegger, O.; Carlier, P.G.; Reyes, M.; Marty, B. Spatially Regularized Parametric Map Reconstruction for Fast Magnetic Resonance Fingerprinting. *Med. Image Anal.* **2020**, *64*, 101741. [CrossRef]
12. Fang, Z.; Chen, Y.; Liu, M.; Xiang, L.; Zhang, Q.; Wang, Q.; Lin, W.; Shen, D. Deep Learning for Fast and Spatially Constrained Tissue Quantification From Highly Accelerated Data in Magnetic Resonance Fingerprinting. *IEEE Trans. Med. Imaging* **2019**, *38*, 2364–2374. [CrossRef]
13. Hermann, I.; Martinez-Heras, E.; Rieger, B.; Schmidt, R.; Golla, A.K.; Hong, J.S.; Lee, W.K.; Yu-Te, W.; Nagtegaal, M.; Solana, E.; et al. Accelerated White Matter Lesion Analysis Based on Simultaneous T1 and T2* Quantification Using Magnetic Resonance Fingerprinting and Deep Learning. *Magn. Reson. Med.* **2021**, *86*, 471–486. [CrossRef]
14. Yang, M.; Jiang, Y.; Ma, D.; Mehta, B.B.; Griswold, M.A. Game of Learning Bloch Equation Simulations for MR Fingerprinting. *arXiv* **2020**, arXiv:2004.02270.
15. Chen, D.; Golbabaee, M.; Gomez, P.A.; Menzel, M.I.; Davies, M.E. A Fully Convolutional Network for MR Fingerprinting. *arXiv* **2019**, arXiv:1911.09846.
16. Li, G.; Zrimec, J.; Ji, B.; Geng, J.; Larsbrink, J.; Zelezniak, A.; Nielsen, J.; Engqvist, M.K. Performance of Regression Models as a Function of Experiment Noise. *Bioinform. Biol. Insights* **2021**, *15*, 1–10. [CrossRef]
17. Zhang, K.; Zuo, W.; Chen, Y.; Meng, D.; Zhang, L. Beyond a Gaussian Denoiser: Residual Learning of Deep CNN for Image Denoising. *IEEE Trans Image Process* **2017**, *26*, 3142–3155. [CrossRef]
18. Bloch, F. Nuclear Induction. *Phys. Rev.* **1946**, *70*, 460. [CrossRef]
19. Lin, T.-Y.; Dollár, P.; Girshick, R.; He, K.; Hariharan, B.; Belongie, S. Feature Pyramid Networks for Object Detection. In Proceedings of the IEEE Computer Society, London, UK, 1 July 2017; pp. 936–944.
20. Vaswani, A.; Shazeer, N.; Parmar, N.; Uszkoreit, J.; Jones, L.; Gomez, A.N.; Kaiser, Ł.; Polosukhin, I. Attention Is All You Need. In Proceedings of the 31st International Conference on Neural Information Processing Systems, Long Beach, CA, USA, 4–9 December 2017; Curran Associates Inc.: Red Hook, NY, USA, 2017; pp. 6000–6010.
21. Wang, X.; Girshick, R.; Gupta, A.; He, K. Non-Local Neural Networks. In Proceedings of the 2018 IEEE/CVF Conference on Computer Vision and Pattern Recognition, Salt Lake City, UT, USA, 18–23 June 2018; pp. 7794–7803.
22. Rieger, B.; Zimmer, F.; Zapp, J.; Weingartner, S.; Schad, L.R. Magnetic Resonance Fingerprinting Using Echo-Planar Imaging: Joint Quantification of T1 and T2* Relaxation Times. *Magn. Reson. Med.* **2017**, *78*, 1724–1733. [CrossRef]
23. Rieger, B.; Akcakaya, M.; Pariente, J.C.; Llufriu, S.; Martinez-Heras, E.; Weingartner, S.; Schad, L.R. Time Efficient Whole-Brain Coverage with MR Fingerprinting Using Slice-Interleaved Echo-Planar-Imaging. *Sci. Rep.* **2018**, *8*, 6667. [CrossRef]
24. Hermann, I.; Chacon-Caldera, J.; Brumer, I.; Rieger, B.; Weingartner, S.; Schad, L.R.; Zollner, F.G. Magnetic Resonance Fingerprinting for Simultaneous Renal T1 and T2* Mapping in a Single Breath-Hold. *Magn. Reson. Med.* **2020**, *83*, 1940–1948. [CrossRef]
25. Ashburner, J.; Barnes, G.; Chen, C.-C.; Daunizeau, J.; Flandin, G.; Friston, K.; Kiebel, S.; Kilner, J.; Litvak, V.; Moran, R. *SPM12 Manual*; Wellcome Trust Centre for Neuroimaging: London, UK, 2014; Volume 2464.
26. Zhang, H.; Goodfellow, I.; Metaxas, D.; Odena, A. Self-Attention Generative Adversarial Networks. *arXiv* **2019**, arXiv:1805.08318.
27. Mohan, J.; Krishnaveni, V.; Guo, Y. A Survey on the Magnetic Resonance Image Denoising Methods. *Biomed. Signal Process. Control* **2014**, *9*, 56–69. [CrossRef]

28. Larsson, E.-M.; Nilsson, H.; Holtås, S.; Ståhlberg, F. Coil Selection for Magnetic Resonance Imaging of the Cervical and Thoracic Spine Using a Vertical Magnetic Field. *Acta Radiol.* **1989**, *30*, 141–146. [CrossRef]
29. Reiss-Zimmermann, M.; Gutberlet, M.; Köstler, H.; Fritzsch, D.; Hoffmann, K.-T. Improvement of SNR and Acquisition Acceleration Using a 32-Channel Head Coil Compared to a 12-Channel Head Coil at 3T. *Acta Radiol.* **2013**, *54*, 702–708. [CrossRef]
30. Gudbjartsson, H.; Patz, S. The Rician Distribution of Noisy MRI Data. *Magn. Reson. Med.* **1995**, *34*, 910–914. [CrossRef]
31. Nowak, R.D. Wavelet-Based Rician Noise Removal for Magnetic Resonance Imaging. *IEEE Trans. Image Processing* **1999**, *8*, 1408–1419. [CrossRef]
32. Soviany, P.; Ionescu, R.T. Optimizing the Trade-Off between Single-Stage and Two-Stage Deep Object Detectors Using Image Difficulty Prediction. In Proceedings of the 2018 20th International Symposium on Symbolic and Numeric Algorithms for Scientific Computing (SYNASC), Timisoara, Romania, 20–23 September 2018; pp. 209–214.
33. Blystad, I.; Håkansson, I.; Tisell, A.; Ernerudh, J.; Smedby, Ö.; Lundberg, P.; Larsson, E.-M. Quantitative MRI for Analysis of Active Multiple Sclerosis Lesions without Gadolinium-Based Contrast Agent. *Am. J. Neuroradiol.* **2016**, *37*, 94–100. [CrossRef]
34. Krüger, G.; Glover, G.H. Physiological Noise in Oxygenation-Sensitive Magnetic Resonance Imaging. *Magn. Reson. Med.* **2001**, *46*, 631–637. [CrossRef]
35. Péran, P.; Hagberg, G.; Luccichenti, G.; Cherubini, A.; Brainovich, V.; Celsis, P.; Caltagirone, C.; Sabatini, U. Voxel-Based Analysis of R2* Maps in the Healthy Human Brain. *J. Magn. Reson. Imaging* **2007**, *26*, 1413–1420. [CrossRef]
36. Wansapura, J.P.; Holland, S.K.; Dunn, R.S.; Ball, W.S. NMR Relaxation Times in the Human Brain at 3.0 Tesla. *J. Magn. Reson. Imaging* **1999**, *9*, 531–538. [CrossRef]
37. Jiang, Y.; Ma, D.; Keenan, K.E.; Stupic, K.F.; Gulani, V.; Griswold, M.A. Repeatability of Magnetic Resonance Fingerprinting T1 and T2 Estimates Assessed Using the ISMRM/NIST MRI System Phantom. *Magn. Reson. Med.* **2017**, *78*, 1452–1457. [CrossRef]
38. Körzdörfer, G.; Kirsch, R.; Liu, K.; Pfeuffer, J.; Hensel, B.; Jiang, Y.; Ma, D.; Gratz, M.; Bär, P.; Bogner, W.; et al. Reproducibility and Repeatability of MR Fingerprinting Relaxometry in the Human Brain. *Radiology* **2019**, *292*, 429–437. [CrossRef] [PubMed]
39. Andica, C.; Hagiwara, A.; Hori, M.; Nakazawa, M.; Goto, M.; Koshino, S.; Kamagata, K.; Kumamaru, K.K.; Aoki, S. Automated Brain Tissue and Myelin Volumetry Based on Quantitative MR Imaging with Various In-Plane Resolutions. *J. Neuroradiol.* **2018**, *45*, 164–168. [CrossRef] [PubMed]
40. Lee, W.-H.; Ozger, M.; Challita, U.; Sung, K.W. Noise Learning Based Denoising Autoencoder. *IEEE Commun. Lett.* **2021**, *25*, 2983–2987. [CrossRef]
41. Wang, S.; Peterson, D.J.; Gatenby, J.C.; Li, W.; Grabowski, T.J.; Madhyastha, T.M. Evaluation of Field Map and Nonlinear Registration Methods for Correction of Susceptibility Artifacts in Diffusion MRI. *Front. Neuroinformatics* **2017**, *11*, 17. [CrossRef] [PubMed]
42. Giorgio, A.; De Stefano, N. Clinical Use of Brain Volumetry. *J. Magn. Reson. Imaging* **2013**, *37*, 1–14. [CrossRef]
43. Balakrishnan, G.; Zhao, A.; Sabuncu, M.R.; Guttag, J.; Dalca, A.V. VoxelMorph: A Learning Framework for Deformable Medical Image Registration. *IEEE Trans. Med. Imaging* **2019**, *38*, 1788–1800. [CrossRef]
44. Ronneberger, O.; Fischer, P.; Brox, T. U-Net: Convolutional Networks for Biomedical Image Segmentation. *arXiv* **2015**, arXiv:1505.04597.

*Article*

# Learning a Metric for Multimodal Medical Image Registration without Supervision Based on Cycle Constraints

Hanna Siebert *, Lasse Hansen and Mattias P. Heinrich

Institute of Medical Informatics, Universität zu Lübeck, 23538 Lübeck, Germany;
hansen@imi.uni-luebeck.de (L.H.); heinrich@imi.uni-luebeck.de (M.P.H.)
* Correspondence: siebert@imi.uni-luebeck.de

**Abstract:** Deep learning based medical image registration remains very difficult and often fails to improve over its classical counterparts where comprehensive supervision is not available, in particular for large transformations—including rigid alignment. The use of unsupervised, metric-based registration networks has become popular, but so far no universally applicable similarity metric is available for multimodal medical registration, requiring a trade-off between local contrast-invariant edge features or more global statistical metrics. In this work, we aim to improve over the use of handcrafted metric-based losses. We propose to use synthetic three-way (triangular) cycles that for each pair of images comprise two multimodal transformations to be estimated and one known synthetic monomodal transform. Additionally, we present a robust method for estimating large rigid transformations that is differentiable in end-to-end learning. By minimising the cycle discrepancy and adapting the synthetic transformation to be close to the real geometric difference of the image pairs during training, we successfully tackle intra-patient abdominal CT-MRI registration and reach performance on par with state-of-the-art metric-supervision and classic methods. Cyclic constraints enable the learning of cross-modality features that excel at accurate anatomical alignment of abdominal CT and MRI scans.

**Keywords:** image registration; cycle constraint; multimodal features; self-supervision; rigid alignment

**Citation:** Siebert, H.; Hansen, L.; Heinrich, M.P. Learning a Metric for Multimodal Medical Image Registration without Supervision Based on Cycle Constraints. *Sensors* **2022**, *22*, 1107. https://doi.org/10.3390/s22031107

Academic Editors: Vahid Abolghasemi, Hossein Anisi and Saideh Ferdowsi

Received: 28 December 2021
Accepted: 27 January 2022
Published: 1 February 2022

**Publisher's Note:** MDPI stays neutral with regard to jurisdictional claims in published maps and institutional affiliations.

## 1. Introduction

Medical image registration based on deep learning methods has gathered great interest over the last few years. Yet, certain challenges, especially in multimodal registration, need to be addressed for learning based approaches, as evident from the recent MICCAI challenge *Learn2Reg* [1]. In order to avoid an elaborate comprehensive annotation of all relevant anatomies and to avoid label bias, unsupervised, metric-based registration networks are widely used for intramodal deep learning based registration [2,3].

However, this poses an additional challenge for multimodal registration problems, as currently no universal metric has been developed and a trade-off has to be made between using local contrast-invariant edge features such as NGF, LCC, and MIND or more global statistical metrics like mutual information. Metric-based methods also entail the difficulty of tuning hyperparameters that balance similarity weights (ensuring similarity between fixed image and warped moving image) and regularisation weights (ensuring plausible deformations).

Ground truth deformations for direct supervision are only available when using synthetic deformation fields. The now very popular FlowNet [4] estimates deformation fields between pairs of input images from a synthetically generated dataset that has been obtained by applying affine transformations to images. However, for medical applications, synthetic deformations have been deployed for monomodal image registration [5–7]. Alternatively, label supervision that primarily maximises the alignment of known structures with expert annotations could be employed [2,8,9]. This leads to improved registration of anatomies that are well represented, but can introduce a bias and deteriorating performance for unseen labels.

On the one hand, the focus of supervised approaches on a limited set of labelled structures may be particularly inadequate for diagnosis of a pathology that cannot be represented sufficiently in the training data. Using metric supervision, on the other hand, has little potential to improve upon classical algorithms that employ the same metric as similarity terms during optimisation. With efficient (parallelised) implementations, adequate runtimes of less than a minute have recently been achieved for classical algorithms.

Learning completely without metric or label supervision, self-supervision, would remedy the aforementioned problems and enable the development of completely new registration methods and multimodal feature descriptors without introducing annotation or engineering biases.

Self-supervision approaches have been used in medical and non-medical learning based image processing tasks. Recently, a self-supervised approach for learning pretext-invariant representations for object detection has outperformed supervised pre-training in [10]. By minimising a contrastive loss function, the authors construct image representations that are invariant to image patch pertubation, similar to the representation of transformed versions of the same image and differ from representations of other images. In [11], semantic features have been learned with self-supervision in order to recognise the rotation that has been applied to an image given four possible transformations as multiples of 90 degrees. The learned features have been useful for various visual perception tasks. For rigid registration between point clouds, an iterative self-supervised method has been proposed in [12]. Here, partial-to-partial registration problems have been addressed by learning geometric priors directly from data. The method comprises a keypoint detection module which identifies points that match in the input point clouds based on co-contextual information and aligns common keypoints. For monomodal medical image registration, in [13] spatial transformations between image pairs have been estimated in a self-supervised learning procedure. Therefore, an image-wise similarity metric between fixed and warped moving images is maximised in a multi-resolution framework while the deformation fields are regularised for smoothness.

In [14], cycle-consistency in time is used for learning visual correspondence from unlabelled video data for self-supervision. Their idea is to obtain supervision for correspondence by tracking backward and then forward, i.e., along a cycle in time, and use the inconsistency between the start and end points as the loss function. For image-to-image translation, a cycle-consistent adversarial network approach is introduced in [15]. The authors use a cycle consistency loss that induces the assumption that forward and backward translation should be bijective and inverse of each other. Another approach that addresses inconsistency is introduced in [16] for medical image registration. It uses information from a complete set of pairwise registrations, aggregates inconsistency, and minimizes the group-wise inconsistency of all pairwise image registrations by using a regularized least-squares algorithm. The idea to measure consistency via registration cycles for monomodal medical image data has been used in [17] that estimates forward and reverse transformation jointly in a non-deep-learning approach and [18] using registration circuits to correct registration errors. In [19], a monomodal unsupervised medical image registration method that trains deep neural network for deformable registration is presented using CNNs with cycle-consistency. This approach uses two registration networks that process the two input images as fixed and moving images inversely to each other and gives the deformed volumes to the networks again to re-deform the images to impose cycle-consistency.

Previous deep learning based registration work has often omitted the step of rigid or affine registration, despite its immense challenges due to often large initial misalignments. Image registration challenges such as [1] provide data that has been pre-aligned with help of non-deep-learning-based methods, whereas the challenge's image registration tasks are then often addressed with deep learning based methods. Rigid transformation is often the inital step before performing deformable image registration, and only few works [20] investigate deep learning techniques for this step. As evident from the CuRIOUS challenge [21], so far no CNN approach was able to learn a rigid or affine mapping between multimodal scan pairs (MRI and ultrasound of neurosurgery) with an adequate robustness. Besides that, no label bias can occur with rigid alignment. Hence, a learning model for large linear transformations is of great importance.

*Contributions*

In order to avoid the difficulty of choosing a metric for multimodal image registration, we propose a completely new concept. For learning multimodal features for image registration, our learning method requires neither label supervision nor handcrafted metrics. It extends upon research that successfully learned monomodal alignment through synthetic deformations, but transforms this concept to multimodal tasks without resorting to complex modality synthesis.

The basic idea of our novel learning based approach is illustrated in Figure 1. It relies on geometric instead of metric supervision. In this work

- We introduce a cycle based approach including cycles that for each pair of CT and MRI scans comprise two multimodal transformations to be estimated and one known synthetic monomodal transformation.
- We restrict ourselves to rigid registration and aim to learn multimodal registration between CT and MRI without metric supervision by minimising the cycle discrepancy.
- We use a CNN for feature extraction with initially separate encoder blocks for each modality followed by shared weights within the last layers.
- We use a correlation layer without trainable weights and a differentiable least squares fitting procedure to find an optimal 3D rigid transformation.
- We created to the best of our knowledge the first annotated MRI/CT dataset with paired patient data that are made publicly available with manual segmentations for liver, spleen, left and right kidney.

Our extensive experimental validation on 3D rigid registration demonstrates the high accuracy that can be achieved and the simplicity of training such networks.

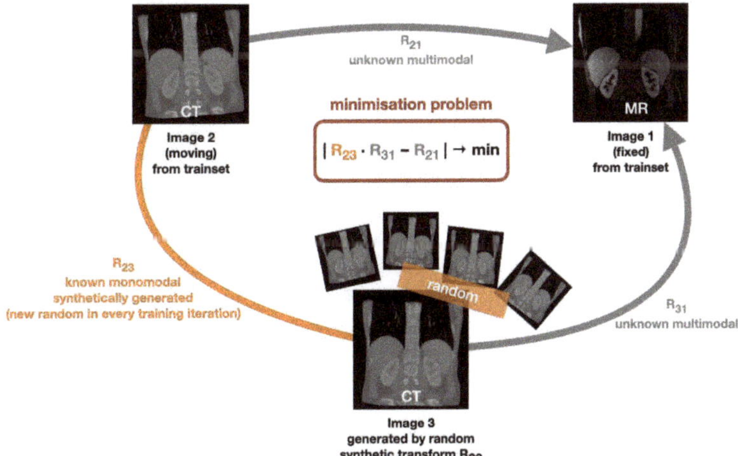


**Figure 1.** Our proposed self-supervised learning concept for multimodal image registration aiming to minimise a cycle discrepancy. In every training iteration, another (known) random transformation matrix $R_{23}$ is used to generate a synthetic image. Like this, a cycle consisting of two unknown multimodal transformations (with the transformation matrices $R_{21}$ and $R_{31}$) and a known monomodal transformation (with the transformation matrix $R_{31}$) is obtained, leading to the minimisation problem of $|R_{23} \cdot R_{31} - R_{21}| \rightarrow$ min that is used for learning.

## 2. Materials and Methods

We introduce a learning concept for multimodal image registration that learns without metric supervision. Therefore, we propose a method to learn with the help of a self-supervised learning procedure using three-way cycles. For our registration models, the architectural design consists of modules for feature extraction, correlation, and registration. Implementation details, open source code and trained models can be found at github.com/multimodallearning/learning_without_metric (accessed on 8 March 2021).

### 2.1. Self-Supervised Learning Strategy

Our deep learning based method learns multimodal registration without using metric supervision. Instead, it is based on geometric self-supervision by minimising the cycle discrepancy created through a cycle consisting of two multimodal transformation and one monomodal transformation. The basic cycle idea is illustrated in Figure 1: Initially, a fixed image ( Image 1) and a moving image ( Image 2) exist. The transformation $R_{21}$ is unknown and is to be learned by our method. In each training iteration, we randomly deform the moving image ( Image 2) by applying a known random transformation $R_{23}$ and hereby obtain a synthetic image ( Image 3). By bringing the individual transformations into a cycle, the minimisation problem of

$$|R_{23} \cdot R_{31} - R_{21}| \rightarrow \min \tag{1}$$

can be derived. We chose to minimise the discrepancy as given in Equation (1) instead of minimising the difference between the transformation combination $R_{23} \cdot R_{31} \cdot R_{12}$ and the identity transformation Id with $|R_{23} \cdot R_{31} \cdot R_{12} - \text{Id}| \rightarrow \min$ in order to avoid that our method only learns identity warping. For optimisation, we use the mean squared error loss function to minimise the cycle discrepancy between the two flow fields generated by the transformation matrices $R_{21}$ and $R_{23,31} = R_{23} \cdot R_{31}$.

As we restrict our model to rigid registration, we create the synthetic transformations $R_{23}$ by randomly initialising rigid transformation matrices with values that are assumed to be realistic from an anatomical point of view.

The advantages of our learning concept are manifold. First, in comparison to supervising the learning with a known similarity metric and regularisation term, the need for balancing a weighting term is removed and the method is applicable to new datasets without domain knowledge. Second, it enables multimodal learning, which is not feasible using synthetic deformations in conjunction with image-based loss terms (cf. [6]). Third, it avoids the use of domain discriminators as used, e.g., in the CycleGAN approach [15,22], which usually requires a large set of training scans with comparable contrast in each modality and may be sensitive to hyper-parameter choices.

On first sight, it might seem daring to use such a weak guidance. While it is clear that once suitable features are learned the loss term enables convergence, since the cycle constraint is fulfilled. Yet to initiate training towards improved features, we primarily rely on the power of randomness (by drawing multiple large synthetic deformations) and explorative learning. In addition, the architecture contains a number of stabilising elements: a patch-based correlation layer computation, outlier rejection and least squares fitting, that are described in detail below in Sections 2.3 and 2.4.

### 2.2. Training Pipeline

We apply our self-supervised learning strategy in the training procedure by going through the same steps in each training iteration as visualised in Figure 2: First, a random transformation matrix $R_{23}$ is generated and applied on the moving image in order to obtain the synthetic image. Then, moving and fixed image are passed through feature extraction, correlation layer and transformation computation module to obtain the transformation matrix $R_{21}$. The same step is also performed for fixed and synthetic image to obtain $R_{31}$. After this, $R_{23}$ and $R_{31}$ are combined to obtain $R_{23,31}$. Finally, the mean squared error of the deformations calculated with help of $R_{21}$ and $R_{23,31}$ is determined. The individual modules for this training pipeline are described in more detail in the following Sections 2.3 and 2.4.

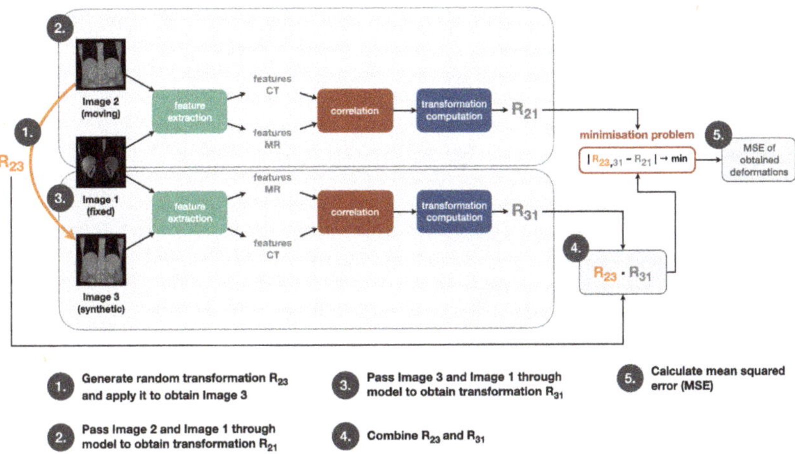


**Figure 2.** Pipeline to train our registration model: A random transformation matrix $R_{23}$ is generated and used to obtain the synthetic image. The pair of moving and fixed image as well as the pair of synthetic and fixed image are passed through feature extraction, correlation layer and transformation computation module (see following Sections 2.3 and 2.4) to obtain the transformation matrices $R_{21}$ and $R_{31}$. Then, $R_{23}$ and $R_{31}$ are combined to obtain $R_{23,31}$. As a final step, the mean squared error (MSE) of the deformations calculated with help of $R_{21}$ and $R_{23,31}$ is determined.

### 2.3. Architecture

The architecture used for our registration method comprises three main components for feature extraction, correlation, and transformation computation.

We chose to use a CNN for feature extraction with initially separate encoder blocks for each modality and shared weights within the last few layers. These features are subsequently fed into the correlation layer, which has no trainable weights and whose output could be directly converted into displacement probabilities. Our method employs a robust and differentiable least squares fitting to find an optimal 3D rigid transformation subject to outlier rejection. Figure 3 visualises the procedure for feature extraction, correlation, and computation of the rigid transformation matrix that is used for registration.

For our feature extraction CNN, we use a Y-shaped architecture (cf. Figure 3) [9] starting with a separate network part for each of the two modalities (ModalityNet), which takes the respective input and passes it through two sequences with a structure of $2\times$,

- (Strided) 3D convolution with a kernel size of three and padding of one;
- 3D instance normalisation;
- leaky ReLU.

The two convolutions of the first sequence are non-strided and output eight feature channels. The first convolution of the second sequence has a stride of two and doubles the number of feature channels to 16, whereas the second convolution of the second sequence is non-strided and keeps the number of 16 feature channels. Whereas the size of the input dimensions are preserved within the first convolution sequence, the strided convolution within the second sequence leads to a halving of each feature map dimension. The output of the ModalityNets are passed into a final module with shared weights (SharedNet), which finalises the feature extraction by applying two sequences of the same structure as used for the separate ModalityNets. Here, the first sequence comprises non-strided convolutions that output 16 feature channels while keeping the spatial dimensions as output by the ModalityNets. The first convolution of the second sequence has a stride of two leading to another halving of the spatial dimension's sizes and doubles the number of feature channels to 32. The second convolution of the second sequence is non-strided and keeps the number of 32 feature channels. The output of the SharedNet is given to a $1 \times 1 \times 1$-convolution providing the final number of 16 feature channels followed by a Sigmoid activation function. As we use correlation and transformation estimation techniques without trainable weights, our model only comprises 80k trainable parameters within the feature extraction part.

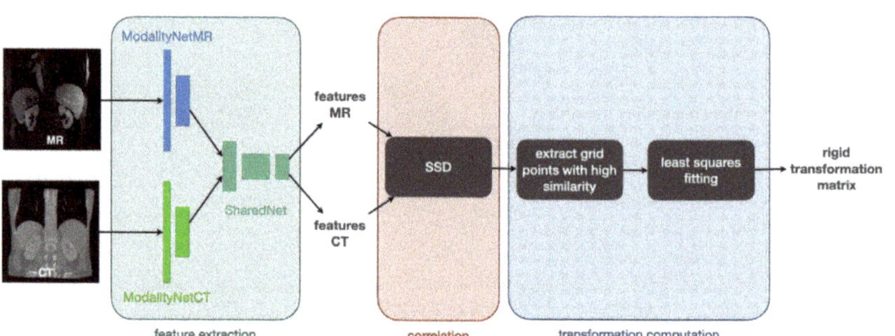


**Figure 3.** The process of feature extraction, correlation, and computation of the rigid transformation matrix: A CNN is used for feature extraction starting with a separate network part for each modality (ModalitiyNetMR and ModalityNetCT) followed by a module with shared weights (Shared-Net). The obtained features are correlated by calculating patch-wise the sum of squared differences (SSD). Subsequently, grid points with high similarity are extracted and used to define point-wise correspondences to calculate the rigid transformation matrix with a least squares fitting.

### 2.4. Correlation and Transformation Computation

As suggested in previous research [3,4], the use of a dense correlation layer that explores a large number of discretised displacements at once is employed to capture larger deformations robustly. This way the learned features are used to define a sum of squared differences cost function akin to metric learning [23].

Similar to [24], which operates directly on input image pairs and uses normalised cross correlations (NCC), we use a block-matching technique to find correspondences between the fixed features and a set of transformed moving features. We correlate the obtained features by calculating patch-wise the sum of squared differences (SSD) and extract points with high similarity of a coarse grid with a spacing of 12 voxel. The extracted grid points are used to define point-wise correspondences to calculate the rigid transformation matrix with a robust (trimmed) least squares fitting procedure.

For the correlation layer, we choose a set of $11 \times 11 \times 11$ discrete displacements with a capture range of approx. 40 voxel in the original volumes. After calculating the sum-of-squared-differences cost volume, we sort the obtained SSD costs and reject the 50% of the displacement choices that entail the highest similarity costs. We apply the Softmax function on the remaining displacement choices to obtain differentiable soft correspondences. While we use this differentiable approach to estimate regularised transformations within a framework that comprises trainable CNN parameters, the learned features could also be used for other optimisation frameworks [9].

The displacement candidates output by the Softmax function are added to the coarse moving grid points. In a least squares fitting procedure comprising five iterations, the final rigid transformation matrix that serves for transformation of the moving image is determined. The best-fitting rigid transformation can be found by computing the singular value decomposition $S = U\Sigma V^T$ with the matrix $S = X^T Y^T$ ($X$: centered fixed grid points $x_i$, $Y$:

centered moving grid points with added displacement candidates $y_i$) and the orthogonal matrices $U$ and $V$ obtained by the singular value decomposition. This leads to the rotation

$$Q = V \begin{pmatrix} 1 & & & & \\ & 1 & & & \\ & & \ldots & & \\ & & & 1 & \\ & & & & \det(VU^T) \end{pmatrix} U^T \tag{2}$$

and the translation

$$t = \bar{y} - Q\bar{x} \tag{3}$$

with $\bar{x}$ being the mean values for fixed grid points and $\bar{y}$ the mean moving grid points with added displacement candidates.

This way, the rigid transformation matrices $R_{21}$ and $R_{31}$ are determined. To combine the synthetic transformation $R_{23}$ and the predicted transformation $R_{31}$ matrix multiplication is used yielding $R_{23,31}$. The transformation matrices $R_{21}$ and $R_{23,31}$ are used to compute the affine grids that are then given to the MSE loss function during training and the affine grid computed by $R_{21}$ is used for warping during inference to align the moving image to the fixed image.

This approach has the advantage of being very compact with only 80k parameters ensuring memory efficiency and fast convergence of training. The multimodal features learned by our model are generally usable for image alignment and can be given to various optimisation methods for image registration once trained with our method.

## 3. Experiments and Results

Our experiments are performed on 16 paired abdominal CT and MR scans from collections of The Cancer Imaging Archive (TCIA) project [25–28]. We have manually created labels for four abdominal organs (liver, spleen, left kidney, right kidney), which we use for the evaluation of our methods. Apart from a withheld test set, they are publicly released for other researchers to train and compare their multimodal registration models. The pre-processing comprises reorientation, resampling to an isotropic resolution of 2 mm and cropping/padding to volume dimensions of $192 \times 160 \times 192$.

To increase the number of training and testing pairs and model realistic variations in initial misalignment we augment the scans with 8 random rigid transformations each that on average reflect the same Dice overlap (of approx. 43%) as the raw data. All models are trained for 100 epochs with a mini-batch size of 4 in less than 45 min each using $\approx 8$ GByte GPU memory.

The weights of the CNN used for feature extraction (FeatCNN) are trained for 100 epochs using the Adam optimiser with an initial learning rate of 0.001 and an cosine annealing scheduling.

### 3.1. Comparison of Training Strategies

We compare three different strategies to train our FeatCNN in a two-fold cross-validation:

1. FeatCNN + Cycle Discrepancy (ours): Our proposed self-supervised cycle learning strategy;
2. FeatCNN + MI Loss: Learning with metric-supervision using Mutual Information (MI) as implemented by [29];
3. FeatCNN + NCC$^2$ Loss: Learning with metric-supervision using squared local normalised cross correlations (NCC$^2$) [24,30];
4. FeatCNN + Label Loss: Supervised learning with label supervision.

All methods share the same settings for the correlation layer and a trimmed least square transform fitting (with five iterations and 50% outlier rejection). Hyperparameters were determined on a single validation scan (#15) for cyclic training and left unaltered for all other experiments. The same trainable FeatCNN comprising the layers as described in Section 2.3

is used to train with our Cycle Discrepancy Loss, MI, NCC$^2$, and Label Loss. For correlation, we chose to extract corresponding grid points within a grid with a spacing of 12 voxels and use patches with a radius of 2 to patch-wise calculate the SSD. We use a displacement radius of 4 and discretise the set of displacements possibilities for the correlation layer with a displacement step (resp. voxel spacing) of 5. To adjust the smoothness of the soft correspondences, the costs obtained by SSD computation are multiplied by a factor of 150 when given to the Softmax function. As the soft-correspondences are needed for differentiability only during training, we increase this factor to 750 for inference.

For our cycle discrepancy method, we create the synthetic transformation matrices $R_{23}$ by randomly initialising them with values that are assumed to be realistic from an anatomical point of view. Therefore, the maximum rotation is $\pm 23°$ and the maximum translation $\pm 42$ voxel (which equals 84 mm for our experiments) in every image dimension.

The results demonstrate a clear advantage of our proposed self-supervised learning procedure with an average Dice of 72.3% compared to the state-of-the-art MI metric loss with 68.14% and NCC$^2$ Loss with 68.1%, which is suitable for multimodal registration due to its computation involving small local neighbourhoods [24] (see Table 1 for qualitative and Figure 4 for quantitative results). This result comes close to the theoretical upper bound of our model trained with full label supervision with 79.55%.

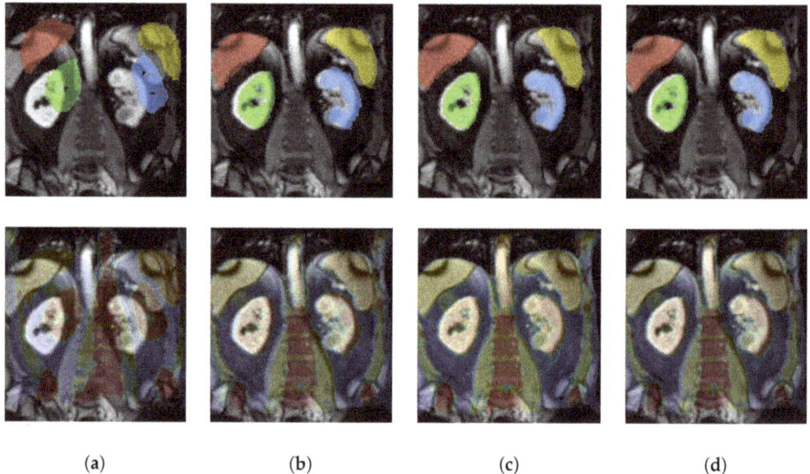

(a)    (b)    (c)    (d)

**Figure 4.** Qualitative results of our proposed cycle discrepancy approach FeatCNN + Cycle Discrepancy (**c**). We visualise the comparison to initial (**a**) before warping as well as to the methods FeatCNN + MI Loss (**b**) and FeatCNN + Label Loss (**d**) (coronal slices). The **top row** shows the fixed MRI and (warped) moving labels. The **bottom row** visualizes the (warped) moving CT and a jet colourmap overlay of the fixed MRI scan.

**Table 1.** Results for our cross-validation experiments: Dice scores listed by anatomical structures of our 3D experiments using FeatCNN for feature extraction and MI Loss, $NCC^2$ Loss, Label Loss or our Cycle Discrepancy for training.

|  | Liver | Spleen | Lkidney | Rkidney | Mean |
|---|---|---|---|---|---|
| initial | 59.32 ± 14.03 | 36.90 ± 19.49 | 36.59 ± 19.53 | 37.02 ± 22.08 | 42.46 ± 18.78 |
| FeatCNN + MI Loss | 75.07 ± 9.38 | 63.17 ± 22.13 | 69.86 ± 26.34 | 64.46 ± 29.45 | 68.14 ± 21.92 |
| FeatCNN + $NCC^2$ Loss | 75.08 ± 12.22 | 61.09 ± 23.69 | 72.19 ± 27.51 | 64.04 ± 31.76 | 68.10 ± 23.80 |
| FeatCNN + Cycle Discrepancy | 77.95 ± 8.16 | 69.89 ± 16.00 | 70.18 ± 24.34 | 71.85 ± 34.40 | 72.30 ± 20.75 |
| FeatCNN + Label Loss | 81.24 ± 8.75 | 73.84 ± 18.32 | 83.15 ± 26.62 | 79.97 ± 33.59 | 79.55 ± 21.82 |

### 3.2. Comparison of Inference Strategies and Increased Trainset

To further enhance our method, we extend it by a two-level warping approach during inference. Therefore, we present our model the input moving and fixed image to warp the moving image and then apply our model to the resulting warped moving image and the fixed image again. For both warping steps, we set a displacement radius of 7 voxel and a grid spacing of 8 voxel. For the first warping step, we use a displacement discretisation of 4 voxel and refine this hyperparameter to 2 voxel for the second warping step.

Moreover, as our dataset is quite small and our method does not require labels, when considering an application scenario where a number of MR/CT scan pairs have to be aligned offline, a fine-tuning of the networks on this test data would be feasible. Therefore, we aim to further increase the performance of our method with training on all available paired CT and MR scans without splitting the dataset.

In Table 2 we compare the results of single-level and two-level warping as well as the cross-validation results and the results when training on the whole available image data. We compare the results achieved by our method with the results achieved using the rigid image registration tool *reg_aladin* of NiftyReg [24] applied to the image pairs used without the symmetric version and one registration level.

**Table 2.** Results for our experiments comparing single-level and two-level warping approach as well as cross-validation and training without withheld data: Dice scores listed by anatomical structures of our experiments using Cycle Discrepancy for training.

|  | Liver | Spleen | Lkidney | Rkidney | Mean |
|---|---|---|---|---|---|
| initial | 59.32 ± 14.03 | 36.90 ± 19.49 | 36.59 ± 19.53 | 37.02 ± 22.08 | 42.46 ± 18.78 |
| cross-validation 1 warp | 77.95 ± 8.16 | 69.89 ± 16.00 | 70.18 ± 24.34 | 71.85 ± 34.40 | 72.30 ± 20.75 |
| cross-validation 2 warps | 80.71 ± 9.33 | 72.12 ± 17.08 | 79.33 ± 26.06 | 74.65 ± 36.91 | 76.68 ± 22.34 |
| trained without withheld data 1 warp | 81.04 ± 8.22 | 71.11 ± 18.03 | 76.27 ± 24.25 | 76.49 ± 32.64 | 76.23 ±20.88 |
| trained without withheld data 2 warps | 81.85 ± 0.58 | 76.77 ± 13.64 | 79.81 ± 24.52 | 80.17 ± 34.65 | 79.65 ± 20.25 |
| NiftyReg *reg_aladin* | 83.97 ± 6.19 | 76.55 ± 12.00 | 79.83 ± 7.12 | 79.26 ± 37.55 | 79.90 ± 15.15 |

Introducing a second warping step increased our cross-validation results by more than 4% points. When training without a withheld testset, we achieved further improvements by another 3% points. These results are on a par with the results of state-of-the-art classic method NiftyReg- *reg_aladin*.

## 4. Discussion

In this work, we presented a completely new concept for multimodal feature learning with application to 3D image registration without supervision of labels or handcrafted metrics. We introduced a new supervision strategy that is based on synthetic random transformations (two across modality and one within) that form a triangular cycle. Minimising the two multimodal transformations in such a cycle constraint avoids singular solutions (predicting identity transforms) and enables the learning of large rigid deformations. Through explorative learning, we are able to successfully train modality independent feature extractors that enable highly accurate and fast multimodal medical image alignment by minimising a cycle discrepancy in training. We also created the first public multimodal 3D MRI/CT abdominal dataset with manual segmentations for validation. To the best of our knowledge our work is also the first deep learning model for robustly estimating large misalignments of multimodal scans.

Despite the very promising results, there are a number of potential extensions that could further improve our concepts. The idea of incremental learning and predicting more useful synthetic transformations to improve detail alignment could be considered and has already shown potential in preliminary 2D experiments.

While the gap between training and test accuracy is relatively small due to the robust architectural design, further fine-tuning would be applicable at test time (since no supervision is required) with moderate computational effort. Combining hand-crafted domain knowledge with self-supervised learning might further boost accuracy. Similarly, domain adaptation through adversarial training could be incorporated to explicitly model the differences of modalities. While the gap between training and test accuracy is relatively small due to the robust architectural design, further fine-tuning would be applicable at test time (since no supervision is required) with moderate computational effort.

## 5. Conclusions

With our method, we were able to improve over the use of handcrafted metric-based losses by using synthetic three-way cycles. By minimising the cycle discrepancy, we are able to learn multimodal registration between CT and MRI without metric supervision. We created a robust method to estimate large rigid transformations that is differentiable in end-to-end learning. Our method is able to successfully perform intra-patient abdominal CT-MRI registration that outperforms state-of-the-art metric-supervision.

**Author Contributions:** Conceptualization, M.P.H., L.H. and H.S.; methodology, M.P.H., L.H. and H.S.; software, M.P.H., L.H. and H.S.; validation, M.P.H., L.H. and H.S.; formal analysis, M.P.H., L.H. and H.S.; investigation, M.P.H., L.H. and H.S.; resources, M.P.H.; data curation, M.P.H. and H.S.; writing—original draft preparation, H.S.; writing—review and editing, M.P.H., L.H. and H.S.; visualization, H.S.; supervision, M.P.H.; project administration, M.P.H.; funding acquisition, M.P.H. All authors have read and agreed to the published version of the manuscript.

**Funding:** This research was funded by the Federal Ministry of Education and Research (BMBF) grant number 16DHBQP052.

**Institutional Review Board Statement:** Not applicable.

**Informed Consent Statement:** Not applicable.

**Data Availability Statement:** Our experiments are performed on abdominal CT and MR scans from collections of The Cancer Imaging Archive (TCIA) project [25–28]. Material is available under TCIA Data Usage Policy and Creative Commons Attribution 3.0 Unported License. Material has been modified for direct usage in registration and deep learning algorithms: We have reorientated the data, resampled it to an isotropic resolution of 2 mm, and used cropping and padding to achieve voxel dimensions of $192 \times 160 \times 192$. We have also manually created segmentations for liver, spleen, left kidney, and right kidney. The results shown here are in whole or part based upon data generated by the TCGA Research Network: http://cancergenome.nih.gov/ (accessed on 7 January 2021).

**Conflicts of Interest:** The authors declare no conflict of interest.


# References

1. Hering, A.; Hansen, L.; Mok, T.C.W.; Chung, A.C.S.; Siebert, H.; Häger, S.; Lange, A.; Kuckertz, S.; Heldmann, S.; Shao, W.; et al. Learn2Reg: Comprehensive multi-task medical image registration challenge, dataset and evaluation in the era of deep learning. *arXiv* **2021**, arXiv:2112.04489.
2. Balakrishnan, G.; Zhao, A.; Sabuncu, M.R.; Guttag, J.; Dalca, A.V. An unsupervised learning model for deformable medical image registration. In Proceedings of the IEEE Conference on Computer Vision and Pattern Recognition, Salt Lake City, UT, USA, 18–23 June 2018; pp. 9252–9260.
3. Heinrich, M.P. Closing the gap between deep and conventional image registration using probabilistic dense displacement networks. In Proceedings of the International Conference on Medical Image Computing and Computer-Assisted Intervention, Shenzhen, China, 13–17 October 2019; Springer: Berlin/Heidelberg, Germany, 2019; pp. 50–58.
4. Dosovitskiy, A.; Fischer, P.; Ilg, E.; Hausser, P.; Hazirbas, C.; Golkov, V.; Van Der Smagt, P.; Cremers, D.; Brox, T. Flownet: Learning optical flow with convolutional networks. In Proceedings of the IEEE International Conference on Computer Vision, Santiago, Chile, 7–13 December 2015; pp. 2758–2766.
5. Eppenhof, K.A.; Pluim, J.P. Pulmonary CT registration through supervised learning with convolutional neural networks. *IEEE Trans. Med. Imaging* **2018**, *38*, 1097–1105. [CrossRef]
6. Eppenhof, K.A.; Lafarge, M.W.; Moeskops, P.; Veta, M.; Pluim, J.P. Deformable image registration using convolutional neural networks. In Proceedings of the Medical Imaging 2018: Image Processing, Houston, TX, USA, 10–15 February 2018; International Society for Optics and Photonics: Bellingham, WA, USA, Volume 10574, p. 105740S.
7. Krebs, J.; Mansi, T.; Delingette, H.; Zhang, L.; Ghesu, F.C.; Miao, S.; Maier, A.K.; Ayache, N.; Liao, R.; Kamen, A. Robust non-rigid registration through agent-based action learning. In Proceedings of the International Conference on Medical Image Computing and Computer-Assisted Intervention, Quebec City, QC, Canada, 11–13 September 2017; Springer: Berlin/Heidelberg, Germany, 2017; pp. 344–352.
8. Hu, Y.; Modat, M.; Gibson, E.; Li, W.; Ghavami, N.; Bonmati, E.; Wang, G.; Bandula, S.; Moore, C.M.; Emberton, M.; et al. Weakly-supervised convolutional neural networks for multimodal image registration. *Med. Image Anal.* **2018**, *49*, 1–13. [CrossRef] [PubMed]
9. Blendowski, M.; Hansen, L.; Heinrich, M.P. Weakly-supervised learning of multi-modal features for regularised iterative descent in 3D image registration. *Med. Image Anal.* **2021**, *67*, 101822. [CrossRef]
10. Misra, I.; van der Maaten, L. Self-supervised learning of pretext-invariant representations. In Proceedings of the IEEE/CVF Conference on Computer Vision and Pattern Recognition, Seattle, WA, USA, 13–19 June 2020; pp. 6707–6717.
11. Komodakis, N.; Gidaris, S. Unsupervised representation learning by predicting image rotations. In Proceedings of the International Conference on Learning Representations (ICLR), Vancouver, BC, Canada, 30 April–3 May 2018.
12. Wang, Y.; Solomon, J.M. PRNet: Self-Supervised Learning for Partial-to-Partial Registration. In *Advances in Neural Information Processing Systems*; Wallach, H., Larochelle, H., Beygelzimer, A., d'Alché-Buc, F., Fox, E., Garnett, R., Eds.; Curran Associates, Inc.: Vancouver, BC, Canada, 2019; Volume 32.
13. Li, H.; Fan, Y. Non-rigid image registration using self-supervised fully convolutional networks without training data. In Proceedings of the 2018 IEEE 15th International Symposium on Biomedical Imaging (ISBI 2018), Washington, DC, USA, 4–7 April 2018; pp. 1075–1078.
14. Wang, X.; Jabri, A.; Efros, A.A. Learning correspondence from the cycle-consistency of time. In Proceedings of the IEEE/CVF Conference on Computer Vision and Pattern Recognition, Long Beach, CA, USA, 16–20 June 2019; pp. 2566–2576.
15. Zhu, J.Y.; Park, T.; Isola, P.; Efros, A.A. Unpaired image-to-image translation using cycle-consistent adversarial networks. In Proceedings of the IEEE International Conference on Computer Vision, Venice, Italy, 22–29 October 2017; pp. 2223–2232.
16. Gass, T.; Székely, G.; Goksel, O. Consistency-based rectification of nonrigid registrations. *J. Med. Imaging* **2015**, *2*, 014005. [CrossRef] [PubMed]
17. Christensen, G.E.; Johnson, H.J. Consistent image registration. *IEEE Trans. Med. Imaging* **2001**, *20*, 568–582. [CrossRef] [PubMed]
18. Datteri, R.D.; Dawant, B.M. Automatic detection of the magnitude and spatial location of error in non-rigid registration. In Proceedings of the International Workshop on Biomedical Image Registration, Nashville, TN, USA, 7–8 July 2012; Springer: Berlin/Heidelberg, Germany, 2012; pp. 21–30.
19. Kim, B.; Kim, J.; Lee, J.G.; Kim, D.H.; Park, S.H.; Ye, J.C. Unsupervised deformable image registration using cycle-consistent cnn. In Proceedings of the International Conference on Medical Image Computing and Computer-Assisted Intervention, Shenzhen, China, 13–17 October 2019; Springer: Berlin/Heidelberg, Germany, 2019; pp. 166–174.
20. de Vos, B.D.; Berendsen, F.F.; Viergever, M.A.; Sokooti, H.; Staring, M.; Išgum, I. A deep learning framework for unsupervised affine and deformable image registration. *Med. Image Anal.* **2019**, *52*, 128–143. [CrossRef] [PubMed]

21. Xiao, Y.; Rivaz, H.; Chabanas, M.; Fortin, M.; Machado, I.; Ou, Y.; Heinrich, M.P.; Schnabel, J.A.; Zhong, X.; Maier, A.; et al. Evaluation of MRI to ultrasound registration methods for brain shift correction: The CuRIOUS2018 challenge. *IEEE Trans. Med. Imaging* **2019**, *39*, 777–786. [CrossRef] [PubMed]
22. Xu, Z.; Luo, J.; Yan, J.; Pulya, R.; Li, X.; Wells, W.; Jagadeesan, J. Adversarial uni-and multi-modal stream networks for multimodal image registration. In Proceedings of the International Conference on Medical Image Computing and Computer-Assisted Intervention, Lima, Peru, 4–8 October 2020; Springer: Berlin/Heidelberg, Germany, 2020; pp. 222–232.
23. Simonovsky, M.; Gutiérrez-Becker, B.; Mateus, D.; Navab, N.; Komodakis, N. A deep metric for multimodal registration. In Proceedings of the International Conference on Medical Image Computing and Computer-Assisted Intervention, Athens, Greece, 17–21 October 2016; Springer: Berlin/Heidelberg, Germany, 2016; pp. 10–18.
24. Modat, M.; Cash, D.M.; Daga, P.; Winston, G.P.; Duncan, J.S.; Ourselin, S. Global image registration using a symmetric block-matching approach. *J. Med. Imaging* **2014**, *1*, 024003. [CrossRef] [PubMed]
25. Clark, K.; Vendt, B.; Smith, K.; Freymann, J.; Kirby, J.; Koppel, P.; Moore, S.; Phillips, S.; Maffitt, D.; Pringle, M.; et al. The Cancer Imaging Archive (TCIA): Maintaining and operating a public information repository. *J. Digit. Imaging* **2013**, *26*, 1045–1057. [CrossRef] [PubMed]
26. Akin, O.; Elnajjar, P.; Heller, M.; Jarosz, R.; Erickson, B.; Kirk, S.; Filippini, J. Radiology data from the cancer genome atlas kidney renal clear cell carcinoma [TCGA-KIRC] collection. *Cancer Imaging Arch.* **2016**. [CrossRef]
27. Linehan, M.; Gautam, R.; Kirk, S.; Lee, Y.; Roche, C.; Bonaccio, E.; Jarosz, R. Radiology data from the cancer genome atlas cervical kidney renal papillary cell carcinoma [KIRP] collection. *Cancer Imaging Arch.* **2016**. [CrossRef]
28. Erickson, B.; Kirk, S.; Lee, Y.; Bathe, O.; Kearns, M.; Gerdes, C.; Lemmerman, J. Radiology Data from The Cancer Genome Atlas Liver Hepatocellular Carcinoma [TCGA-LIHC] collection. *Cancer Imaging Arch.* **2016**. [CrossRef]
29. Sandkühler, R.; Jud, C.; Andermatt, S.; Cattin, P.C. AirLab: Autograd image registration laboratory. *arXiv* **2018**, arXiv:1806.09907.
30. Balakrishnan, G.; Zhao, A.; Sabuncu, M.R.; Guttag, J.; Dalca, A.V. Voxelmorph: A learning framework for deformable medical image registration. *IEEE Trans. Med. Imaging* **2019**, *38*, 1788–1800. [CrossRef] [PubMed]

*Article*

# Effect of Auditory Discrimination Therapy on Attentional Processes of Tinnitus Patients

Ingrid G. Rodríguez-León [1], Luz María Alonso-Valerdi [2], Ricardo A. Salido-Ruiz [1], Israel Román-Godínez [1], David I. Ibarra-Zarate [2] and Sulema Torres-Ramos [1,*]

[1] Division of Cyber-Human Interaction Technologies, University of Guadalajara (UdG), Guadalajara 44100, Jalisco, Mexico; ingrid.rleon@alumnos.udg.mx (I.G.R.-L.); ricardo.salido@academicos.udg.mx (R.A.S.-R.); israel.roman@academicos.udg.mx (I.R.-G.)

[2] Tecnologico de Monterrey, Escuela de Ingenieria y Ciencias, Monterrey 64849, Nuevo Leon, Mexico; lm.aloval@tec.mx (L.M.A.-V.); david.ibarra@tec.mx (D.I.I.-Z.)

[*] Correspondence: sulema.torres@academicos.udg.mx

**Abstract:** Tinnitus is an auditory condition that causes humans to hear a sound anytime, anywhere. Chronic and refractory tinnitus is caused by an over synchronization of neurons. Sound has been applied as an alternative treatment to resynchronize neuronal activity. To date, various acoustic therapies have been proposed to treat tinnitus. However, the effect is not yet well understood. Therefore, the objective of this study is to establish an objective methodology using electroencephalography (EEG) signals to measure changes in attentional processes in patients with tinnitus treated with auditory discrimination therapy (ADT). To this aim, first, event-related (de-) synchronization (ERD/ERS) responses were mapped to extract the levels of synchronization related to the auditory recognition event. Second, the deep representations of the scalograms were extracted using a previously trained Convolutional Neural Network (CNN) architecture (MobileNet v2). Third, the deep spectrum features corresponding to the study datasets were analyzed to investigate performance in terms of attention and memory changes. The results proved strong evidence of the feasibility of ADT to treat tinnitus, which is possibly due to attentional redirection.

**Keywords:** tinnitus; auditory discrimination therapy; EEG evaluation; event-related synchronization; event-related desynchronization; convolutional neural network

**Citation:** Rodríguez-León, I.G.; Alonso-Valerdi, L.M.; Salido-Ruiz, R.A.; Román-Godínez, I.; Ibarra-Zarate, D.I.; Torres-Ramos, S. Effect of Auditory Discrimination Therapy on Attentional Processes of Tinnitus Patients. *Sensors* **2022**, *22*, 937. https://doi.org/10.3390/s22030937

Academic Editors: Vahid Abolghasemi, Hossein Anisi and Saideh Ferdowsi

Received: 22 December 2021
Accepted: 20 January 2022
Published: 26 January 2022

**Publisher's Note:** MDPI stays neutral with regard to jurisdictional claims in published maps and institutional affiliations.

## 1. Introduction

Tinnitus is the perception of sound in the absence of an external source [1]. It affects between 5 and 15% of the world population [2]. Tinnitus is caused by exposure to loud noise, fever, ototoxicity, or a transient disturbance in the middle ear [1]. Tinnitus can be perceived by people of all ages, either those with normal hearing or those with hearing loss [3]. Lenhardt classified tinnitus into objective and subjective [4]. Objective tinnitus is associated with peripheral vascular abnormalities detectable by stethoscopic inspection, whereas subjective tinnitus is determined as an acoustic perception merely experienced by the patient [5]. The tinnitus of interest for the present investigation is the subjective one.

Subjective tinnitus can become chronic and refractory, and it may be caused by the over synchronization of neurons, which affects cognitive, attentional, emotional, and even motor processes [1]. Cognitive impairment has been frequently reported in patients with tinnitus over the last few years [6]. Particularly, working memory and attentional processes that are affected include deficits in (1) executive control of attention [7], (2) attentional changes [6], and (3) selective and divided attention [8]. Furthermore, tinnitus differs across patients in its perceptual characteristics (e.g., frequency and intensity), in its time course (constant, fluctuating, and intermittent), response to interventions (e.g., masking sounds and somatic maneuvers), etiologic factors, and comorbidities [9]. This heterogeneity of tinnitus is reflected by a substantial variability in tinnitus pathophysiology [10], which causes a high variability in the treatment outcome. Therefore, a major challenge in clinical tinnitus research is the identification of relevant criteria for subtyping patients [11,12].

The attentional neurophysiological mechanisms altered by the presence of tinnitus can be recorded over the human scalp using the electroencephalography (EEG) technique [13]. EEG allows monitoring neural oscillations and ongoing electrical activity, which is made up of several simultaneous oscillations at different frequencies [14–16]. Neural oscillations have traditionally been studied based on event-related experiments, where event-related potentials and (de-) synchronization levels have been estimated [5]. Specifically, event-related neural oscillatory responses at different frequency bands reflect different stages of neural information processing [14,15,17]. Event-related oscillations are typically studied as (1) event-related desynchronization (ERD), which refers to the phasic relative power decrease of a certain frequency band, and (2) event-related synchronization (ERS), which implies a relative power increase. As the term indicates, both ERD and ERS are neural patterns occurring in relation to emotional, cognitive, motor, sensory, and/or perceptual events [18–20]. In tinnitus patients, power changes in various frequency bands reflects changes in neural synchrony [5]. The levels of synchronization related to auditory stimuli are carried out here to evaluate the effect of auditory discrimination therapy (ADT).

It is well established that sound brings about physiological, cognitive, and psychological changes, which is why sound-based therapies have become seven of the twenty-five most widely used treatments for tinnitus according to [12]. ADT is an acoustic therapy based on the oddball paradigm principle. This therapy is designed to reduce attention toward tinnitus, thereby reducing its perception [21]. The oddball paradigm consists of a pair of stimuli: standard and deviant pulses, which are randomly presented. The patient must identify deviant (40%) from standard (60%) pulses. This therapy intends to redirect the patient attention toward other sensorial events different from tinnitus so as to reduce its perception. It requires the attention of the patient on the therapy by presenting a composed sound of standard and deviant pulses in a random way. The patient must identify which type of pulse is presented, either standard or deviant. The standard pulse is the same tone that the tinnitus is, and the deviant pulse is 10% more than the standard one. Auditory discrimination has shown an improvement in tinnitus symptoms attributed to the rehabilitation of auditory processing frequencies of the auditory cortex damaged due to tinnitus [22] and prevention of auditory cortex reorganization [23]. Training at tones that differed from the dominant tinnitus pitch is beneficial due to the effect of lateral inhibition. Furthermore, stimulating specific frequency regions close to but not within the tinnitus frequency region will likely promote or strengthen lateral inhibitory activity, thus disrupting the pathological synchronous activity of the tinnitus-generating region [24]. There are currently several areas of opportunity suggested by the scientific community to study [25]. A distinctive niche refers to finding objective measures to evaluate the effect of treatments in patients with tinnitus, since there are conventional clinical protocols based on a trial-and-error procedure, and there is no formal and adequate follow-up of the treatment. At present, the most used way to evaluate acoustic therapies is through subjective methods such as the visual analogue scale and ad hoc questionnaires [3]. For instance, [26] evaluated the effectiveness of using sound generators with individual adjustments to relieve tinnitus in patients unresponsive to previous treatments and according to the Tinnitus Handicap Inventory (THI) test. The authors found improvement in quality of life, with good response to sound therapy. Not only subjective but also objective evaluation has been recently undertaken. The investigation presented by [27] compared sound therapies based on music, retraining, neuromodulation (e.g., ADT), and binaural sounds using neuro-audiology assessments and psychological evaluations. The first assessment revealed that the whole frequency structure of the neural networks showed a higher level of activeness in tinnitus sufferers than in control individuals. According to the psychological evaluation, the retraining treatment was the most effective sound-based therapy to reduce tinnitus perception and to release stress and anxiety after 60 days of treatment. Nonetheless, binaural sounds and ADT produced very similar effects. Furthermore, ADT showed to exert less side effects. Secondly, [28] evaluated the feasibility of Binaural Sound Therapy (BST) for tinnitus treatment by comparing its effect with Music Therapy (MT) effect. According to the THI questionnaire outcomes, BST reduced tinnitus perception. On the other hand, slightly major neural synchronicity over the right frontal lobe was reflected after two-month treatment.

In the light of the above discussion, the present work aims to establish a methodology based on EEG analysis to evaluate objectively the effectiveness of ADT to redirect the attention of patients with tinnitus. For this purpose, the database "Acoustic therapies for tinnitus treatment: An EEG database" [29] was used. From the database, only control and ADT groups were selected. Afterwards, ERD and ERS responses were mapped for two study cases: (1) before and (2) after applying the ADT. For ERD–ERS maps, Continuous Wavelet Transform (CWT) related to auditory material recognition was computed. Thereafter, deep representations from the resulting scalograms images using pre-trained Convolutional Neural Networks (CNNs) were extracted. Finally, deep spectrum features were analyzed to investigate the performance in terms of cognitive changes, specifically those related to attention and memory. The foregoing may provide solid evidence of the feasibility of ADT to treat subjective, chronic, and refractory tinnitus. The conduction of the investigation is described below.

## 2. Materials and Methods

The methodology for this work was undertaken into four steps: (1) to analyze and select the EEG signals of interest from the aforementioned database, (2) to estimate the ERD/ERS maps based on CWT, (3) to extract deep features based on CNN, and (4) to analyze statistically data based on centroids and Euclidean distances. This methodology is shown in Figure 1 and described in detailed in the following paragraphs.

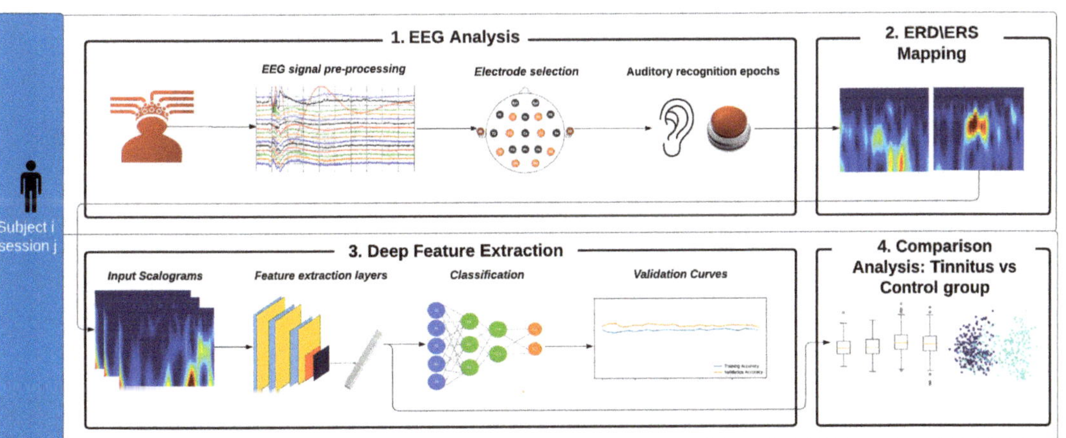


**Figure 1.** Four-step based methodology followed for the current research study: (1) EEG Analysis, (2) ERD/ERS Mapping, (3) Deep Feature Extraction, and (4) Comparison Analysis: Tinnitus vs. Control group.

### 2.1. EEG Database

The database for this research is available at Mendeley Data under the title "Acoustic therapies for tinnitus treatment: An EEG database" [29]. This database was created by following a protocol formerly approved by the Ethical Committee of the National School of Medicine of the Tecnologico de Monterrey, described, published, and registered under the trial number: ISRCTN14553550.

From the cohort, two groups were selected: tinnitus patients treated with ADT and controls. There were eleven participants per group. Both groups were treated for 8 weeks and were instructed to use the sound-based therapy for one hour every day at any time of the day. Note that controls were acoustically stimulated with relaxing music. In both cases, the sound therapy was monitored by psychometric and electroencephalographic evaluations before and after the 8-week treatment. For the EEG monitoring, four auditory stimulation conditions were found: (1) 3 min at resting state, (2) 3 min at listening to the corresponding therapy, (3) 2.5 min at listening to intermittent stimuli, and (4) 5 min at listening to everyday soundscapes where individuals had to identify 5 different sounds. The last case was the only one analyzed for this research. As this research aimed to evaluate objectively the effectiveness of ADT to redirect the patient's attention, the EEG analysis of tinnitus patients when recognizing everyday sounds (e.g., mobile ring, car horn) at common soundscapes could reveal whether the tinnitus attention had been reduced, and they were able to identify those sounds.

Two different soundscapes were played, while five associated auditory stimuli were randomly played. Whenever participants identified auditory stimuli, they pressed a keyboard button. The soundscapes and their related auditory stimuli to be identified for each monitoring session were: (1) *construction in progress*: (i) human sound (yelling), (ii) police siren, (iii) mobile dialing, (iv) bang, and (v) hit; and (2) *restaurant*: (i) human sound (tasting food), (ii) microwave sound, (iii) glass breaking, (iv) door closing, and (v) soda can being opened. All the stimuli lasted 1 s and were repeated 50 times at a random rate. Participants kept their eyes closed during the stimulation. Every monitoring session was around 60 min long [3]. The experimental timing protocol is illustrated in Figure 2.

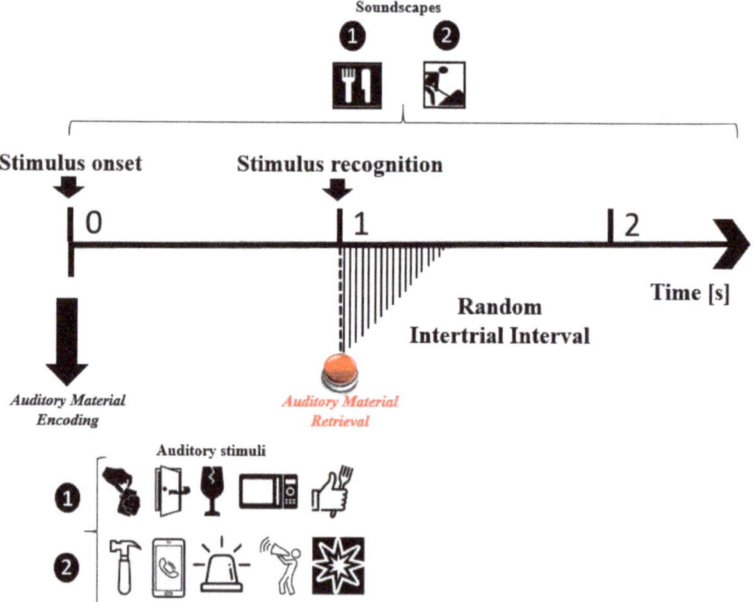


**Figure 2.** Timing protocol for EEG data in use. Each trial was around 60 min long. In each trial, participants listened to a soundscape and identified five randomly played auditory stimuli by pressing a button on the keyboard. There were two types of induced events: (1) auditory material encoding and (2) auditory material retrieval.

To record the EEG data, a g.USBamp amplifier was used, which was configured as stated in Table 1. Furthermore, clinical (level of hearing loss and frequency, intensity, and laterality of tinnitus) and demographic (gender, age) characteristics from the cohort selected were registered.

**Table 1.** EEG recording system configuration.

| Sampling rate | 256 Hz |
| --- | --- |
| Number of channels | 16 |
| Channels used by region | Prefrontal (FP1, FP2), Frontal (F7, F3, Fz, F4, F8), Temporal (T3, T4, T5, T6), Central (C3, C4), Parietal (Pz), Occipital (O1, O2) |
| Reference method | Monopolar @ Cz |
| Electrode placement system | International 10–20 system |

### 2.2. EEG Signal Pre-Processing

The EEG signals were pre-processed as follows. Firstly, the low-frequency components were eliminated by applying a Butterworth-type Band Pass digital filter with order 6 of zero phase, and with cutoff frequencies between 0.1 and 30 Hz. Secondly, channels were removed according to the criteria reported in [30]: flat for more than 5 s, maximum acceptable high-frequency noise standard deviation of 4, minimum acceptable correlation with nearby channels of 0.8. Thirdly, Artifact Subspace Reconstruction (ASR) bad burst correction was performed in order to remove bad data periods with transient or large-amplitude artifacts that exceeded 20 times the standard deviation of the calibrated data [30]. Fourthly, Independent Component Analysis (ICA) was applied with RunICA function. Finally, the independent components (ICs) distinguished as non-brain sources were rejected by the ICLabel classifier. The probability range for components flagged for rejection was set between 0.6 and 1. There were five non-brain source categories: (1) muscular, (2) ocular, and (3) electrocardiographic artifacts, (4) line noise, and (5) channel noise.

Due to the previous pre-processing stage alongside with some missing material recognition responses in the initial monitoring session, there was a significant loss of auditory material retrieval events; therefore, the sample of interest had to be reduced to 5 tinnitus patients composed of four adults aged 30–59 years old and one elderly aged 60–85 years old: 3 males and 2 females.

Table A1 (located in Appendix A) shows up the rejected channels, the percentage of bad data periods with transient or large-amplitude artifacts, and the independent components distinguished as non-brain sources.

### 2.3. ERD/ERS Maps

To begin this process, EEG signals over the frontal lobe and middle line (Fz) were carried out to monitor the ADT effect on tinnitus sufferers. Channel Fz was selected to analyze EEG information, since it is the recording site for clinical diagnosis of tinnitus.

Secondly, the epochs were extracted 500 ms before and 1 s after the keyboard button press; i.e., the recognition of the familiar sound played randomly during the everyday soundscape (Figure 2). This event refers to the auditory material retrieval. A negative window was proposed as a reference to measure changes in potential prior to the event whilst the positive window is aligned with the timing protocol corresponding to the time of appearance of ERD/ERS responses associated with the auditory memory and attentional mechanisms involved [31].

Thirdly, the CWT was the time-frequency analysis applied to each of 50 epochs per stimulus (5 stimuli in total). Wavelet of the Complex Gaussian family (Equation (1)) was selected, since they are based on complex-valued sinusoids constituting an analytic signal, possessing the shift invariance property. The sampling frequency was 256 Hz. The frequency range oscillated between 0.1 and 30 Hz.

$$f(x) = C_p e^{-ix} e^{-x^2} \tag{1}$$

The integer $p$ is the parameter of this family built from the complex Gaussian function. $C_p$ is such that $\| f^p \|^2 = 1$ where $f^p$ is the $p^{th}$ derivative of $f$.

Fourthly, the baseline correction (*BC*) was carried out using the subtraction method based on Equation (2).

$$BC = \left(P(t, f) - \overline{R}(f)\right) \tag{2}$$

where $P(t,f)$ is the power value given a time-frequency point subtracted by the average value of the baseline values from $-400$ to $-100$ ms at each frequency range prior to the appearance of an auditory recognition event [32].

Finally, the coefficient matrices resulting from the CWT per epoch were averaged, and the absolute value was carried out to obtain only real estimations. CWT scalograms were plotted as a function of time windows from $-500$ ms to 1 s and a frequency ranging from 0.1 to 30 Hz, for the purpose of representing the auditory synchronization and desynchronization activity over the Fz area before and after the ADT-based procedure.

*2.4. Deep Feature Extraction*

The CNN is often used in disease detection and classification [33,34]. Nonetheless, in this paper, it was executed with the aim of extracting a distributed vector representation of the scalograms images resulted from training a model to classify tinnitus from control patients. From now on, such vector representations will be known as *deep spectrum features*. The premise with such deep spectrum features is that images from tinnitus patients result in vector representations that are closer among them and, at the same time, distant from vector representations corresponding to control participants. The CNN utilized was the MobileNet V2, which is based on a streamlined architecture that uses depth-wise separable convolutions, a form of factorized convolutions, with the aim to build lightweight deep neural networks. MobileNet uses $3 \times 3$ depth wise separable convolutions, which uses between 8 and 9 times less computation, and it is extremely efficient relative to standard convolutions. Furthermore, the model has the effect of drastically reducing model size and computational cost [35]. This feature helps face the high computing capability and the large memory requirements characterized in a CNN method [33]. The pre-trained CNN was transferred to our recognition of auditory material task for extracting the deep spectrum features from the scalogram images carried out in the previous section.

The dataset used was 2468 scalogram images, divided into four classes, tinnitus patients before (801 images) and after (667 images) the treatment and control subjects before (500 images) and after (500 images) the treatment. There is a significantly larger number of tinnitus samples compared to the control ones (approximately 59% against 41%, respectively).

The pixel values in the images were into the range [0, 255]. So, as part of the model expectation, the pre-processing method included with the CNN model was executed to rescale the pixel values in $[-1, 1]$. Furthermore, the scalograms were resized from $1200 \times 900$ to $160 \times 160$.

To start with, the base model from the MobileNet-V2, which is pre-trained on the ImageNet dataset model, was executed to classify between controls and tinnitus patients before the corresponding sound-based treatment.

Secondly, the feature extractor converted each $160 \times 160 \times 3$ image into a $5 \times 5 \times 1280$ block of features. Hence, a classifier was added on top of it so the top-level classifier can be trained accordingly.

Thirdly, in order to generate predictions from the block of features, a GlobalAverage-Pooling2D layer was used to average over the spatial $5 \times 5$ spatial locations with the aim to convert the features to a single 1280-element vector per image. In addition, a Dense layer was applied to convert these features into a single prediction per image. Positive numbers predicted class 1 (Control participants), and negative numbers predicted class 0 (Tinnitus patients). There were 1.2K trainable parameters in the Dense layer, which were divided in 2 variable objects: the weights and biases.

Fourthly, the model was compiled. An Adam optimizer was used with a learning rate of $1 \times 10^{-4}$, dropout value of 0.2, and a batch size of 32. The architecture of the model executed is shown in Figure 3. An exhaustive search was executed to find optimal learning, epochs, batch size rate, and dropout values hyper parameters in the classifier block; learning rates from $1 \times 10^{-3}$ to $1 \times 10^{-6}$, dropout values from 0.1 to 0.5, epochs from 15 to 100, and batch size from 25 to 45 were explored.

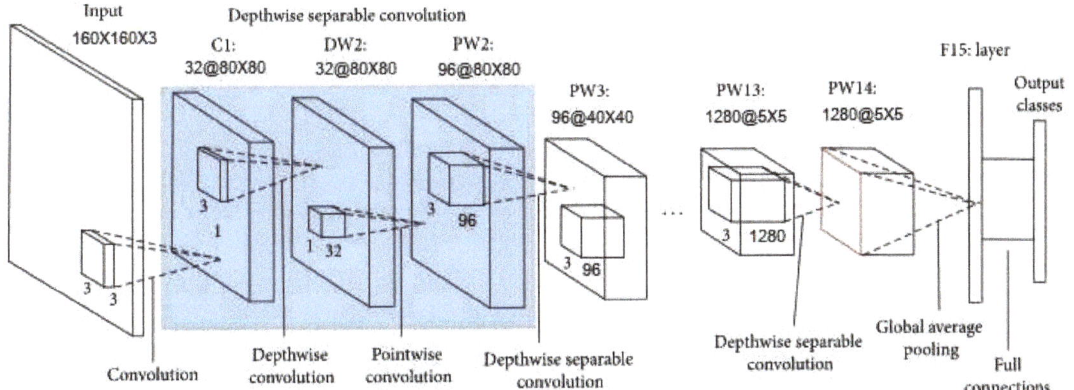


**Figure 3.** MobileNet-V2 architecture.

Fifthly, the MobileNet-V2 base model was trained by using 25 epochs. Learning curves of the training and validation accuracies were plotted (Figure A1 located in Appendix A), getting 69% accuracy on the validation set. An 80/20 validation was applied: 80% of data was used for model construction, and 20% of the data was used for model validation. The validation metrics were evaluated after the corresponding epochs.

Finally, the convolutional base, pre-loaded with weights trained on ImageNet without the classification layers, was applied for the feature extraction of scalogram images related to the auditory material recognition task carried out from tinnitus patients and controls during the two monitoring sessions: before and after the corresponding sound-based treatment.

### 2.5. Comparison Analysis: Tinnitus vs. Control Group

Once deep spectrum features were extracted per scalogram, in order to analyze tinnitus and control groups, a statistical evaluation was performed to acquire the significant differences among all the study datasets. Furthermore, an estimator was calculated to evaluate the effect of the sound-based therapy, and finally, centroids and distances were obtained to measure the closeness between the instances of the tinnitus group and control group.

#### 2.5.1. Statistical Evaluation

The statistical analyses were conducted separately for each dataset: tinnitus patients and controls before and after the treatment considering the recognition of auditory material.

The Lilliefors test was used to assess data distribution between-tinnitus subjects, within-tinnitus subjects, and within-control subjects before and after the sound-based treatments. After achieving a normal distribution, the statistical significance of any differences among the groups stated in Table 2 was evaluated with the Student's $t$-test. $p$-values were stated at 5% for both statistical processes. $p$-values greater than 0.05 will represent a statistically significant relationship in ERD/ERS responses between the indicated study data sets, whilst $p$-values less than 0.05 will show significant differences. Significant relationship responses between the tinnitus group after the sound-based treatment versus control group could help point out whether ADT was a reliable treatment. Additionally, box plots were created.

**Table 2.** Study groups. Tinnitus vs. Control group.

| | | | Tinnitus | | | |
|---|---|---|---|---|---|---|
| | | | Intra-Subject Comparison | | Inter-Subject Comparison | |
| | | | **Before** | **After** | **Before** | **After** |
| Control | Intra-subject comparison | Before | X | X | X | X |
| | | After | X | X | X | X |

### 2.5.2. The Differences in Differences (DID) Estimator

The DID estimator was estimated to analyze the differential effect of the sound-based treatment on the tinnitus group versus the control group in both experimental designs: between subjects and within subjects. The DID model is based on Equation (3).

$$Y = \beta_0 + \beta_1 Time + \beta_2 Intervention + \beta_3 (Time \cdot Intervention) + \varepsilon \tag{3}$$

where $\beta_0$ is the baseline average, $\beta_1$ is the time trend in the control group, $\beta_2$ is the difference between two groups pre-intervention, and $\beta_3$ is the difference in changes over time.

DID is a quasi-experimental design that makes use of longitudinal data from treatment and control groups to estimate a causal effect of a specific intervention or treatment by comparing the changes in outcomes over time. DID requires data from pre-/post-intervention, such as cohort or repeated cross-sectional data. The approach gets rid of biases in post-intervention period comparisons between the treatment and control group and from comparisons over time in the treatment group [36].

### 2.5.3. Centroid and Distance Measures

Firstly, there were calculated centroid values based on the mean values of the coordinates of all the data instances from control and tinnitus groups before and after the treatment (Equation (4)).

$$C_i = \frac{1}{p} \sum_{j=1}^{p} x_i^j \tag{4}$$

$x^u$ is the $u$-th deep spectrum feature vector where $x^u \in \mathbb{R}^{1280}$, $u \in \{1, 2, \ldots, p\}$ ($p$ is the number of scalograms for a given group). Additionally, $i \in \{1, 2, \ldots, 1280\}$ where $i$ is the $i$-th component of the vector $x$.

Secondly, Euclidian distance was calculated between each data instance and the corresponding centroids (Equation (5)). Media (Equation (6)) and standard deviations (Equation (7)) were reported. By applying the present criteria, it was possible to measure the closeness between the instances of the tinnitus group after receiving the therapy with respect to the control centroids. Analysis based on centroids and distances offered a novel multidimensional approach for identifying tinnitus groups already treated that exhibited similarities in ERD/ERS responses compared with control groups. If the mean Euclidian distance between the instances of the tinnitus group after treatment and the centroids of the control group is shorter than the corresponding between the instances of the tinnitus group before treatment and the centroids of the control group, this could indicate the existence of neural similarities, which could support the effectiveness of treatment in some scenarios.

$$D = \sqrt{\left(x_1^u - C_1^k\right)^2 + \left(x_2^u - C_2^k\right)^2 + \ldots + \left(x_{1280}^u - C_{1280}^k\right)^2} \tag{5}$$

where $x^u$ is a deep spectrum feature vector and $C^k$ is the $k$-th centroid.

$$\bar{x} = \frac{\sum_{i=1}^{N} D}{N} \tag{6}$$

$$s = \sqrt{\frac{\sum_{i=1}^{N}(D_i - \bar{x})^2}{N}} \tag{7}$$

In summary, the pipeline of the EEG analysis undertaken for this research was followed in four stages: (1) EEG Analysis, (2) ERD/ERS Mapping, (3) Deep Feature Extraction, and (4) Comparison Analysis. Figure 4 presents in detail the whole pipeline.

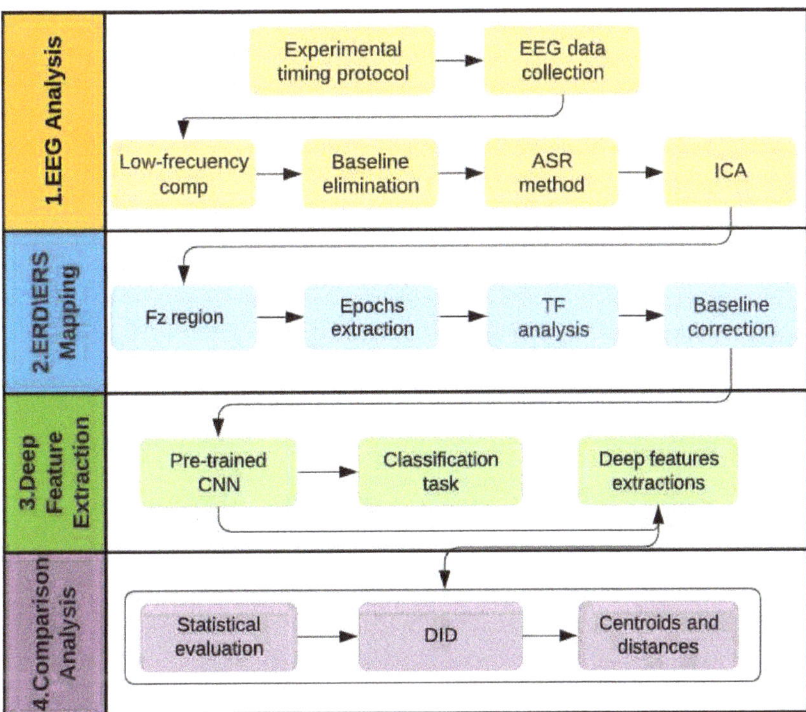


**Figure 4.** Pipeline of the EEG analysis to evaluate the effectiveness of ADT to treat subjective, chronic, and refractory tinnitus.

## 3. Results

Table 3 shows the training and validation accuracies of the MobileNet-V2 model used in the current research study. Although the classification metric is not the main purpose of the work, the classification percentage was reported to obtain a reference of the model performance used for the extraction of deep features.

Table 4 shows the clinical (laterality, frequency, and intensity of tinnitus, heart rate, and hearing loss) and demographic (age, sex) characteristics of the study sample of tinnitus patients.

From the 11 participants, five were selected. The rest of them were rejected for any of the following two reasons: there were no auditory material recognition responses in the initial monitoring session during the acoustic therapy or during the pre-processing stage due to segment rejection for artifacts, and/or the channel Fz was eliminated due to the transient or large amplitude artifacts.

**Table 3.** Training and validation accuracies of the MobileNet-V2 model used in the current research study.

| Epochs | Training Accuracy | Validation Accuracy |
|---|---|---|
| 1 | 0.6321 | 0.642 |
| 2 | 0.6324 | 0.6307 |
| 3 | 0.6331 | 0.625 |
| 4 | 0.6341 | 0.6364 |
| 5 | 0.6345 | 0.6335 |
| 6 | 0.635 | 0.6349 |
| 7 | 0.637 | 0.6359 |
| 8 | 0.6477 | 0.6392 |
| 9 | 0.6511 | 0.6449 |
| 10 | 0.6623 | 0.6492 |
| 11 | 0.681 | 0.6392 |
| 12 | 0.682 | 0.644 |
| 13 | 0.6874 | 0.6392 |
| 14 | 0.681 | 0.6477 |
| 15 | 0.682 | 0.6591 |
| 16 | 0.681 | 0.66 |
| 17 | 0.6825 | 0.672 |
| 18 | 0.6835 | 0.6899 |
| 19 | 0.6855 | 0.6899 |
| 20 | 0.6817 | 0.6909 |
| 21 | 0.682 | 0.6591 |
| 22 | 0.681 | 0.66 |
| 23 | 0.6825 | 0.672 |
| 24 | 0.6835 | 0.6821 |
| 25 | 0.6855 | 0.6899 |
| Average | 0.6758 | 0.661684211 |

**Table 4.** Clinical and demographic characteristics of the study sample: Tinnitus patients.

| Subjects | Age | Sex * | Laterality ** | Frequency [Hz] | Intensity [dB] | BPM *** | HL ****-L | HL-R |
|---|---|---|---|---|---|---|---|---|
| 1 | Adult | M | R | 125 | 90 | 75 | 96 | 20 |
| 2 | Elderly | M | R | 6000 | 70 | 79 | 56 | 52 |
| 3 | Adult | M | L | 8000 | 50 | 69 | 29 | 30 |
| 4 | Adult | F | B | 2000 | 87.5 | 86 | 63 | 70 |
| 5 | Adult | F | B | 6000 | 20 | * | 13 | 10 |

* M: male, F: female, ** R: right, L: left, B: both. *** BPM: beats per minute. **** HL: hearing loss → L: left and R: right.

Event-related (de) synchronizations maps extracted during the auditory recognition task before and after the sound-based treatment are shown in Figures 5 and 6.

In Table 5, we can see $p$-values as a result of the Student's $t$-test to statistically assess all tinnitus patients and control participants before and after the sound-based treatment under the experimental condition related to the recognition of acoustic material. Estimations indicated with a plus sign refer to those $p$-values above 0.05. These represent a statistically significant relationship in the ERD/ERS responses between the two study conditions. On the other hand, in Table 6, we can see $p$-values as a result of the Student's $t$-test to statistically assess each tinnitus patient and all control participants before and after the sound based treatment under the experimental condition of recognition of acoustic material. Estimations indicated with a plus sign refer to those $p$-values above 0.05. These represent a statistically significant relationship in the ERD/ERS responses between the two stated study datasets.

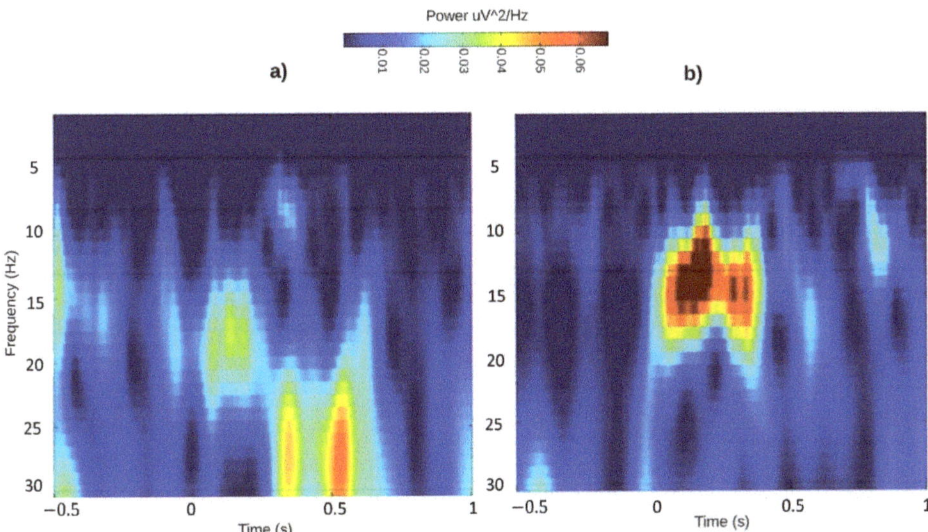

**Figure 5.** Tinnitus group. ERD/ERS responses over Fz before (**a**) and after (**b**) ADT-based treatment during the auditory recognition event. Fz was selected to illustrated central tendencies since it is the clinical recording site to diagnose tinnitus.

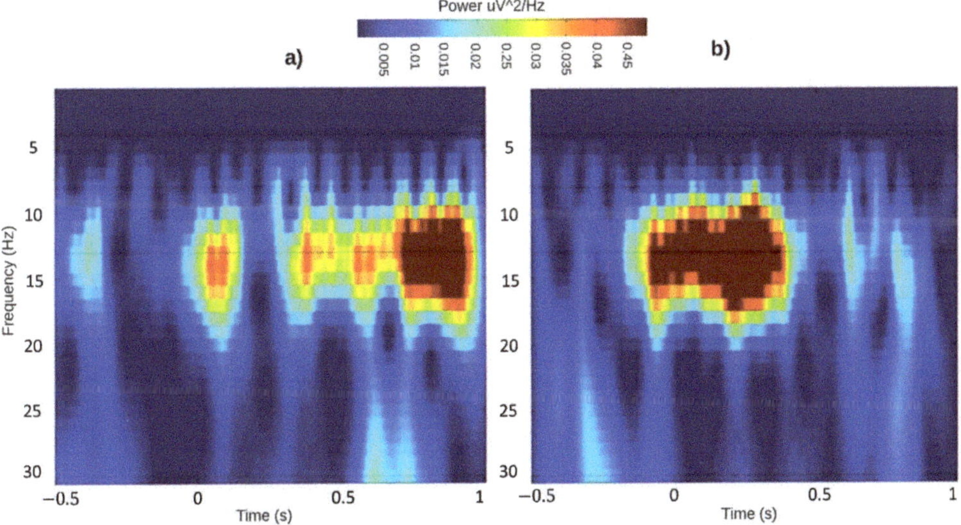

**Figure 6.** Control group. ERD/ERS responses over Fz before (**a**) and after (**b**) the sound-based treatment during the auditory recognition event. Fz was selected to illustrated central tendencies, since it is the clinical recording site to diagnose tinnitus.

**Table 5.** $p$-values as a result of within-subjects design where the Student's $t$-test was applied in tinnitus subjects versus control participants in different sessions undertaken before and after the sound-based treatment.

| | Tinnitus S1 *–Control S1 | Tinnitus S1–Control S2 ** | Tinnitus S2–Control S1 | Tinnitus S2–Control S2 | Tinnitus S1–Tinnitus S2 | Control S1–Control S2 |
|---|---|---|---|---|---|---|
| Tinnitus Patients | - | - | + | - | + | |
| Control Patients | | | | | | - |

* S1: before the sound-based treatment. ** S2: after the sound-based treatment. -: significant differences ($p < 0.05$). +: significant relationship ($p > 0.05$).

**Table 6.** $p$-values as a result of between-subjects design where the Student's $t$-test was applied in each tinnitus subject versus the control participants in different sessions undertaken before and after the sound-based treatment.

| Tinnitus Patients | Tinnitus S1 *–Control S1 | Tinnitus S1–Control S2 ** | Tinnitus S2–Control S1 | Tinnitus S2–Control S2 | Tinnitus S1–Tinnitus S2 |
|---|---|---|---|---|---|
| 1 | + | - | + | - | + |
| 2 | - | - | - | - | + |
| 3 | - | - | + | - | + |
| 4 | - | - | - | - | + |
| 5 | - | + | + | + | + |

* S1: before the sound-based treatment. ** S2: after the sound-based treatment. -: significant differences ($p < 0.05$). +: significant relationship ($p > 0.05$).

In Figure 7, boxplots display the distribution of the different study datasets: tinnitus and control groups in two monitoring sessions: before and after the sound-based treatment.

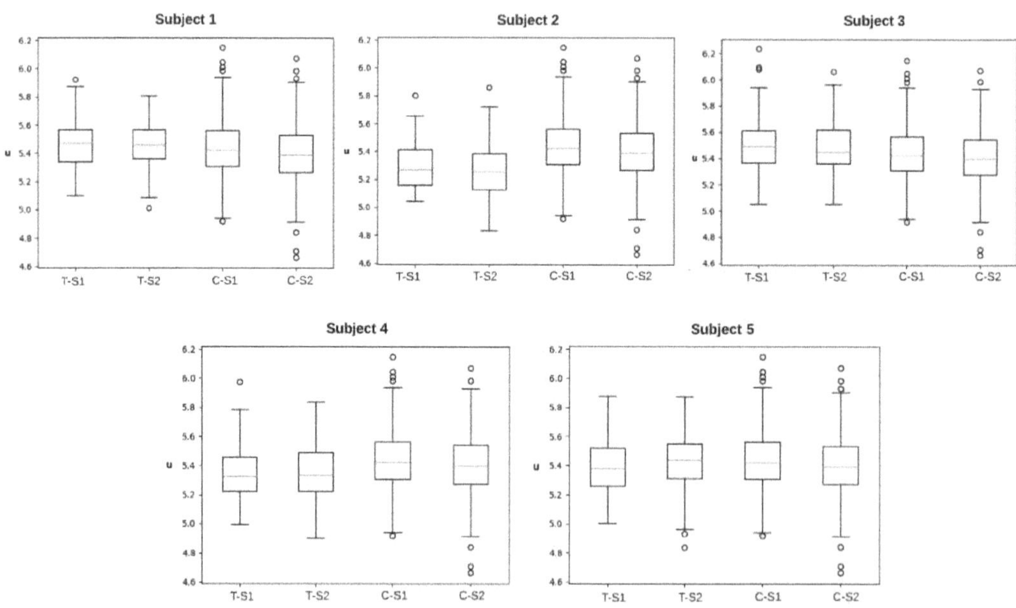


**Figure 7.** Box plots of five subjects as a result of between-subjects design to obtain the statistical distribution between-tinnitus subjects versus control participants in two monitoring sessions. T-S1: tinnitus group before the sound-based treatment, T-S2: tinnitus group after the sound-based treatment, C-S1: control group before the sound-based treatment, C-S2: control group after the sound-based treatment.

Table 7 shows the differential effect of the sound-based treatment on the 'tinnitus group' versus the 'control group' in both experimental designs: between-subjects and within-subjects. The DID negative refers to a negative therapy effect, whilst positive estimators have to do with a positive treatment effect.

**Table 7.** DID between-subjects and within-subjects.

| Subjects | DID | ADT-Based Treatment Effect |
| --- | --- | --- |
| 1 | 0.0327 | Positive effect |
| 2 | −0.0018 | Negative effect |
| 3 | 0.0152 | Positive effect |
| 4 | 0.0464 | Positive effect |
| 5 | 0.0741 | Positive effect |
| All tinnitus patients | 0.0225 | Positive effect |

On the other side, in Table 8, we can see the means and standard deviations of Euclidian distances between each data instance of tinnitus and control groups before and after the treatment with regard to the corresponding control centroids with the aim to measure the closeness among the different study groups.

**Table 8.** Distance measures among data instances of control and tinnitus groups and control centroids.

| Instances-Centroids | | Subject 1 | Subject 2 | Subject 3 | Subject 4 | Subject 5 |
| --- | --- | --- | --- | --- | --- | --- |
| Tinnitus S1 *— | Mean | 2.0867 | 1.9623 | 1.9567 | 1.8840 | 1.9407 |
| Control S1 ** | STD | 0.2664 | 0.2350 | 0.2798 | 0.2002 | 0.2553 |
| Tinnitus S2— | Mean | 2.1166 | 1.9433 | 1.9745 | 1.8803 | 1.9214 |
| Control S1 | STD | 0.2747 | 0.2520 | 0.2733 | 0.2349 | 0.2537 |
| Tinnitus S1— | Mean | 2.0166 | 1.9092 | 2.0065 | 1.8570 | 1.9025 |
| Control S2 | STD | 0.2738 | 0.2379 | 0.2830 | 0.2081 | 0.2623 |
| Tinnitus S2— | Mean | 2.0432 | 1.8968 | 2.0307 | 1.8560 | 1.8884 |
| Control S2 | STD | 0.2821 | 0.2640 | 0.2852 | 0.2404 | 0.2568 |
| Control S1— | Mean | 1.7758 | 1.7562 | 1.8997 | 1.7025 | 1.8513 |
| Control S2 | STD | 0.2827 | 0.2345 | 0.2969 | 0.2298 | 0.3289 |

* S1: before the sound-based treatment. ** S2: after the sound-based treatment.

## 4. Discussion

The aim of this study was to establish an objective methodology based on EEG analysis to measure changes in attentional processes in tinnitus patients treated with ADT.

Regarding the ERD/ERS responses of the tinnitus group (Figure 5), the absence of ERS response during the initial monitoring session (before ADT) and the increase in 4–13 Hz ERS during the final monitoring session (after ADT) could indicate increased cognitive demands such as semantic memory (cognitive processes responsible for accessing and/or bringing back information from long-term memory) and attentional processes [37] during the performance of the experimental task. Moreover, regarding [1,38], the alpha power increase in the final session may indicate that the ADT-based treatment had increased attention to everyday acoustic environments, and tinnitus sufferers were able to identify typical related auditory stimulus. Furthermore, during the first session, high-frequency energy is observed between 25 and 30 Hz after 500 ms of the stimulus onset. This could mean that tinnitus patients were able to identify the auditory stimuli at high frequencies as they perceived the task with a high complexity level because alongside the tinnitus sounds, they heard their own tinnitus causing a division in their attention. Nonetheless, during the final monitoring session, the responses are observed as normal. In addition, there was a notable decrease in the reaction time from 0 to 500 ms, and there was a frequency decrease in the neurons communication with the aim to meet the task.

On the other hand, ERD/ERS responses of the control group (Figure 6) kept high levels of synchronization within the alpha band in both monitoring sessions, which could indicate that the semantic memory was maintained throughout the sound-based therapy. However, the reaction time was changed as well. During the first monitoring session, there was a dispersed reaction time from 0 to 1 s as the experimental paradigm is new for the subjects. Therefore, the reaction times were more diverse. Even so, the central tendency is tardy, closed to the second 1. On the other side, during the final monitoring session, such variability decreases considerably downsizing the reaction time range from 1 to 500 ms.

One recurring problem with tinnitus research is that there is no objective way of assessing whether treatments counteract tinnitus. A recent systematic review examined the work to date on trying to find suitable objective measures of tinnitus [39]. The authors identified 21 articles, studying objective tests that included blood tests, electrophysiological measures, radiological measures, and balance tests. They concluded that the quality of evidence was generally poor and had failed to identify any reliable or reproducible objective measures of tinnitus. According to a subjective comparison among several acoustic therapies with the aim to evaluate the effect in tinnitus patients through a psychological evaluation [27], the re-training treatment was the most effective sound-based therapy to reduce tinnitus perception and to release stress and anxiety after 60 days of treatment. Nonetheless, binaural sounds and ADT produced very similar effects. Furthermore, ADT showed to exert less side effects. Nonetheless, nothing has yet been shown to offer the necessary specificity and sensitivity to be used as a biomarker in tinnitus treatment. As findings have shown, considerable variability and lack of consistency among studies suggest that further work in this area is needed [25]. Unlike the current research study, we herein proposed a quantitative approach based on EEG analysis and deep feature extraction to objectively measure ADT-based treatment comparing the tinnitus group with a control group to ensure reproducibility and sensibility measurement. A recent study by [28] combined objective and subjective measures to evaluate the effect of BST in tinnitus patients. The THI questionnaire reported that BST increased tinnitus perception in 15% of the patients. Furthermore, according to EEG monitoring, BST did not tend to reduce tinnitus perception but instead appeared to reduce tinnitus distress due to the slightly major neural synchronicity over the right frontal lobe found after the treatment. Unlike the current research, a new methodology was herein proposed as a first approach to evaluate the effect of the ADT-based treatment by EEG analysis.

In contrast to evoked activity, induced response refers to modulations of ongoing neural activity commonly quantified by event-related oscillations (EROs). As EROs reflect the coupling and uncoupling of neural networks, these EEG parameters give an insight into the functional neural network dynamics [5]. As far as it is known, ERD/ERS has not been undertaken to monitor electrophysiological changes in tinnitus sufferers during an acoustic therapy, it had been exemplified above the versatility of ERD/ERS estimation to capture the dynamics of neural oscillations related to emotional, cognitive, perceptual, and motor events [5]. Based on the previous statement, ERD/ERS maps were extracted so that deep features can be carried out to quantify the level of synchrony of the EEG signals by performing a cross-sectional study, comparing the tinnitus patients with control subjects at the end of the ADT-based treatment.

Based on [12], we supported the notion that tinnitus heterogeneity influences the observed variability in treatment response after an analysis of collected data of 5017 tinnitus bearers where participants reported which treatments they tried, the duration and the outcome of the given treatment, alongside with the demographic and tinnitus characteristics. Sound therapy can effectively suppress tinnitus, at least in some patients [40], but there is still a lack of research on the efficacy of sound therapy. It is necessary to analyze the characteristics of individual tinnitus patients and to unify the assessment criteria of tinnitus [24]. In Tables 6 and 7, *p*-values above 0.05 and DID results suggest all the adult patients had a positive effect after the ADT-based treatment, whilst the elderly patient, under the same experimental conditions, had a negative effect. Furthermore, the subject who faced a significant improvement having the highest DID estimator and a similar statistical distribution to the control groups before and after the sound-based treatment is the one with the lowest tinnitus intensity registered, alongside with low hearing loss in both ears.

Regarding treatment duration, it should be interpreted with caution, as it is well-known that certain treatments require some time for adaptation, whereas other treatments require longer periods to be effective [12]. There is still uncertainty about the duration of treatment that may be required to achieve an improvement [25]. During this study, ADT-based treatments lasted 8 weeks. However, they were not applied for all patients even though 2 months is the minimum necessary time that has been empirically reported to find changes [12].

Tinnitus impairment can be quantified by various validated questionnaires such as THI. However, a recent analysis revealed a high variability in the outcome instruments used in clinical trials, indicating the need to standardize outcome measurement [9]. Furthermore, the outcome measures carried out through the THI in [12] were retrospective and subjective, which could have biased the results. This is why questionnaires are considered a subjective metric. According to [25], a further limitation of the current tools for assessing tinnitus impact is the reliability and repeatability of such measures: self-report measures of tinnitus have an associated risk of variability, as they supply a momentary snapshot, whereas the experience of tinnitus changes with time and context. Based on the previous evidence, it was proposed a first quantitative approach to objectively measure and evaluate the effects of ADT using ERD/ERS techniques along with the extraction of deep spectrum features. Significant relationship responses between the 'tinnitus group' after the sound-based treatment versus the 'control group' (Tables 5 and 6), positive DID estimators (Table 7), and close distance measures (Table 8) indicate the existence of neural modifications, which could explain why this treatment is so effective in some scenarios. Results from this research might help point out ADT as a potential solution for certain patients, but it is not a viable treatment for many others.

According to [24], patients with more severe initial tinnitus respond better to sound therapy; however, in the current study, the opposite results were observed. In Tables 6 and 7, $p$-values above 0.05 and positive DID estimators suggest that the subject who faced a better performance is the one with the lowest tinnitus intensity registered, alongside with low hearing loss in both ears. The elderly patient who did not benefit from acoustic therapy was due to the time he had suffered from tinnitus: around 30 years.

Our study comes with some inherent limitations. First, although we started analyzing 11 tinnitus patients, this number was reduced to 5 tinnitus subjects due to one of the following reasons: the rest did not show auditory material recognition responses in the initial monitoring session before receiving the ADT-based treatment or during the preprocessing stage, and the channel Fz was eliminated due to the transient or large amplitude artifacts. The final sample was insufficient, so it might not be representative of all patients with tinnitus. Second, the improvement trend is inevitable; however, it would be interesting to carry out a deep spectrum features analysis by theta, alpha, and beta bands to know exactly which cognitive demands are increasing or decreasing in terms of semantic, working memory, and attentional processes in each tinnitus subject compared with control subjects.

## 5. Conclusions

In conclusion, a new methodology based on ERD/ERS analysis and deep spectrum features extraction was successfully implemented to measure changes in attentional processes in tinnitus patients treated with ADT. Based on the previous implementation, our results pointed out that tinnitus attention was significantly reduced after the ninth week of an ADT-based treatment in adult patients. Furthermore, the therapy reported significant improvements in the patients with the lowest intensity recorded of tinnitus, alongside with low hearing loss in both ears. It is worth mentioning that this acoustic therapy is based on redirecting the attention that the patient has his tinnitus, this attention is focused on the deviant pulse of the oddball paradigm that is different from the frequency of the tinnitus. After eight weeks of treatment, the patient reports a reduction in the perception, but beyond the reduction in the level of tinnitus perception, there is a reduction in the attention level, which results in the improvement of the patient.

Future work will entail measuring the EEG signals over the whole frontal lobe (Fp1, Fp2, F7, F3, Fz, F4, and F8). Furthermore, different neural network architectures could be applied to ensure the increase of the accuracy percentage in the classification stage to make the deep feature extraction stage more reliable.

**Author Contributions:** Conceptualization, I.G.R.-L., L.M.A.-V. and S.T.-R.; methodology, I.G.R.-L., L.M.A.-V., I.R.-G., S.T.-R. and R.A.S.-R.; validation, I.G.R.-L., L.M.A.-V., S.T.-R., I.R.-G., R.A.S.-R. and D.I.I.-Z.; formal analysis, I.G.R.-L., L.M.A.-V., S.T.-R., I.R.-G., R.A.S.-R. and D.I.I.-Z.; investigation, I.G.R.-L., L.M.A.-V., S.T.-R., I.R.-G. and R.A.S.-R.; data curation, I.G.R.-L., L.M.A.-V., S.T.-R., I.R.-G. and R.A.S.-R.; writing—original draft preparation, I.G.R.-L., L.M.A.-V., S.T.-R., I.R.-G. and R.A.S.-R.; writing—review and editing, I.G.R.-L., L.M.A.-V., S.T.-R., I.R.-G., R.A.S.-R. and D.I.I.-Z.; visualization, I.G.R.-L., L.M.A.-V., S.T.-R., I.R.-G., R.A.S.-R. and D.I.I.-Z.; supervision, I.G.R.-L., L.M.A.-V., S.T.-R., I.R.-G., R.A.S.-R. and D.I.I.-Z. All authors have read and agreed to the published version of the manuscript.

**Funding:** This research received no external funding.

**Institutional Review Board Statement:** The database used was created by following a protocol formerly approved by the Ethics Committee of the National School of Medicine of the Tecnólogico de Monterrey, described, published, and registered under the trial number: ISRCTN14553550.

**Informed Consent Statement:** Informed consent was obtained from all subjects involved in the study.

**Data Availability Statement:** The database used in this study is available at https://data.mendeley.com/datasets/kj443jc4yc/1 (accessed on 20 November 2021).

**Conflicts of Interest:** The authors declare no conflict of interest.

## Appendix A

**Table A1.** Rejected channels, the percentage of bad data periods with transient or large-amplitude artifacts, and the independent components distinguished as non-brain sources.

| Subjects | Sessions | Channels Rejected | Percentage of Bad Data Periods | Components Flagged for Rejection |
|---|---|---|---|---|
| 1 | 1 | 0 | 17.5% | 3 |
| | 2 | 1 | 0.4% | 3 |
| 2 | 1 | 2 | 0.0% | 2 |
| | 2 | 1 | 0.0% | 7 |
| 3 | 1 | 2 | 0.0% | 1 |
| | 2 | 2 | 0.0% | 0 |
| 4 | 1 | 4 | 46.5% | 4 |
| | 2 | 1 | 20.0% | 3 |
| 5 | 1 | 2 | 13.0% | 4 |
| | 2 | 4 | 15.1% | 3 |
| 6 | 1 | 0 | 3.4% | 3 |
| | 2 | 2 | 1.3% | 3 |
| 7 | 1 | 12 | 1.9% | 0 |
| | 2 | 3 | 1.4% | 2 |
| 8 | 1 | 4 | 12.4% | 2 |
| | 2 | 3 | 1.7% | 4 |
| 9 | 1 | 1 | 5.3% | 3 |
| | 2 | 0 | 0.7% | 2 |
| 10 | 1 | 0 | 1.1% | 2 |
| | 2 | 0 | 0.7% | 3 |
| 11 | 1 | 1 | 0.7% | 3 |
| | 2 | 4 | 1.1% | 2 |

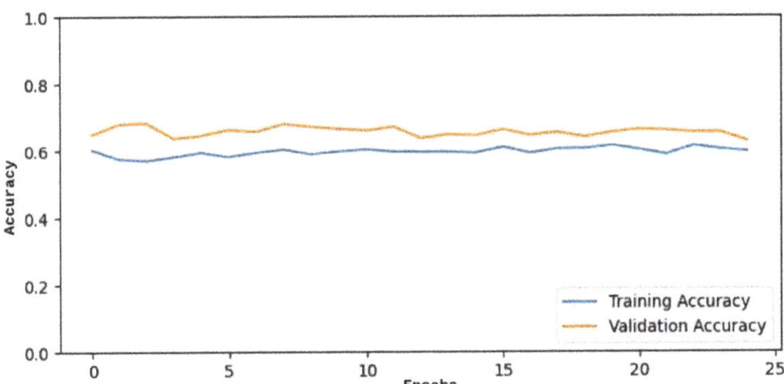

**Figure A1.** Training and validation curves to display the classification performance of the MobileNet-V2 model used in the current research study.

# References


1. Eggermont, J.J.; Roberts, L.E. The neuroscience of tinnitus: Understanding abnormal and normal auditory perception. *Front. Syst. Neurosci.* **2012**, *6*, 53. [CrossRef] [PubMed]
2. Meyer, M.; Luethi, M.S.; Neff, P.; Langer, N.; Buchi, S. Disentangling tinnitus distress and tinnitus presence by means of EEG power analysis. *Neural Plast.* **2014**, *2014*, 468546. [CrossRef]
3. Alonso-Valerdi, L.M.; Ibarra-Zarate, D.I.; Tavira-Sanchez, F.J.; Ramirez-Mendoza, R.A.; Recuero, M. Electroencephalographic evaluation of acoustic therapies for the treatment of chronic and refractory tinnitus. *BMC Ear Nose Throat Disord.* **2017**, *17*, 9. [CrossRef] [PubMed]
4. Lenhardt, M. Home medical device for tinnitus treatment. In *Up to Date on Tinnitus*, 1st ed.; Fayez, B., Ed.; IntechOpen: Rijeca, Croatia, 2011; Volume 1, pp. 137–152. Available online: https://www.intechopen.com/books/up-to-date-on-tinnitus/home-medical-device-for-tinnitus-treatment (accessed on 12 December 2021).
5. Sanchez, F.J.T. Evaluation of Acoustic Therapies as a Treatment for Chronic and Refractory Tinnitus. Ph.D. Thesis, Polytechnic University of Madrid, Madrid, Spain, 2019. Available online: http://oa.upm.es/57107/1/FRANCISCO_JOSE_TAVIRA_SANCHEZ.pdf (accessed on 14 December 2021).
6. Hallam, R.S.; McKenna, L.; Shurlock, L. Tinnitus impairs cognitive efficiency. *Int. J. Audiol.* **2004**, *43*, 218–226. [CrossRef] [PubMed]
7. Heeren, A.; Maurage, P.; Perrot, H.; De Volder, A.; Renier, L.; Araneda, R.; Lacroix, E.; Decat, M.; Deggouj, N.; Philippot, P. Tinnitus specifically alters the top-down executive control sub-component of attention: Evidence from the Attention Network Task. *Behav. Brain Res.* **2014**, *269*, 147–154. [CrossRef] [PubMed]
8. Rossiter, S.; Stevens, C.; Walker, G. Tinnitus and its effect on working memory and attention. *J. Speech Lang. Hear. Res.* **2006**, *49*, 150–160. [CrossRef]
9. Kleinjung, T.; Langguth, B. Avenue for Future Tinnitus Treatments. *Otolaryngol. Clin. N. Am.* **2020**, *53*, 667–683. [CrossRef]
10. Elgoyhen, A.B.; Langguth, B.; De Ridder, D.; Vanneste, S. Tinnitus: Perspectives from human neuroimaging. *Nat. Rev. Neurosci.* **2015**, *16*, 632–642. [CrossRef] [PubMed]
11. Cederroth, C.R.; Gallus, S.; Hall, D.A.; Kleinjung, T.; Langguth, B.; Maruotti, A.; Meyer, M.; Norena, A.; Probst, T.; Pryss, R.; et al. Editorial: Towards an Understanding of Tinnitus Heterogeneity. *Front. Aging Neurosci.* **2019**, *11*, 53. [CrossRef] [PubMed]
12. Simoes, J.; Neff, P.; Schoisswohl, S.; Bulla, J.; Schecklmann, M.; Harrison, S.; Vesala, M.; Langguth, B.; Schlee, W. Toward Personalized Tinnitus Treatment: An Exploratory Study Based on Internet Crowdsensing. *Front. Public Health* **2019**, *7*, 157. [CrossRef]
13. Duncan, C.C.; Barry, R.J.; Connolly, J.F.; Fischer, C.; Michie, P.T.; Naatanen, R.; Polich, J.; Reinvang, I.; Van Petten, C. Event-related potentials in clinical research: Guidelines for eliciting, recording, and quantifying mismatch negativity, P300, and N400. *Clin. Neurophysiol.* **2009**, *120*, 1883–1908. [CrossRef] [PubMed]
14. Basar, E.; Basar-Eroglu, C.; Karakas, S.; Schurmann, M. Oscillatory brain theory: A new trend in neuroscience. *IEEE Eng. Med. Biol. Mag.* **1999**, *18*, 56–66. [CrossRef] [PubMed]
15. Basar, E.; Basar-Eroglu, C.; Karakas, S.; Schurmann, M. Brain oscillations in perception and memory. *Int. J. Psychophysiol.* **2000**, *35*, 95–124. [CrossRef]
16. Krause, C.M. Brain electric oscillations and cognitive processes. In *Experimental Methods in Neuropsychology*, 1st ed.; Kenneth, H., Ed.; Springer: Boston, MA, USA, 2003; Volume 1, pp. 111–130. [CrossRef]
17. Basar, E.; Basar-Eroglu, C.; Karakas, S.; Schurmann, M. Are cognitive processes manifested in event-related gamma, alpha, theta and delta oscillations in the EEG? *Neurosci. Lett.* **1999**, *259*, 165–168. [CrossRef]

18. Aranibar, A.; Pfurtscheller, G. On and off effects in the background EEG activity during one-second photic stimulation. *Electroencephalogr. Clin. Neurophysiol.* **1978**, *44*, 307–316. [CrossRef]
19. Pfurtscheller, G. Graphical display and statistical evaluation of event-related desynchronization (ERD). *Electroencephalogr. Clin. Neurophysiol.* **1977**, *43*, 757–760. [CrossRef]
20. Pfurtscheller, G.; Aranibar, A. Evaluation of event-related desynchronization (ERD) preceding and following voluntary self-paced movement. *Electroencephalogr. Clin. Neurophysiol.* **1979**, *46*, 138–146. [CrossRef]
21. Herraiz, C.; Diges, I.; Cobo, P. Auditory discrimination therapy (ADT) for tinnitus management. *Prog. Brain Res.* **2007**, *166*, 467–471. [CrossRef]
22. Herraiz, C.; Diges, I.; Cobo, P.; Aparicio, J.M.; Toledano, A. Auditory discrimination training for tinnitus treatment: The effect of different paradigms. *Eur. Arch. Otorhinolaryngol.* **2010**, *267*, 1067–1074. [CrossRef] [PubMed]
23. Flor, H.; Hoffmann, D.; Struve, M.; Diesch, E. Auditory discrimination training for the treatment of tinnitus. *Appl. Psychophysiol. Biofeedback* **2004**, *29*, 113–120. [CrossRef] [PubMed]
24. Wang, H.; Tang, D.; Wu, Y.; Zhou, L.; Sun, S. The state of the art of sound therapy for subjective tinnitus in adults. *Ther. Adv. Chronic Dis.* **2020**, *11*, 1–22. [CrossRef] [PubMed]
25. McFerran, D.J.; Stockdale, D.; Holme, R.; Large, C.H.; Baguley, D.M. Why Is There No Cure for Tinnitus? *Front. Neurosci.* **2019**, *13*, 802. [CrossRef] [PubMed]
26. Barros, S.; Akira, F.; Kaouru, F.; Tsuneo, E.; Oliveira, N. Effectiveness of sound therapy in patients with tinnitus resistant to previous treatments: Importance of adjustments. *Braz. J. Otorhinolaryngol.* **2016**, *82*, 297–303. [CrossRef] [PubMed]
27. Alonso, L.; Gonzalez, J.; Ibarra, D. Neuropsychological monitoring of current acoustic therapies as alternative treatment of chronic tinnitus. *Am. J. Otolaryngol.* **2021**, *42*, 1–10. [CrossRef]
28. Ibarra, D.; Naal, N.; Alonso, L. Binaural sound therapy for tinnitus treatment: A psychometric and neurophysiological evaluation. *Am. J. Otolaryngol.* **2022**, *43*, 103248. [CrossRef]
29. Ibarra, D.; Alonso, L.; Cuevas, A.; Intriago, L. Acoustic Therapies for Tinnitus Treatment: An EEG Database. *Mendeley Data* **2021**. [CrossRef]
30. Chang, C.Y.; Hsu, S.H.; Pion-Tonachini, L.; Jung, T.P. Evaluation of Artifact Subspace Reconstruction for Automatic Artifact Components Removal in Multi-Channel EEG Recordings. *IEEE Trans. Biomed. Eng.* **2020**, *67*, 1114–1121. [CrossRef]
31. Krause, C.M. Cognition- and memory-related ERD/ERS responses in the auditory stimulus modality. *Prog. Brain Res.* **2006**, *159*, 197–207. [CrossRef] [PubMed]
32. Zhang, Z.; Li, H. *EEG Signal Processing and Feature Extraction*, 1st ed.; Springer: Singapore, 2019; pp. 23–116. [CrossRef]
33. Cahyo, A. Development of Mobile Skin Cancer Detection using Faster R-CNN and MobileNet V2 Model. In Proceedings of the ICITACEE, Semarang, Indonesia, 24–25 September 2020. [CrossRef]
34. Velasco, J.; Pascion, C.; Wilmar, J.; Apuang, J.; Cruz, J.; Gomez, M.; Molina, B.; Tuala, L.; Thio, A.; Jorda, R. A Smartphone-Based Skin Disease Classification Using MobileNet CNN. *Int. J. Adv. Trends Comput. Sci. Eng.* **2019**, *8*, 2632–2637. [CrossRef]
35. Howard, A.; Zhu, M.; Chen, B.; Kalenichenko, D.; Wang, W.; Weyand, T.; Adam, H. Mobilenets: Efficient convolutional neural networks for mobile vision applications. *arXiv* **2017**, arXiv:1704.04861. Available online: https://arxiv.org/abs/1704.04861 (accessed on 12 December 2021).
36. Wing, C.; Simon, K.; Bello, R. Designing Difference in Difference Studies: Best Practices for Public Health Policy Research. *Annu. Rev. Public Health* **2018**, *39*, 453–469. [CrossRef] [PubMed]
37. Klimesch, W. EEG alpha and theta oscillations reflect cognitive and memory performance: A review and analysis. *Brain Res. Rev.* **1999**, *29*, 169–195. [CrossRef]
38. Weisz, N.; Moratti, S.; Meinzer, M.; Dohrmann, K.; Elbert, T. Tinnitus perception and distress is related to abnormal spontaneous brain activity as measured by magnetoencephalography. *PLoS Med.* **2005**, *2*, e20153. [CrossRef] [PubMed]
39. Jackson, R.; Vijendren, A.; Phillips, J. Objective Measures of Tinnitus: A Systematic Review. *Otol. Neurotol.* **2019**, *40*, 154–163. [CrossRef]
40. Tyler, R.S.; Perreau, A.; Powers, T.; Watts, A.; Owen, R.; Ji, H.; Mancini, P.C. Tinnitus Sound Therapy Trial Shows Effectiveness for Those with Tinnitus. *J. Am. Acad. Audiol.* **2020**, *31*, 6–16. [CrossRef] [PubMed]

*Article*

# Differences in Physiological Signals Due to Age and Exercise Habits of Subjects during Cycling Exercise

Szu-Yu Lin [1,†], Chi-Wen Jao [1,2,†], Po-Shan Wang [1,3], Michelle Liou [4], Jun-Liang Wu [5], Hsiao Chun [1], Ching-Ting Tseng [1] and Yu-Te Wu [1,6,*]

1    Institution of Biophotonics, National Yang Ming Chiao Tung University, Taipei 112, Taiwan; betty810720@nycu.edu.tw (S.-Y.L.); c3665810@ms24.hinet.net (C.-W.J.); b8001071@yahoo.com.tw (P.-S.W.); apply91122@gmail.com (H.C.); shps961421@gmail.com (C.-T.T.)
2    Department of Research, Shin Kong Wu Ho-Su Memorial Hospital, Taipei 111, Taiwan
3    Department of Neurology, Municipal Gandau Hospital, Taipei 112, Taiwan
4    Institute of Statistical Science, Academia Sinica, Taipei 115, Taiwan; mliou@stat.sinica.edu.tw
5    Department of Health of Beitou District, Taipei City Government, Taipei 112, Taiwan; wclwclwcl@health.gov.tw
6    Brain Research Center, National Yang Ming Chiao Tung University, Taipei 112, Taiwan
*    Correspondence: ytwu@ym.edu.tw
†    These authors contributed equally to this paper.

**Abstract:** Numerous studies indicated the physical benefits of regular exercise, but the neurophysiological mechanisms of regular exercise in elders were less investigated. We aimed to compare changes in brain activity during exercise in elderly people and in young adults with and without regular exercise habits. A total of 36 healthy young adults (M/F:18/18) and 35 healthy elderly adults (M/F:20/15) participated in this study. According to exercise habits, each age group were classified into regular and occasional exerciser groups. ECG, EEG, and EMG signals were recorded using V-AMP with a 1-kHz sampling rate. The participants were instructed to perform three 5-min bicycle rides with different exercise loads. The EEG spectral power of elders who exercised regularly revealed the strongest positive correlation with their exercise intensity by using Pearson correlation analysis. The results demonstrate that exercise-induced significant cortical activation in the elderly participants who exercised regularly, and most of the $p$-values are less than 0.001. No significant correlation was observed between spectral power and exercise intensity in the elders who exercised occasionally. The young participants who exercised regularly had greater cardiac and neurobiological efficiency. Our results may provide a new exercise therapy reference for adult groups with different exercise habits, especially for the elders.

**Keywords:** exercise; EEG; EMG; ECG; brain activity; age; exercise habit

**Citation:** Lin, S.-Y.; Jao, C.-W.; Wang, P.-S.; Liou, M.; Wu, J.-L.; Chun, H.; Tseng, C.-T.; Wu, Y.-T. Differences in Physiological Signals Due to Age and Exercise Habits of Subjects during Cycling Exercise. *Sensors* **2021**, *21*, 7220. https://doi.org/10.3390/s21217220

Academic Editors: Vahid Abolghasemi, Hossein Anisi and Saideh Ferdowsi

Received: 30 September 2021
Accepted: 27 October 2021
Published: 29 October 2021

**Publisher's Note:** MDPI stays neutral with regard to jurisdictional claims in published maps and institutional affiliations.



## 1. Introduction

Regular physical exercise is associated with health benefits and is a crucial element of preventive strategies for promoting health. During exercise, moving the body requires a substantial degree of brain activity, necessitating the activation of numerous neurons to generate, receive, and interpret repeated, rapid-fire messages from the nervous system [1]. However, the neurophysiological mechanisms underlying the effects of exercise are poorly understood and require further investigation. Cycling is a common exercise, and daily cycling can enable a large proportion of the population to meet their recommended physical activity levels [2]. Several studies have reported that cycle ergometers are suitable for measuring physiological signals emitted during exercise. Studies on cycling exercise have reported that such exercise can induce specific changes in cortical activity. These changes are measured through various methods, including electroencephalography (EEG), the aim of which is to study the modulation of brain activity associated with performing cycling tasks [3–7].

Hottenrott et al. reported that cortical brain activation could be measured during cycling exercise; they suggested that higher cortical brain activation is necessary to increase muscle strength at higher cadences [4]. Enders et al. recently revealed that EEG power increased significantly in the frontal cortex and parietal cortex as fatigue accumulated throughout high-intensity cycling exercise activities. Notably, they observed a broadband increase in EEG power, in contrast to other studies that investigated various exercise conditions and observed changes that were limited to the alpha and beta bands [5]. Brummer et al. localized the exercise-induced changes in brain cortical activity by using the active-EEG/low-resolution electromagnetic tomography analysis and demonstrated that motor cortex activity increased with additional exercise intensity on a cycle ergometer [6]. Although Brummer et al. used different methodologies from other, earlier research, all of the aforementioned studies have focused on the effects of exercise intensity on cortical activity in young people or athletes [7]. Few studies have examined the activity of the cerebral cortex during exercise in other segments of the population, especially in older adults. Moreover, few studies have investigated the neurobiological differences between regular and occasional exercisers during physical exercise.

Accordingly, the aim of the present study was to investigate the changes in brain activity during exercise in elderly people and young adults. Previous studies have proposed the use of heart rate as a measure of exercise intensity [8]. They have described a positive linear correlation between increasing exercise intensity and changes in heart rate. However, because of age-related factors, the heart rate should not be directly used as an index for measuring exercise intensity. Santos reported that the aging process significantly alters the mean heart rate, which decreases with advancing age [9]. Therefore, the mean heart rates of young adults and elderly people at rest differ. In the present study, we used the average maximum heart rate ratio (AMHRR) [10,11], which can reduce the effect of age on the resting heart rate and maximum heart rate in response to exercise, and hypothesized that the AMHRR would facilitate the comparison of EEG and electromyography (EMG) readings between elderly people and young adults at the same exercise intensity.

We measured cardiac, cerebral, and muscular activity levels in elderly people and young adults in response to cycling exercise and investigated the differences between physiological signals obtained from four study groups: regularly exercising elderly people, occasionally exercising elderly people, regularly exercising young people, and occasionally exercising young people. In general, under a constant cycling period and intensity, regularly exercising young adults could achieve higher exercise efficiency with lower brain activation compared with the other participants. However, occasionally exercising young adults and elderly people may need to recruit more muscle units and increase the activation of the motor cortex during cycling compared with regularly exercising young adults. We hypothesized that physiological signal patterns would be similar between the occasionally exercising young adults and elderly people. We also anticipated that as age increases, the significant differences of physiological signals between occasional and regular exercisers may be more obvious in elderly adults than in young adults.

## 2. Materials and Methods

### 2.1. Participants and Data Acquisition

This study included 36 healthy young adults (18 men and 18 women aged 22.39 ± 3.56 years) and 35 elderly people (20 men and 15 women aged 64.65 ± 2.21 years) as participants. The elderly participants as well as the young participants were subdivided into 2 groups according to the time spent on exercise per week; specifically, participants who exercised for a total time of more than 3 h every week were considered as regularly exercising individuals, and those exercised for a total time of less than 3 h every week were regarded as occasionally exercising individuals [12]. All participants provided informed consent after receiving a detailed explanation of the purpose and potential benefits, and risks involved in the study. This study was conducted according to the guidelines of the Declaration of Helsinki and approved by the Institutional Review Board of National Yang Ming Chiao

Tung University (YM106115E-1, 7 March 2019). Moreover, all participants were confirmed by physicians that their body mass index (BMI) was not overweight and without any lower limb or pelvic injuries, and had no brain-related diseases such as stroke, epilepsy, neurodegenerative diseases, orthopedic, or cardiovascular diseases. ECG, EMG, and EEG signals were recorded using V-AMP (Brain Products GmbH, Munich, Germany) with a 1 kHz sampling rate. The EEG channels included 10 wired wet electrodes, namely F3, F4, Fz, C3, C4, Cz, P3, P4, Pz, and A1, and were used according to the international 10/20 system (Figure 1a) [13]. The ground electrode was positioned at FPz. The EEG impedance level was maintained at <20 kΩ during the recording. The A1 channel was used as the reference for all electrodes.

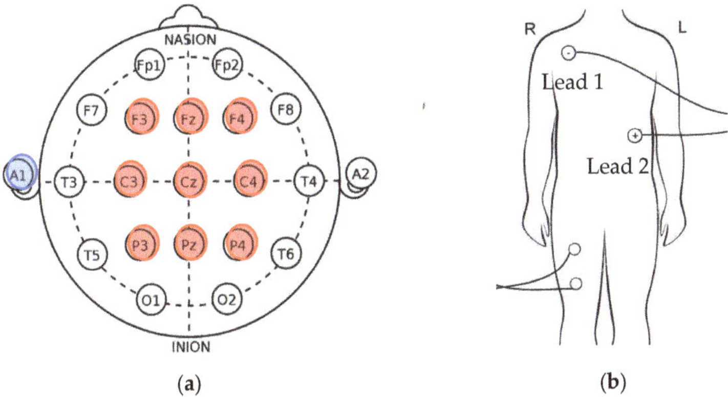


**Figure 1.** Locations of electrodes for EMG, ECG, and EEG. (a) location of electrodes for EEG (b) location of electrodes for ECG and EMG.

Electrocardiography (ECG) signals were recorded using 2 bipolar lead electrodes. The lead 1 (negative) electrode was situated below the right clavicle, on the mid-clavicular line within the rib cage frame; the lead 2 (positive) electrode was placed on the lower left abdomen, also within the rib cage frame. The surface EMG (sEMG) electrodes were placed on the quadriceps muscle (Figure 1b).

*2.2. Experimental Protocol*

We conducted an experiment to record EEG, ECG, and EMG signals while the participants performed the cycling exercise. Each of the participants sat on an electronically braked cycle ergometer in the upright position, with electrodes attached to their body. The study involved a pretest session and an experimental session. The pretest session involved 10 40 s stages of increasing workload with 20 s of rest between stages. For every participant, the workload ranged from 1 to 10. After the pretest session, the participants took a 5 min rest. The root mean square (RMS) amplitudes of EMG signals recorded for each stage were calculated, and the maximum RMS amplitude was considered the subject-specific maximum workload. For a participant, a workload corresponding to 40% of the maximum RMS amplitude was defined as the suitable workload for this participant. For safety, we assigned lighter exercise loads to the elderly participants to avoid injury or muscle damage due to over-load, considering the effects of declining physiological function with aging. Hence, the pretest session was considered to be excessively strenuous for the elderly participants, their suitable workload was set to 3.

In the experimental session, the participants were asked to ride the bicycle in 3 5-min exercise stages, resting for 30 s between stages. These 3 stages corresponded to relatively light, suitable, and relatively heavy workloads. EEG, ECG, and EMG signals were recorded simultaneously while the participants performed the exercise. Signals were also recorded for 5 min before the exercise (pre-exercise period) and for another 5 min after the exercise (post-exercise period). In this study, we required subjects to minimize their head and upper body movement as much as possible during the experiment. The participants were also asked not to move during the resting period. Figure 2 illustrates the overall experimental protocol.

**pretest session**

| Rest 20s | **WL1** 40s | Rest 20s | **WL2** 40s | Rest 20s | **WL3** 40s | Rest 20s | **WL4** 40s | Rest 20s | **WL5** 40s | Rest 20s | **WL6** 40s | Rest 20s | **WL7** 40s | Rest 20s | **WL8** 40s | Rest 20s | **WL9** 40s | Rest 20s | **WL10** 40s |
|---|---|---|---|---|---|---|---|---|---|---|---|---|---|---|---|---|---|---|---|

**experiment session**

| Pre-exercise rest 5 min | Lighter WL 5 min | Rest 30s | Suitable WL 5 min | Rest 30s | heavier WL 5 min | Post-exercise rest 5 min |
|---|---|---|---|---|---|---|

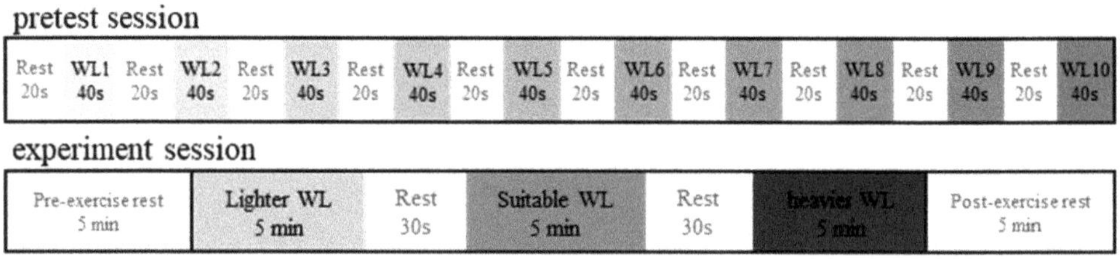

**Figure 2.** Overall experimental protocol.

*2.3. Data Analysis*

2.3.1. ECG Analysis

ECG signals were detrended to remove low-frequency shifts, and the peak-to-peak R waves were identified to calculate RR intervals. The RR interval is the time elapsed between 2 successive R waves of the QRS signal on the ECG. We further used the AMHRR to monitor the status of each participant during the experiment [14]. The AMHRR can be defined as follows:

$$\text{AMHRR} = \frac{\text{averaged heart rate in each stage} - \text{RHR}}{\text{predicted maximal heart rate}(220 - \text{age} - \text{RHR})} \times 100\% \tag{1}$$

where RHR is the average heart rate during rest [10,11].

2.3.2. EEG Analysis

For each participant, EEG signals recorded during the 5 min rest and during the exercise sessions were subjected to band-pass filtering between 1 and 45 Hz. Although participants were advised not to blink their eyes, clench their teeth, tense their muscles, or move their heads, these activities occasionally occurred and introduced artifacts into the EEG data. All signals with these artifacts were discarded during offline data processing. We further applied a moving average to the remaining signals for artifact suppression. Subsequently, each signal was divided into non-overlapping 1 min segments and then subjected to the wavelet transform [15].

The wavelet transform is based on small wavelets with a limited duration. The wavelet transform of a continuous-time signal $x(t)$ can be defined as follows:

$$\text{WT}(a, b) = \int_{-\infty}^{\infty} x(t)\psi_{(a,b)}(t)dt \tag{2}$$

where

$$\psi_{(a,b)}(t) = \frac{1}{\sqrt{|a|}}\psi\left(\frac{t - b}{a}\right) \tag{3}$$

is called the mother wavelet. The notations $a$ and $b$ denote the dimensionless frequency scale variable and time-like translation variable, respectively. The Wavelet transform enables the achievement of excellent localization both in the time domain through translations of the mother wavelet and in the scale (frequency) domain through dilations.

In this study, we used the Morlet wavelet [15] to transform each 1 min non-overlapping segment of an EEG signal (Figure 3b) in the 9 channels into temporal-spectral maps (Figure 3c). Each of these maps had 60,000 samples on the horizontal axis and 7 passbands—namely 1–4 (delta), 4–8 (theta), 8–10 (low alpha), 10–12 (high alpha), 13–21 (low beta), 21–30 (high beta), and 31–45 (gamma) Hz—on the vertical axis (Figure 3d). The spectral power levels in each frequency band were averaged to obtain a frequency-averaged temporal power curve, which was again averaged across time to derive a frequency-time-averaged value. Thus, the average power per minute per frequency band was calculated. Each exercise stage was 5 min. Thus, the average power was calculated for 3 different workloads (Figure 3e). Subsequently, to normalize the average power for each exercise stage, this power was divided by the power at rest before exercise, thus yielding the normalized power (Figure 3f).

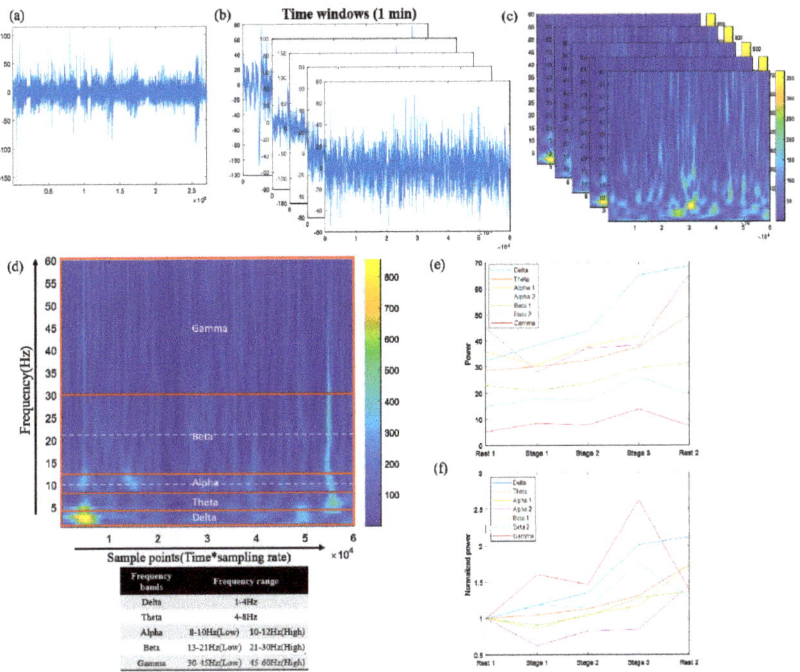

**Figure 3.** EEG signal analysis procedure. (**a**) Five-minute EEG signals during exercise. (**b**) Five-minute signals divided into 1-min segments. (**c**) Temporal–spectral maps after the application of the Morlet wavelet transform on 1-min segments. (**d**) Temporal–spectral map divided into seven bands: delta, theta, low-alpha, high-alpha, low-beta, high-beta, and gamma bands. (**e**) Average power of each frequency band in each exercise stage. (**f**) Normalized average power of each frequency band in each exercise stage.

### 2.3.3. EMG Analysis

The EMG signals were detrended to remove low-frequency shifts caused by the position fluctuations produced during the cycling exercise. The EMG signals were then subjected to band-stop filtering between 55 and 65 Hz for the removal of noise effects. After preprocessing, the signals were divided into 5 s segments (5000 sample points). RMS is usually used to predict muscle activity. Generally, a higher RMS value means higher muscle activity. RMS can be derived as follows:

$$\text{RMS} = \sqrt{\frac{1}{N}\sum_{n=1}^{N} x_n^2} \tag{4}$$

where $x_n^2$ represents the EMG signal and $N$ represents the length of the signal.

2.3.4. Statistical Analysis

Pearson correlation analysis was used to evaluate the linear relationships between normalized power of EEG and the AMHRR or RMS of EMG. The AMHRR was considered an indicator of heart load for the various exercise stages. Thus, we could observe EEG and EMG changes with different exercise loads. In addition, paired-sample *t*-tests were used to examine for significant within-group changes before and after exercise (stage 3 and rest 2) to determine the post-exercise recovery status. In this study, MATLAB R2013b software (Mathworks, Natick, MA, USA) was applied for data analysis. Figure 4 illustrates a summary of the analysis procedures of EEG, ECG and EMG used in this study.

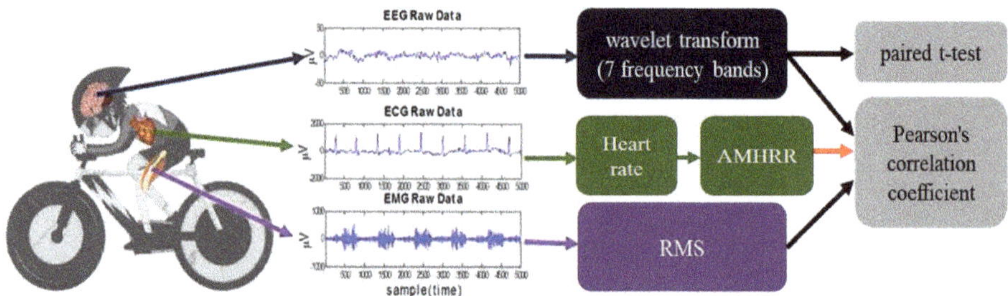


**Figure 4.** EEG, ECG, and EMG procedures in this study.

### 3. Results

*3.1. Changes in Heart Rate and AMHRR with Exercise Stages*

The mean heart rate and the AMHRR of the young and elderly participants during the different exercise stages are presented in Tables 1 and 2. Figure 5 illustrates the ECG analysis results for mean heart rate and AMHRR.The results revealed that in all groups, the heart rate and the AMHRR increased gradually with each exercise stage. The heart rates of the elderly participants were lower than those of the young participants. However, the AMHRR values of the elderly participants were not significantly different from those of the young participants, indicating that the cardiac load conditions of both the young and elderly participants were similar. The AMHRR was derived by normalizing the heart rate and excluding the effects of basal heart rate and age. Therefore, the AMHRR was suitable for observing the physiological state of the heart. We used Pearson correlation coefficient analysis to estimate the association between normalized EEG power and AMHRR per minute.

**Table 1.** ANOVA results for heart rate in young and elderly participants during different exercise stages. * Indicates *p*-value < 0.05.

| | | Heart Rate (BPm) | | | |
| --- | --- | --- | --- | --- | --- |
| | | Mean | SD | F | Post Hoc Test ($p < 0.05$) |
| Rest 1 | Youth Regular | 87.26 | 10.39 | 14.97 * | Elderly Regular, Occasional |
| | Youth Occasional | 91.67 | 13.86 | | Elderly Regular, Occasional |
| | Elderly Regular | 72.57 | 8.18 | | Youth Regular, Occasional |
| | Elderly Occasional | 73.07 | 9.27 | | Youth Regular, Occasional |
| Stage 1 | Youth Regular | 127.33 | 16.11 | 11.65 * | Elderly Regular, Occasional |
| | Youth Occasional | 122.19 | 13.39 | | Elderly Regular, Occasional |
| | Elderly Regular | 103.95 | 16.63 | | Youth Regular, Occasional |
| | Elderly Occasional | 103.65 | 14.23 | | Youth Regular, Occasional |

**Table 1.** *Cont.*

| | | Heart Rate (BPm) | | | |
|---|---|---|---|---|---|
| | | Mean | SD | F | Post Hoc Test ($p < 0.05$) |
| Stage 2 | Youth Regular | 140.34 | 19.90 | 9.35 * | Elderly Regular, Occasional |
| | Youth Occasional | 134.19 | 16.38 | | Elderly Regular, Occasional |
| | Elderly Regular | 115.43 | 19.18 | | Youth Regular, Occasional |
| | Elderly Occasional | 114.12 | 17.14 | | Youth Regular, Occasional |
| Stage 3 | Youth Regular | 152.12 | 21.66 | 10.81 * | Elderly Regular, Occasional |
| | Youth Occasional | 147.16 | 17.20 | | Elderly Regular, Occasional |
| | Elderly Regular | 124.22 | 20.69 | | Youth Regular, Occasional |
| | Elderly Occasional | 122.78 | 18.06 | | Youth Regular, Occasional |
| Rest 2 | Youth Regular | 110.19 | 16.87 | 5.65 | |
| | Youth Occasional | 109.42 | 18.10 | | |
| | Elderly Regular | 94.03 | 15.45 | | |
| | Elderly Occasional | 94.20 | 13.32 | | |

**Table 2.** ANOVA results for AMHRR in young and elderly participants during different exercise stages.

| | | AMHRR (%) | | | |
|---|---|---|---|---|---|
| | | Mean | SD | F | Post Hoc Test ($p < 0.05$) |
| Rest 1 | Youth Regular | | | | |
| | Youth Occasional | | | | |
| | Elderly Regular | | | | |
| | Elderly Occasional | | | | |
| Stage 1 | Youth Regular | 36.33 | 11.79 | 1.47 | |
| | Youth Occasional | 28.29 | 10.36 | | |
| | Elderly Regular | 37.78 | 17.69 | | |
| | Elderly Occasional | 36.89 | 16.17 | | |
| Stage 2 | Youth Regular | 48.33 | 14.58 | 1.41 | |
| | Youth Occasional | 40.72 | 12.84 | | |
| | Elderly Regular | 51.98 | 20.87 | | |
| | Elderly Occasional | 49.79 | 20.29 | | |
| Stage 3 | Youth Regular | 58.96 | 16.26 | 0.71 | |
| | Youth Occasional | 53.69 | 12.83 | | |
| | Elderly Regular | 62.54 | 24.02 | | |
| | Elderly Occasional | 60.56 | 21.28 | | |
| Rest 2 | Youth Regular | 20.95 | 9.60 | 1.96 | |
| | Youth Occasional | 17.59 | 8.66 | | |
| | Elderly Regular | 26.08 | 16.50 | | |
| | Elderly Occasional | 25.86 | 13.06 | | |

### 3.2. Changes in EEG during Exercise in Young Participants with and without Exercise Habits

We used Pearson correlation analysis to estimate the correlation between normalized EEG power and the AMHRR. Table 3 presents a summary of the regression coefficients of normalized EEG power and the AMHRR for all frequency bands. According to this table, a moderately strong correlation was observed, with the normalized coefficient ranging from 0.4 to 0.6. The results demonstrated that changes in EEG at the most frequency bands at C3, C4, and Cz were significantly and positively correlated with the AMHRR in both the young and elderly participants ($p < 0.001$). Moreover, the effect of exercise on EEG was mainly observed in the alpha band.

**Table 3.** Regression coefficients for the correlation between normalized EEG power and AMHRR during exercise. Boldface values represent moderate positive correlation between normalized EEG power and the AMHRR (correlation coefficient > 0.4).

| | | Delta | Theta | L-Alpha | H-Alpha | L-Beta | H-Beta | Gamma |
|---|---|---|---|---|---|---|---|---|
| **C3** | Youth Occasional | 0.1787 (0.1959) | 0.2137 (0.1207) | **0.4083** (0.0022) | **0.4699** (0.0003) | 0.3848 (0.0041) | 0.2407 (0.0796) | 0.1770 (0.2004) |
| | Youth Regular | **0.4426** (0.0008) | **0.4215** (0.0015) | **0.4679** (0.0004) | **0.4493** (0.0007) | 0.3831 (0.0042) | 0.2130 (0.1220) | 0.1404 (0.3113) |
| | Elderly Occasional | 0.3759 (0.0066) | 0.2405 (0.0892) | 0.2583 (0.0673) | 0.2672 (0.0580) | 0.2463 (0.0815) | 0.2088 (0.1414) | 0.2381 (0.0925) |
| | Elderly Regular | **0.7037** (<0.0001) | **0.6519** (<0.0001) | **0.6913** (<0.0001) | **0.6441** (<0.0001) | **0.5516** (<0.0001) | **0.5376** (<0.0001) | **0.5284** (<0.0001) |
| **C4** | Youth Occasional | 0.2498 (0.0685) | 0.2907 (0.0329) | **0.4961** (0.0001) | **0.5641** (<0.0001) | **0.4637** (0.0004) | 0.3110 (0.0221) | 0.2288 (0.0961) |
| | Youth Regular | 0.2103 (0.1268) | 0.2221 (0.1065) | 0.3645 (0.0067) | 0.3643 (0.0068) | 0.2607 (0.0569) | 0.0716 (0.6068) | −0.0280 (0.8406) |
| | Elderly Occasional | 0.3585 (0.0098) | 0.2222 (0.1170) | 0.2012 (0.1569) | 0.2087 (0.1417) | 0.2184 (0.1237) | 0.1866 (0.1899) | 0.2366 (0.0947) |
| | Elderly Regular | **0.6107** (<0.0001) | **0.6164** (<0.0001) | **0.6644** (<0.0001) | **0.6070** (<0.0001) | 0.3623 (0.0071) | 0.2948 (0.0305) | 0.2788 (0.0412) |
| **Cz** | Youth Occasional | 0.1729 (0.2112) | 0.2076 (0.1319) | 0.3869 (0.0038) | **0.4639** (0.0004) | 0.3673 (0.0063) | 0.2385 (0.0824) | 0.1706 (0.2176) |
| | Youth Regular | 0.2310 (0.0929) | 0.2845 (0.0371) | **0.4181** (0.0017) | **0.4113** (0.0020) | 0.3397 (0.0120) | 0.1716 (0.2146) | 0.0650 (0.6404) |
| | Elderly Occasional | 0.3904 (0.0046) | 0.2652 (0.0600) | 0.2624 (0.0629) | 0.2700 (0.0553) | 0.2668 (0.0584) | 0.2342 (0.0981) | 0.2798 (0.0468) |
| | Elderly Regular | **0.4380** (0.0009) | **0.4607** (0.0005) | **0.5498** (<0.0001) | **0.5505** (<0.0001) | **0.4959** (0.0001) | **0.5027** (0.0001) | **0.4986** (0.0001) |

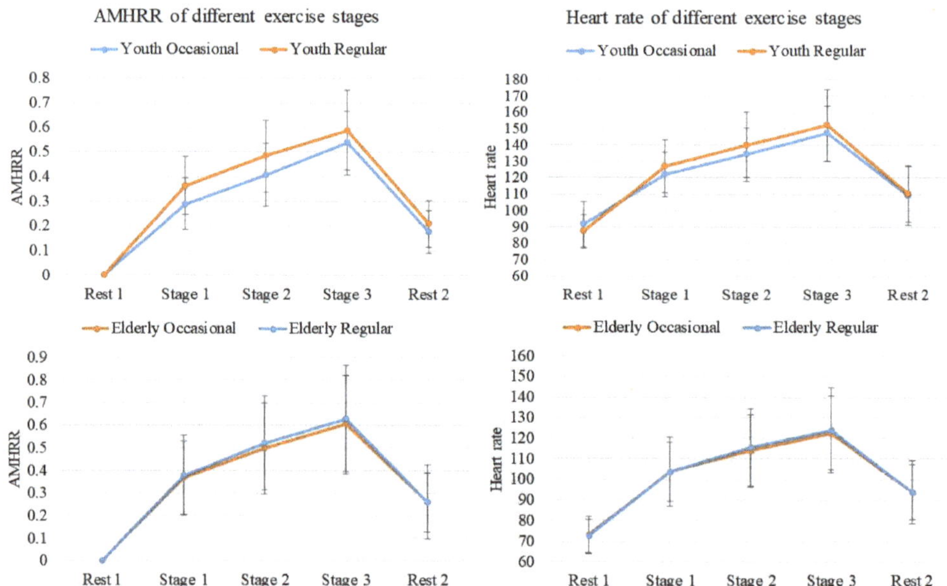



**Figure 5.** ECG analysis results for mean heart rate and AMHRR.

According to Table 3, we could also observe the effect of exercise habits on normalized EEG power in the young participants. The correlation between normalized EEG power at C3 and the AMHRR was higher in young participants who exercised regularly, and the correlation between normalized EEG power at C4 and the AMHRR was higher in young participants who exercised occasionally.

### 3.3. Changes in EEG during Exercise in Elderly Participants

As presented in Table 3, the regression coefficients revealed a moderate or high correlation between normalized EEG power and the AMHRR in the elderly participants who exercised regularly. However, the correlation observed for the elderly participants who exercised occasionally was low and nonsignificant. Combining the results for elderly participants and young participants revealed that maintaining adequate exercise habits was more imperative for older adults than for younger adults. As illustrated in Figure 6, the elderly participants who exercised regularly demonstrated consistent EEG power changes. As the AMHRR increased, the normalized EEG power also increased. By contrast, no clear trend was observed for the elderly participants who exercised occasionally. The changes in EEG power were more dispersed. These results indicated that adequate exercise habits may lead to more stable brain wave changes in elderly people during exercise.

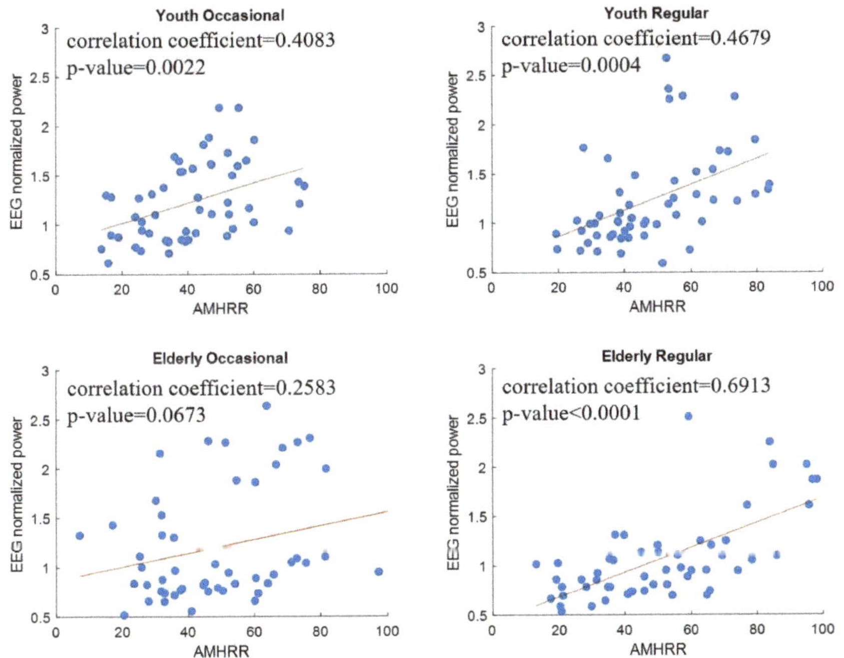


**Figure 6.** Scatter plots of correlation between normalized EEG power and AMHRR in low-alpha band (C3).

### 3.4. Paired t-Test Results Observed during and after Exercise

Figure 7 displays the normalized power values and statistical analysis results observed at C3 (low beta) at stage 3 and during post-exercise rest. Accordingly, the normalized power value during post-exercise rest would decrease to 1 if the power value during the pre- and post-exercise rest periods were identical. According to the plots in Figure 7, we observed the recovery speed of EEG power after exercise. The results revealed a significant difference in the change in normalized power between the exercise stage and post-exercise rest state in the young participants, regardless of their exercise habits. By contrast, in the elderly participants, the difference in the change in normalized power between stage 3 and post-exercise rest states was nonsignificant. Table 4 presents a summary of the results of the paired *t*-test for normalized EEG power in stage 3 and in the post-exercise rest state. In particular, the difference between the young and elderly participants was clearly observed in the beta band. The young participants recovered faster after exercise; therefore, a significant difference in the change in normalized power was observed. By contrast, the elderly participants recovered more slowly after exercise; hence, the difference in the change in normalized power was nonsignificant.

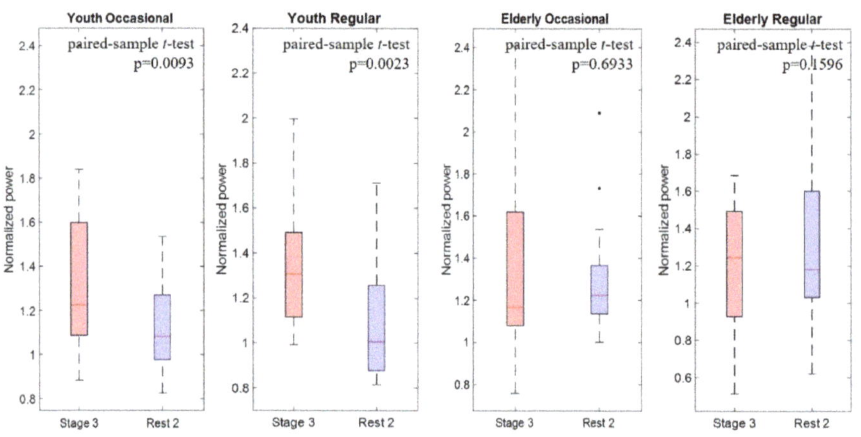

**Figure 7.** Histogram of EEG normalized power in low-beta (C3) band during exercise stage 3 and post-exercise rest.

**Table 4.** *p* values for paired-sample *t*-tests of normalized EEG power during and after exercise.

| | | Delta | Theta | L-Alpha | H-Alpha | L-Beta | H-Beta | Gamma |
|---|---|---|---|---|---|---|---|---|
| C3 | Youth Occasional | $p = 0.0861$ | $p = 0.0872$ | $p = 0.1032$ | $p = 0.0545$ | $p = 0.0093$ | $p = 0.0272$ | $p = 0.0064$ |
| | Youth Regular | $p < 0.001$ | $p = 0.0011$ | $p = 0.6801$ | $p = 0.5305$ | $p = 0.0023$ | $p = 0.0011$ | $p < 0.001$ |
| | Elderly Occasional | $p = 0.0122$ | $p = 0.0621$ | $p = 0.9527$ | $p = 0.9249$ | $p = 0.6933$ | $p = 0.1791$ | $p = 0.0027$ |
| | Elderly Regular | $p = 0.1019$ | $p = 0.6662$ | $p = 0.0180$ | $p = 0.0324$ | $p = 0.1596$ | $p = 0.3657$ | $p = 0.3446$ |
| C4 | Youth Occasional | $p = 0.0658$ | $p = 0.0499$ | $p = 0.0738$ | $p = 0.0348$ | $p = 0.0049$ | $p = 0.0151$ | $p = 0.0034$ |
| | Youth Regular | $p < 0.001$ | $p = 0.0022$ | $p = 0.4626$ | $p = 0.4761$ | $p = 0.0021$ | $p < 0.001$ | $p < 0.001$ |
| | Elderly Occasional | $p = 0.0240$ | $p = 0.0543$ | $p = 0.7284$ | $p = 0.8185$ | $p = 0.5308$ | $p = 0.1379$ | $p = 0.0017$ |
| | Elderly Regular | $p = 0.0122$ | $p = 0.4295$ | $p = 0.0197$ | $p = 0.0569$ | $p = 0.5484$ | $p = 0.9590$ | $p = 0.1634$ |
| Cz | Youth Occasional | $p = 0.0808$ | $p = 0.0902$ | $p = 0.1752$ | $p = 0.0640$ | $p = 0.0106$ | $p = 0.0252$ | $p = 0.0045$ |
| | Youth Regular | $p = 0.0015$ | $p = 0.0211$ | $p = 0.9572$ | $p = 0.8219$ | $p = 0.0109$ | $p = 0.0037$ | $p < 0.001$ |
| | Elderly Occasional | $p = 0.0254$ | $p = 0.0952$ | $p = 0.8433$ | $p = 0.8706$ | $p = 0.9117$ | $p = 0.4567$ | $p = 0.0040$ |
| | Elderly Regular | $p = 0.1706$ | $p = 0.3402$ | $p = 0.6988$ | $p = 0.6412$ | $p = 0.7912$ | $p = 0.6926$ | $p = 0.0321$ |

Overall, the alpha and beta bands could reflect changes in brain wave power during and after exercise. The alpha band can be used to observe changes in brain wave power during exercise, and the beta band can be used to observe recovery in rest states after exercise. However, further understanding of the effect of age and exercise habits on EEG changes is warranted.

*3.5. Relationship between EMG RMS and AMHRR for the Four Test Groups*

Figure 8 displays the results of the linear regression on the differences in EMG RMS between stages 2 and 1 (i.e., ΔEMG RMS) and the difference in AMHRR between stages 2 and 1 (i.e., ΔAMHRR). Because of the extensive individual differences in EMG RMS values and the location of the EMG bipolar electrodes, we normalized the EMG RMS values; that is, we divided the RMS values for stages 2 and 3 by those for stage 1. This can be used to observe the increase in ΔEMG RMS with exercise load. The results revealed a more significant trend of increasing ΔEMG RMS with ΔAMHRR in the young participants than in the elderly participants. Additionally, the regression coefficients for the young participants who exercised occasionally were higher than those for the young participants who exercised regularly. However, for the elderly participants, a low correlation was observed between the ΔEMG RMS values and ΔAMHRR, regardless of their exercise habits, and their ΔEMG RMS values were more clustered. This low correlation may be because in this study, the elderly participants were assigned a fixed cycling load that was lower than those assigned to the young participants.

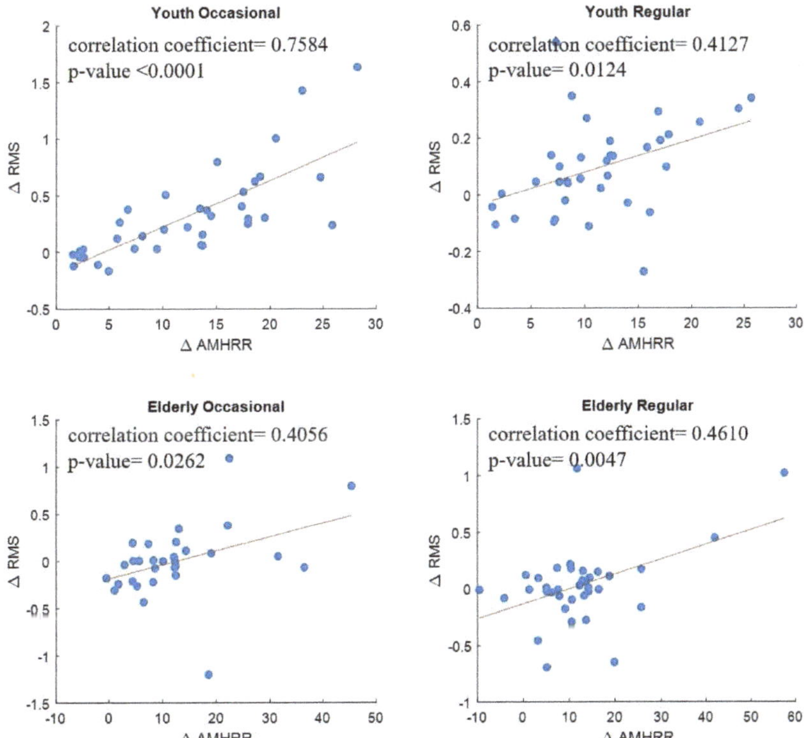


**Figure 8.** Linear regression between △EMG RMS and △AMHRR.

## 4. Discussion

This study used ECG, EMG, and EEG to explore changes in physiological signals transmitted during cycling exercise in young and elderly participants with different exercise habits. We assigned lighter exercise loads to the elderly participants to avoid muscle damage or injury from overload, considering the effects of declining physiological function with aging. According to previous research, exercise intensity (workload) is reflected in the response of many physiological processes, including heart rate [16]. Therefore, we defined the exercise load according to the AMHRR and further observed changes in EEG and EMG with gradually increasing exercise loads.

### 4.1. Spectral Power of EEG Increases with AMHRR during Exercise

We observed that during exercise, the normalized power of each frequency band of the EEG signal was positively and linearly correlated with the AMHRR. We also determined that an increase in normalized EEG power was consistent with an increase in AMHRR. This consistency was observed in most EEG frequency bands, including the delta, theta, low-alpha, high-alpha, low-beta, and high-beta bands. Furthermore, these phenomena were more evident in the low-alpha, high-alpha, low-beta, and high-beta frequency bands. Earlier research reported that cortical activity increased with fatigue during exercise in order to maintain a constant physical output [1]. Schillings et al. also reported that the energy loss associated with fatigue during exercise may cause increased brain activation in the motor cortex [17].

Previous studies determined that during exercise, EEG cortical activation was most affected in the alpha and beta frequency bands [18–21]. Therefore, most experiments and literature reviews on the effects of exercise on EEG cortical activity were limited to these two frequency bands. Several previous studies involving ergometer cycling revealed that incremental graded exercise tests resulted in increased alpha power in the central and parietal regions as well as increased EEG current density in the primary motor region. Bailey et al. showed an increase in alpha and beta power after sustained intensity bicycle ergometer exercise with a progressively increasing workload [3]. Lin et al. reported increased EEG power in the alpha and beta bands in the frontal and central areas during high-resistance pedaling exercise [22]. They further proposed that the fatigue situation would be accompanied by an increase in $\alpha$ and $\beta$ power. However, increased EEG beta activity may be associated with attentional demands and higher levels of arousal. Other studies demonstrated that the effect of exercise on EEG cortical activity was not limited to the alpha and beta bands [3,5]. Our results demonstrated that the alpha band was more suitable for observing changes in brain activation during exercise. However, the beta band was more appropriate for determining the differences between brain activation observed during exercise and that observed during post-exercise rest.

### 4.2. Young People Who Exercise Regularly Have a More Coordinated Use of Their Dominant Leg

Our results reveal that the EEG differences between young participants who exercised regularly and those who exercised occasionally were in the activation of motor cortical areas in the left and right hemispheres (i.e., C3 and C4). A higher correlation was observed between normalized power changes at C3 and exercise load in the young participants who exercised regularly. However, the normalized power at C3 and that at C4 in the young participants who exercised occasionally were moderately correlated. The concept of limb dominance was based on the fact that the two hemispheres of the brain function differently and tend toward activities that use one limb under voluntary control [23]. Bhise et al. observed that when for an inherently manipulative task, most participants used the dominant leg [24]. Young people who exercise regularly have greater coordination in the use of the dominant leg, meaning that they require only the dominant leg to complete the exercise. However, young people who exercise occasionally must use both legs to compensate for the deficiency of the dominant leg [25–27]. The RMS of EMG signals is often used as a concise quantitative indicator of muscle activity; we found that the young participants who exercised occasionally had significantly higher EMG RMS values than did those who exercised regularly. Our results indicate that the dominant legs of young people who exercise occasionally require more force output to perform a given task. However, that the young participants who exercised occasionally had lower EEG activation in the C3 region than did those who exercised regularly.

### 4.3. Regular Exercise in Elderly People Induces Significant Cortical Activation during Exercise

We observed that the highest increase in EEG normalized power occurred when the participants were at their highest AMHRR (exercise workload). This phenomenon was particularly notable in the elderly participants who exercised regularly. The results reveal that the normalized EEG power increased with the AMHRR in the elderly participants who exercised regularly, with the corresponding correlation being moderate to high. The heart rate increases with the delivery of oxygenated blood around the body and into the brain. Muscles require relatively high energy during exercise. Similarly, the brain consumes glucose or other carbohydrates when the body is in motion [28]. Therefore, the brain becomes more active during exercise. This suggests that elderly people who exercise regularly require relatively high exercise performance and muscle strength during exercise, which may induce considerable activation of the cerebral cortex. However, the change in normalized EEG power with respect to exercise load was less consistent in the elderly participants who exercised occasionally. Accordingly, the results reveal no significant correlation between normalized EEG power and the AMHRR. Although this phenomenon could also be observed in the young participants, the results were less pronounced than those observed in the elderly participants. Our results show that the difference in EEG signal changes between the elderly participants who exercised occasionally and those who exercised regularly was more significant than that between the young participants who exercised occasionally and those who exercised regularly. For elderly people, regular exercise can help reduce the functional decline associated with aging.

### 4.4. EEG Recovery after Exercise Is Slower in Elderly People

The paired *t*-test revealed significant beta band activation in the young participants in stage 3 and during post-exercise rest. By contrast, this phenomenon was not observed in the elderly participants. These results indicate that the young participants returned to a resting state more quickly after exercise, whereas the elderly participants required a longer time to recover. Aging affects the post-exercise recovery process. Several studies have revealed a functional decrease in the replenishment of energy supply before and after exercise. Research has presented evidence of differences in acute recovery of physiological parameters after fatiguing exercise between younger and older participants. For similar exercise stimuli, elderly people require a longer recovery period when returning to baseline levels after exercise [29]. Although this study did not reveal a significant difference in exercise recovery between the elderly participants with and without exercise habits, the EEG results demonstrate that the elderly participants who exercised regularly had superior brain regulation of exercise load than did those who exercised occasionally.

However, there are still some limitations in this study. First, the muscle artifacts occurring in the head and neck musculature during cycling exercise may be recorded in EEG signals. In this study, we asked subjects to minimize their head and upper body movement as much as possible during the experiment. Unfortunately, experimental protocols are still sensitive to physiological and non-physiological artifacts, including motion artifacts that may contaminate the EEG recordings. Following the procedure of artifact suppression, we applied a simple cleaning noise method, moving average, to remove the noise caused from motion artifacts in EEG signals. Although these procedures can eliminate motion artifacts but may also decrease the sensitivity in EEG signals. Second, the strength of muscle will decrease with aging, and there exist an individual difference in this aging effect. In this study, we did not take the muscle strength decay of elderly and individual muscle ability into consideration in the experiment setup.

## 5. Conclusions

This study revealed the AMHRR to be a suitable indicator of exercise intensity and that the physiological indicators of ECG and EEG in elderly people are different from those in young people because of aging. We found that the EEG spectral power of elders who exercised regularly revealed the strongest positive correlation with their exercise intensity.

The results demonstrate that exercise-induced significant cortical activation in the elderly participants who exercised regularly, and most of the *p*-values are less than 0.001. No significant correlation was observed between spectral power and exercise intensity in the elders who exercised occasionally. The young participants who exercised regularly had greater cardiac and neurobiological efficiency. Therefore, appropriate exercise habits may benefit brain responsiveness and improve the efficiency of cardiac and neurobiological responses to exercise. Our results may provide a new exercise therapy reference for adult groups with different exercise habits, especially for the elders.

**Author Contributions:** Conceptualization, S.-Y.L., C.-W.J. and Y.-T.W.; methodology, P.-S.W., Y.-T.W. and M.L.; software, S.-Y.L., H.C. and C.-T.T.; validation, C.-W.J., P.-S.W. and Y.-T.W.; formal analysis, S.-Y.L. and C.-W.J.; investigation, P.-S.W. and Y.-T.W.; resources, H.C., C.-T.T. and J.-L.W.; data curation, S.-Y.L., H.C., C.-T.T. and J.-L.W.; writing—original draft preparation, S.-Y.L. and C.-W.J.; writing—review and editing, S.-Y.L., C.-W.J. and Y.-T.W.; visualization, S.-Y.L. and C.-W.J.; supervision, P.-S.W. and Y.-T.W.; project administration, Y.-T.W. and J.-L.W.; funding acquisition, Y.-T.W. and J.-L.W. All authors have read and agreed to the published version of the manuscript.

**Funding:** This research was funded by National Yang Ming Chiao Tung University, grant number 110BRC-B701, Ministry of Education (MOE),Taiwan, grant number 110W219 and the APC was funded by National Yang Ming Chiao Tung University and Ministry of Education, Taiwan.

**Institutional Review Board Statement:** The study was conducted according to the guidelines of the Declaration of Helsinki and approved by the Institutional Review Board of National Yang Ming Chiao Tung University (YM106115E-1, 7 March 2019).

**Informed Consent Statement:** Informed consent was obtained from all subjects involved in the study, and written informed consent has been obtained from the patients to publish this paper.

**Data Availability Statement:** The data are not publicly available due to the privacy concern raised by our IRB.

**Acknowledgments:** The authors thank all the subjects participated in this study and thank Wallace academic editing company for editing this manuscript.

**Conflicts of Interest:** The authors declare no conflict of interest. The funders had no role in the design of the study; in the collection, analyses, or interpretation of data; in the writing of the manuscript, or in the decision to publish the results.

# References

1. Taylor, J.L.; Amann, M.; Duchateau, J.; Meeusen, R.; Rice, C.L. Neural contributions to muscle fatigue: From the brain to the muscle and back again. *Med. Sci. Sports Exerc.* **2016**, *48*, 2294. [CrossRef]
2. Buehler, R.; Pucher, J.; Merom, D.; Bauman, A. Active travel in Germany and the US: Contributions of daily walking and cycling to physical activity. *Am. J. Prev. Med.* **2011**, *41*, 241–250. [CrossRef]
3. Bailey, S.P.; Hall, E.E.; Folger, S.E.; Miller, P.C. Changes in EEG during graded exercise on a recumbent cycle ergometer. *J. Sports Sci. Med.* **2008**, *7*, 505.
4. Hottenrott, K.; Taubert, M.; Gronwald, T. Cortical brain activity is influenced by cadence in cyclists. *Open Sports Sci. J.* **2013**, *6*, 9–14. [CrossRef]
5. Enders, H.; Cortese, F.; Maurer, C.; Baltich, J.; Protzner, A.B.; Nigg, B.M. Changes in cortical activity measured with EEG during a high-intensity cycling exercise. *J. Neurophysiol.* **2016**, *115*, 379–388. [CrossRef]
6. Brümmer, V.; Schneider, S.; Strüder, H.K.; Askew, C.D. Primary motor cortex activity is elevated with incremental exercise intensity. *Neuroscience* **2011**, *181*, 150–162. [CrossRef]
7. Brümmer, V.; Schneider, S.; Abel, T.; Vogt, T.; Strueder, H.K. Brain cortical activity is influenced by exercise mode and intensity. *Med. Sci. Sports Exerc.* **2011**, *43*, 1863–1872. [CrossRef]
8. Karvonen, J.; Vuorimaa, T. Heart rate and exercise intensity during sports activities. *Sports Med.* **1988**, *5*, 303–311. [CrossRef]
9. Santos, M.A.A.; Sousa, A.C.S.; Reis, F.P.; Santos, T.R.; Lima, S.O.; Barreto-Filho, J.A. Does the aging process significantly modify the Mean Heart Rate? *Arq. Bras. Cardiol.* **2013**, *101*, 388–398. [CrossRef] [PubMed]
10. Camarda, S.R.D.A.; Tebexreni, A.S.; Páfaro, C.N.; Sasai, F.B.; Tambeiro, V.L.; Juliano, Y.; Barros Neto, T.L.D. Comparison of maximal heart rate using the prediction equations proposed by Karvonen and Tanaka. *Arq. Bras. Cardiol.* **2008**, *91*, 311–314. [CrossRef]
11. Karvonen, J.J.; Kentala, E.; Mustala, O. The Effects of Training on Heart Rate: A "Longitudinal" Study. *Ann. Med. Exp. Biol. Fenn.* **1957**, *35*, 307–315.

12. Piercy, K.L.; Troiano, R.P.; Ballard, R.M.; Carlson, S.A.; Fulton, J.E.; Galuska, D.A.; Olson, R.D. The physical activity guidelines for Americans. *JAMA* **2018**, *320*, 2020–2028. [CrossRef]
13. Homan, R.W.; Herman, J.; Purdy, P. Cerebral location of international 10–20 system electrode placement. *Electroencephalogr. Clin. Neurophysiol.* **1987**, *66*, 376–382. [CrossRef]
14. Goldberg, L.; Elliot, D.L.; Kuehl, K.S. Assessment of exercise intensity formulas by use of ventilatory threshold. *Chest* **1988**, *94*, 95–98. [CrossRef]
15. Daubechies, I. *The Wavelet Transform, Time-Frequency Localization and Signal Analysis*; Princeton University Press: Princeton, NJ, USA, 2009; pp. 442–486.
16. Michael, S.; Graham, K.S.; Davis, G.M. Cardiac autonomic responses during exercise and post-exercise recovery using heart rate variability and systolic time intervals—A review. *Front. Physiol.* **2017**, *8*, 301. [CrossRef]
17. Schillings, M.L.; Kalkman, J.S.; Van Der Werf, S.P.; Bleijenberg, G.; van Engelen, B.G.M.; Zwarts, M.J. Central adaptations during repetitive contractions assessed by the readiness potential. *Eur. J. Appl. Physiol.* **2006**, *97*, 521–526. [CrossRef]
18. Crabbe, J.B.; Dishman, R.K. Brain electrocortical activity during and after exercise: A quantitative synthesis. *Psychophysiology* **2004**, *41*, 563–574. [CrossRef]
19. Kamijo, K.; Nishihira, Y.; Hatta, A.; Kaneda, T.; Kida, T.; Higashiura, T.; Kuroiwa, K. Changes in arousal level by differential exercise intensity. *Clin. Neurophysiol.* **2004**, *115*, 2693–2698. [CrossRef]
20. Nielsen, B.; Hyldig, T.; Bidstrup, F.; Gonzalez-Alonso, J.; Christoffersen, G.R.J. Brain activity and fatigue during prolonged exercise in the heat. *Pflügers Arch.* **2001**, *442*, 41–48. [CrossRef]
21. Nybo, L.; Nielsen, B. Perceived exertion is associated with an altered brain activity during exercise with progressive hyperthermia. *J. Appl. Physiol.* **2001**, *91*, 2017–2023. [CrossRef]
22. Lin, M.A.; Meng, L.F.; Ouyang, Y.; Chan, H.L.; Chang, Y.J.; Chen, S.W.; Liaw, J.W. Resistance-induced brain activity changes during cycle ergometer exercises. *BMC Sports Sci. Med. Rehabil.* **2021**, *13*, 27. [CrossRef] [PubMed]
23. Velotta, J.; Weyer, J.; Ramirez, A.; Winstead, J.; Bahamonde, R. Relationship between leg dominance tests and type of task. In Proceedings of the ISBS-Conference Proceedings Archive, Porto, Portugal, 27 June–1 July 2011.
24. Bhise, S.A.; Patil, N.K. Dominant and Non dominant Leg Activities in Young Adults. *Int. J. Ther.* **2016**, *5*, 257–264. [CrossRef]
25. Carpes, F.; Rossato, M.; Faria, I.; Mota, C.B. During a simulated 40-km cycling time-trial. *J. Sports Med. Phys. Fit.* **2007**, *47*, 51–57.
26. Lepers, R.O.; Hausswirth, C.H.; Maffiuletti, N.I.; Brisswalter, J.E.; Van Hoecke, J. Evidence of neuromuscular fatigue after prolonged cycling exercise. *Med. Sci. Sports Exerc.* **2000**, *32*, 1880–1886. [CrossRef]
27. Iannetta, D.; Passfield, L.; Qahtani, A.; MacInnis, M.J.; Murias, J.M. Interlimb differences in parameters of aerobic function and local profiles of deoxygenation during double-leg and counterweighted single-leg cycling. *Am. J. Physiol.—Regul. Integr. Comp. Physiol.* **2019**, *317*, R840–R851. [CrossRef]
28. Nybo, L.; Secher, N.H. Cerebral perturbations provoked by prolonged exercise. *Prog. Neurobiol.* **2004**, *72*, 223–261. [CrossRef]
29. Fell, J.; Williams, A.D. The effect of aging on skeletal-muscle recovery from exercise: Possible implications for aging athletes. *J. Aging Phys. Act.* **2008**, *16*, 97–115. [CrossRef]

*Article*

# Safe *Hb* Concentration Measurement during Bladder Irrigation Using Artificial Intelligence

Gerd Reis [1,*], Xiaoying Tan [1], Lea Kraft [2], Mehmet Yilmaz [2], Dominik Stephan Schoeb [2] and Arkadiusz Miernik [2]

[1] Department Augmented Vision, German Research Center for Artificial Intelligence, 67663 Kaiserslautern, Germany; xiaoying.tan@dfki.de

[2] Medical Centre, Department of Urology, Faculty of Medicine, University of Freiburg, 79106 Freiburg, Germany; lea.kraft@uniklinik-freiburg.de (L.K.); mehmet.yilmaz@uniklinik-freiburg.de (M.Y.); dominik.stefan.schoeb@uniklinik-freiburg.de (D.S.S.); arkadiusz.miernik@uniklinik-freiburg.de (A.M.)

* Correspondence: gerd.reis@dfki.de

**Abstract:** We have developed a sensor for monitoring the hemoglobin (*Hb*) concentration in the effluent of a continuous bladder irrigation. The *Hb* concentration measurement is based on light absorption within a fixed measuring distance. The light frequency used is selected so that both arterial and venous *Hb* are equally detected. The sensor allows the measurement of the *Hb* concentration up to a maximum value of 3.2 g/dL (equivalent to $\approx$20% blood concentration). Since bubble formation in the outflow tract cannot be avoided with current irrigation systems, a neural network is implemented that can robustly detect air bubbles within the measurement section. The network considers both optical and temporal features and is able to effectively safeguard the measurement process. The sensor supports the use of different irrigants (salt and electrolyte-free solutions) as well as measurement through glass shielding. The sensor can be used in a non-invasive way with current irrigation systems. The sensor is positively tested in a clinical study.

**Keywords:** hemoglobin sensor; bladder irrigation monitor; absorption near infrared; artificial intelligence; bubble detection

**Citation:** Reis, G.; Tan, X.; Kraft, L.; Yilmaz, M.; Schoeb, D.S.; Miernik, A. Safe *Hb* Concentration Measurement during Bladder Irrigation Using Artificial Intelligence. *Sensors* **2021**, *21*, 5723. https://doi.org/10.3390/s21175723

Academic Editors: Vahid Abolghasemi, Hossein Anisi, Saideh Ferdowsi and Stefano Bettati

Received: 25 June 2021
Accepted: 24 August 2021
Published: 25 August 2021

**Publisher's Note:** MDPI stays neutral with regard to jurisdictional claims in published maps and institutional affiliations.

## 1. Introduction

In urology, continuous bladder irrigation (CBI) is an important standard care procedure [1–4] after transurethral resection of the bladder (TURB) or the prostate (TURP). The dominant goal of CBI is to prevent the formation of blood clots and consecutive bladder tamponade, a medical condition requiring an additional and foremost avoidable follow-up surgery [5]. The purpose of CBI in the given application scenario is, therefore, to keep the blood concentration in the bladder at a very low level. Technically, CBI provides a continuous dilution of the bladder content with fresh irrigation fluid (often saline) and thus, prevents clot formation. Although nearly trivial from a pure technical point of view, CBI is involved when applied in clinical practice. Improper CBI may trigger bladder spasms by irritating the bladder, cause undesired bleeding [5], lead to bladder rupture or perforation [6], and might even become life-threatening [7]. These and other possible complications result primarily from increased pressure, due to a high flow rate. Consequently, there are two optimization goals for an optimally adjusted CBI: On the one hand, the flow should be high enough to dilute the blood sufficiently, and on the other hand, the flow should be as low as possible to avoid pressure-related complications. Since the amount of bleeding after surgery cannot be controlled, the optimal flow speed changes over time. Hence, CBI demands extensive and continuous supervision and management by medical personnel, imposing a heavy burden on nurses responsible for a whole urological station [8,9].

CBI supervision is, on the one hand, comprised of rather technical aspects, such as caring for a filled fluid reservoir and an empty waste reservoir, ensuring a continuous flow of fluids into and out of the bladder. On the other hand, there are aspects with high medical relevance, in particular, choosing the right irrigation flow. Too high a flow will keep the blood concentration low, but may cause severe negative effects to the patient. In the case of the urethral catheter becoming clogged by a clot, continued irrigation may even lead to bladder rupture [5]. Too low a flow will not suffice to dilute the blood and will result in an ineffective procedure.

Today, nurses inspect the coloring of the waste fluid and estimate the flow speed accordingly. Obviously, this approach has significant drawbacks. Estimating the blood concentration in the outflow demands a high level of experience and is observer-dependent. Furthermore, it is affected by external conditions, such as the current illumination. Additionally, nutrition can have a significant effect on urine color, such as betanins leading to beeturia [10], easily confused with hematuria. Other foods known to change urine color include blackberries and rhubarb, which turn urine pink or red, while fava beans and aloe turn it reddish brown. Medications can also affect the color of urine, such as phenazopyridine, a drug used to numb urinary symptoms, the antibiotic rifampin, and laxatives containing senna, which can turn urine reddish-orange. The anti-inflammatory drugs sulfasalazine and phenazopyridine, as well as certain chemotherapy drugs turn urine orange-red, while the antimalarials chloroquine and primaquine, the antibiotics metronidazole and nitrofurantoin, the muscle relaxant methocarbamol, and laxatives containing cascara turn urine reddish-brown. In addition to the drug itself, the food coloring contained in the coating, e.g., in the case of tablets, pills, and dragees, can also discolor the urine. The discoloration of the urine and, thus, of the waste fluid is easily mistaken for an increased amount of blood, which in turn leads to an increased irrigation flow. The color of the excretory fluid is, therefore, not a reliable indicator for the adjustment of the flow. Detection of acute bleeding might become obscured, again putting patients at risk.

In order to better estimate the blood concentration, Hageman et al. [11] developed the Hemostick, a color scale to visually compare blood color. While this device significantly contributes to standardize the estimation of blood concentration between nurses, it still is affected by illumination influences. Furthermore only a limited number of discrete reference colors are available. Therefore, as can be observed in Figure 9, the distinction of blood concentration based on observation or color comparison becomes quickly unfeasible. Ding et al. [12] evaluated a CBI control system able to adjust the irrigation flow automatically based on the estimation of the blood concentration. The system features a color monitor to estimate the blood concentration. It thus resembles an automated version of the approach presented in [11]. In [13], Chan et al. presented a device to measure the blood concentration, using the light absorption principle. In order to estimate the blood concentration, they investigated light emitting diodes (LED) with different colors (i.e., red, green, blue) and finally decided on using green LED. Unfortunately, they did not specify the particular wavelengths of the used LEDs. To measure the transmitted light, a light dependent resistor (LDR) was used. Timm et al. [14] proposed a system based on light absorption for non-invasive estimation of hemoglobin ($Hb$) concentration in human tissue. They used three different, well-defined wavelengths to estimate the $Hb$ concentration. The system was tested in a technical setting. Zhang et al. [15] described a system to estimate blood loss during endoscopic surgery based on $Hb$ measurements. Their system is also based on absorption since they use a photoelectric sensor. Unfortunately, they did not provide any specifics on the sensor.

The new $Hb$ concentration sensor proposed in this article also exploits the light absorption principle and thus, is closely related to the work in [13–15]. The advantage of using the absorption principle compared to color monitoring is the improved accuracy of the estimates. Independence from dietary coloring effects can be achieved through the selection of appropriate light frequencies. In contrast to [13], our approach is far more rigorous. The light absorption properties of blood and other coloring components were taken into account, and an optimal light frequency was selected. In contrast to [14], our system is dedicated to monitor CBI. The measurements take the light absorption of the irrigation tubing into account, as well as influences of the irrigation fluid. In contrast to [15], our system measures the *Hb* concentration in a defined tube in contrast to being integrated in a collecting bucket. In addition, practical aspects, such as gross mismeasurements due to bubble formation in the outflow tubes and measurement through thin glass panes, were taken into account. The system was evaluated in a clinical study.

The proposed sensor is part of a comprehensive mobile CBI monitoring system. Although the overall system is outside the scope of this paper, the main reasons for its developments will be briefly presented. As mentioned earlier, continuous monitoring of CBI is important. In this regard, a single nurse can take care of about two to four patients, provided that they are in the same room and the nurse has no other duties. In reality, however, there are easily twenty concurrent CBI on a urology ward, spread across multiple rooms and cared for by a single nurse. Therefore, continuous monitoring of all patients is not feasible. In addition, the nurse must maintain the CBIs by emptying or changing the waste bags and replacing empty saline bags in a timely manner. Finally, the nurse also has to manage the daily routine of attending to the well-being of patients, dispensing medications, conferring with physicians, and much more. Regardless of what a nurse is doing, he/she always has the pressure in the background that he/she should be monitoring CBIs. This has several negative consequences. Nurses are under constant stress, even when monitoring a particular CBI because other CBIs cannot be monitored at the same time. Nevertheless, acute bleeding in one or more patients can occur at any time. On the other hand, maintenance of the CBIs also places a burden on the nursing staff since overflowing waste bags as well as empty reservoir bags should be avoided as much as possible. The monitoring system relieves the nursing staff of this pressure since the flow rate and *Hb*/blood concentration are constantly monitored by a technical system. The flow and concentration measurements can be used to effectively calculate whether irrigation is being performed optimally. As soon as one of the parameters leaves its predefined limits, alarms can be transmitted to a mobile device and/or to the ward room. In addition, the current status of all CBIs can be visualized simultaneously, giving the nurse an optimal overview of all patients. The use of the monitoring system allows nurses to fully concentrate on their respective tasks, while giving them the security of knowing that they will be informed in time if intervention is required. Overall, the system thus provides the opportunity to improve CBI monitoring and thus patient care, while at the same time reducing the workload of nurses, thus indirectly further improving patient care. The proposed sensor is an extremely important and integral component of such a monitoring system. Additional fields of application are conceivable, but will not be considered further at this point.

## 2. Materials and Methods

### 2.1. Selecting the Measurement Method

A requirement for the sensor development was the seamless integration with existing CBI systems. A refractometer-based approach or the use of other sensors that require direct contact with the medium to measure *Hb* concentration is, therefore, inappropriate. The *Hb* measurement in this work is performed based on light absorption. A known amount of light is exerted by a LED, passed through a tube filled with CBI waste fluid and captured on the opposite side. According to Beer's Law, the concentration is then related to the intensities and measurement section via the following:

$$I_t = I_0 \cdot 10^{-\varepsilon \Delta z c},$$

where $I_t$ and $I_0$ denote the transmitted and incident light intensities, respectively, $\varepsilon$ and $c$ denote the molar absorptivity and concentration of the light absorptive compound, and $\Delta z$ is the optical path length through the drainage fluid. By selecting a proper power supply and wavelength for the LED, we obtain constant $I_0$ and $\varepsilon$. When the outflow tube is full of drainage fluid, $\Delta z$ is also constant and determined by the diameter of the tube. Therefore, we can infer the *Hb* concentration $c$ solely based on $I_t$.

## 2.2. Selecting the Light Source

In this simple form, Beer's Law is only valid for a single light absorbing material. The waste fluid is, however, composed of various materials, i.e., irrigation fluid and potentially additional medications, as well as urine and blood. The latter two materials are compounds on their own. We, therefore, investigated how strong the influence of the additional ingredients would be. In a work from Pegau et al. [16], we found that the influence of salt on the absorption property of water is very small in the near infrared frequency band and that salt might even lower absorptivity. From Palmer et al. [17], we learned that the absorption of water compared to $Hb$ is negligible (approx. ratio 1:40,000). Additionally, urine coloring molecules, such as betanins, have their main absorption in the range of 400–600 nm and absorb only very little light in the near infrared as Gonçalves et al. [18] reported. In summary, we can state that $Hb$ absorption is absolutely dominant, particularly in the near infrared but also in the visible spectral band. Hence, we apply Beer's Law in its simple form without danger of significantly degrading the measurement accuracy.

The absorption curve of $Hb$ and $HbO_2$ is shown in Figure 1. For the formation of blood clots, it is basically irrelevant whether $Hb$ is oxygenated or not. In order to account for both $Hb$ versions at the same time, isobestic points, i.e., points at which two chemical species have the same molar absorptivity, should be used for measurement. For $Hb$ and $HbO_2$ the main isobestic points are at frequencies of 420, 545, 570, and 800 nm. In tests with blood samples and the targeted tubing, it was found that the absorption for the lower frequency points is way too high for a CBI monitoring system. The transmitted amount of light quickly drops below values that can be reliably distinguished from background noise. Additionally, total absorption is reached very quickly, so increasing the light intensity would not help. Combined with knowledge of the absorption properties of other relevant materials, these findings led to the determination of the central measurement frequency at 800 nm.

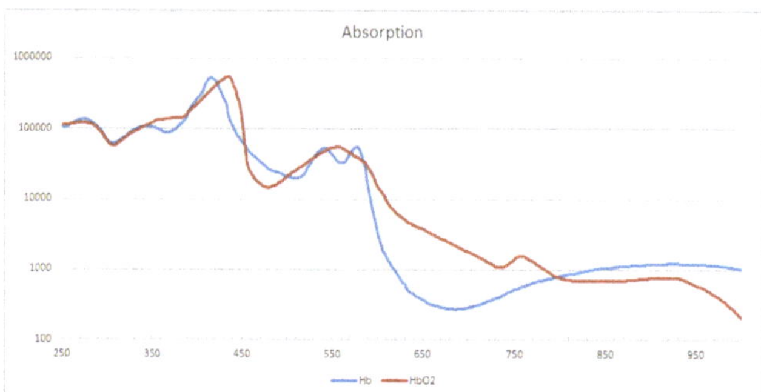


**Figure 1.** Molar extinction coefficient e in $\left[\frac{L}{mol\cdot cm}\right]$ of $Hb$ and $HbO_2$ for wavelengths between 250 and 1000 nm. Major isobestic points are located at 420, 545, 570 and 800 nm.

For an LED, the emitted light intensity as well as the light frequency depend on the applied current. Especially when using batteries in a mobile setting, care must be taken to provide a constant current supply. To this end, we implemented a small electronic component (see Figure 2), that guarantees a constant current over a wide voltage range. Implementation was done on a stripboard, using regular size electronic components.

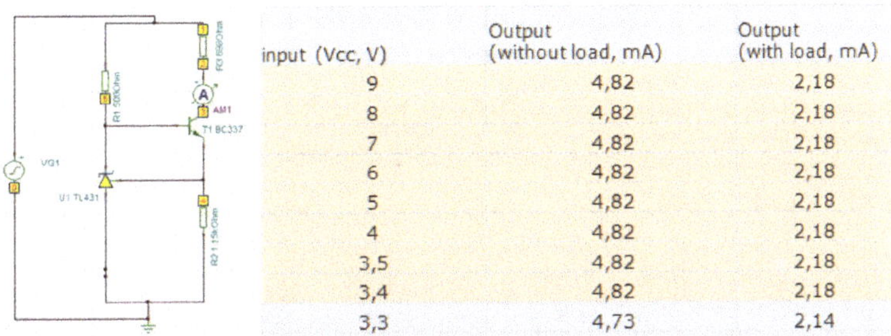

| input (Vcc, V) | Output (without load, mA) | Output (with load, mA) |
|---|---|---|
| 9 | 4,82 | 2,18 |
| 8 | 4,82 | 2,18 |
| 7 | 4,82 | 2,18 |
| 6 | 4,82 | 2,18 |
| 5 | 4,82 | 2,18 |
| 4 | 4,82 | 2,18 |
| 3,5 | 4,82 | 2,18 |
| 3,4 | 4,82 | 2,18 |
| 3,3 | 4,73 | 2,14 |

**Figure 2.** LED needs a constant current to emit a well-defined light. (**Left**): Circuit diagram of the used electronic. (**Right**): Test results with and without load. Constant current can be guaranteed for a wide range of input voltages.

### 2.3. Selecting the Light Sensor

In order to measure the transmitted light, a photo resistor (LDR), or a photo diode (PD) would theoretically be sufficient, as shown in [13]. Taking practical considerations into account, i.e., the fact that the CBI tubes do have reinforcing ribs, a single measurement is not sufficient for a reliable monitoring system. Even more problematic is the fact that bubbles are traveling unpredictably in the outflow tube. Gain ribs appear darker in the detector image and suggest increased $Hb$ values, while the presence of bubbles dramatically decreases light absorption and suggests lower $Hb$ values. While slightly increased $Hb$ readings might be tolerable, significantly lower readings would put patients at risk. As a consequence, measuring the $Hb$ concentration with a single point measurement (LDR/PD) is very dangerous in practice. We propose to use a camera chip instead. At only moderately increased cost, a wealth of measurements can be performed, enabling the application of advanced image processing techniques.

For system demonstration, we chose a monochrome camera without an IR filter, which provides us with 10 bit linear intensities. While the concrete camera model is irrelevant, the three mentioned properties of the camera are important. A monochrome camera lacks a Bayer filter; hence, all pixels generate readings of the same quality. The lack of IR filtering allows for reliable measurements in the near infrared frequency band. The 10 bit color depth increases the dynamic range and provides an increased measurement range.

### 2.4. Extending the Measurement Interval

Considering the exponential relationship between the intensity and $Hb$ concentration, the dynamic range of even a 10 bit sensor is rather limited. Starting from an exposure $E_{Low}$ that just saturates the sensor for pure irrigation fluid, the maximum measurable $Hb$ concentration is less than 0.6 g/dL (equivalent to ≈4% blood). Increasing the exposure time by a factor of 8 ($E_{High}$) allows measurements of more than 3.2 mg/dL $Hb$ (≈20% blood), but leads to overexposure for the $Hb$ concentration below 0.1 g/dL. We, therefore, introduced a dual-exposure setup, which is controlled, using a hysteresis threshold $T_{Low,High}$. We chose $T_{Low}$ (i.e., where we switch from $E_{High}$ to $E_{Low}$) at 0.2 g/dL $Hb$ and $T_{High}$ at 0.6 g/dL $Hb$. In terms of sensor readings, i.e., intensities, the $T_{Low}$ is at 800 and the $T_{High}$ is at 80. The overall measurement range of the proposed sensor, therefore, covers the interval [0–3.23] g/dL hemoglobin concentration or [0–21.5]% blood concentration.

### 2.5. Processing of Sensor Images

We apply three different image processing methods, namely the detection of shadow artifacts caused by the gain ribs, detection of bubbles, and automatic exposure adjustment.

Figure 3 gives an impression of the effect created by the gain ribs. The left subfigure shows a small, opened segment of the tube placed on a white paper with parallel pencil lines. Significant distortion of the lines can be observed near the ribs, which act like plano-convex lenses. The subfigure on the right shows the effect of the ribs on the detector image.

The horizontal position of the artifacts depends on how the tube was inserted into the sensor. Vertical line artifacts with mainly decreased intensity can be observed. In some cases, even a slight intensity increase can occur, depending on the positioning of a rib within the optical pass. Although the positioning of the tube cannot be controlled in practice, the artifacts can be reliably detected. Since the tube does not change its position during the entire CBI process, it is practical to automatically select an artifact-free measurement ROI when setting up the system. This can be realized by inspecting the intensity profile of a horizontal line and comparing it to a known intensity profile. An exemplary profile is depicted in Figure 4. The blue regions are those that are not directly illuminated, while the yellow regions are transitional zones that are partially lit. Both regions are a constant size and determined during the sensor assembly since they solely depend on the actual camera and LED placement. The union of the red and green regions is the potential measurement area. In the example, we detected two artifact regions, marked in red, and an artifact-free region, where we can safely place the measurement ROI.

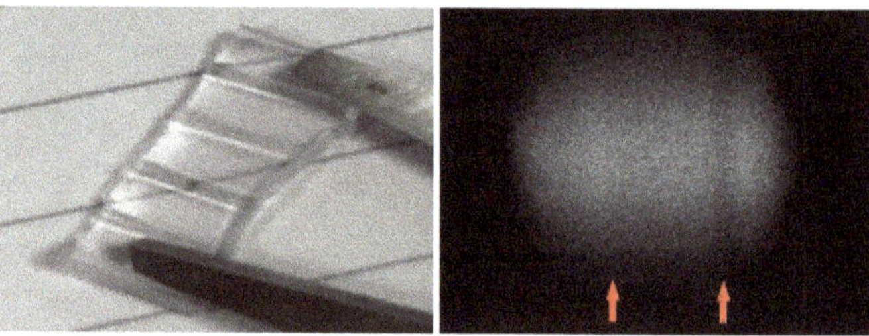

**Figure 3.** (**Left**): A segment of the waste fluid tube. Reinforcement ribs are running along the tube. The ribs significantly change the optical pass of light traveling through the tube. (**Right**): Acquired sensor image with marked artifacts caused by reinforcements. The horizontal location of the shadows depends on the tube placement and cannot be controlled. We note that the right figure was intensity adjusted for better depiction, which amplified intensity variations in the vertical direction as well as the noise level.

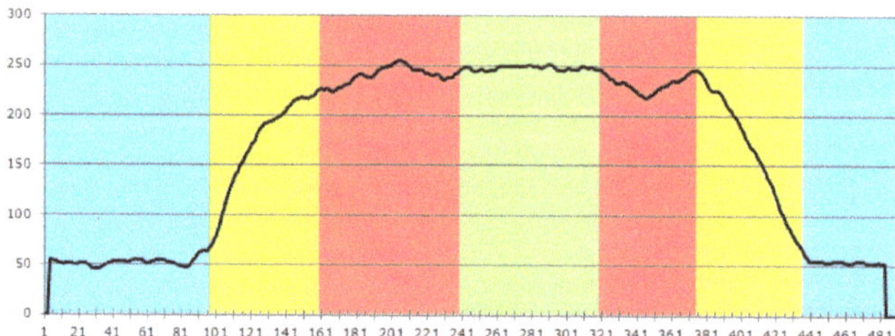

**Figure 4.** Exemplary intensity profile segmented into regions: (blue) background region, (yellow) transitional region, (red) artifact region, (green) measurement region.

As already mentioned earlier, bubbles pose a significant problem to the measurement process. While the gain ribs create static artifacts that only slightly and systematically alter the measurements, bubbles are dynamic by nature and severely impact the measurement. Figure 5 demonstrates the effect of a bubble in the sensor image. Please note that for all image sequences, the same specimen was used, i.e., a small tube section filled with diluted blood and sealed with a small fraction of air, similar to those depicted in Figure 9. Striking is the difference in the image sequences depicted in Figure 5. The only difference between the sequences is the location with respect to the camera and LED, where the bubble travels along the tube. Already, these examples demonstrate that a bubble detection based on classic approaches is not feasible. The situation becomes even more challenging when several small bubbles form a foam-like cluster. Furthermore, we found that also pure image processing of a single image is not sufficient to reliably detect bubbles: in the case that a bubble of a reasonable size remains static within the light pass, the images can be indistinguishable from images without bubbles at a lower *Hb* concentration. It is, therefore, imperative to use a system that can analyze image sequences. To this end, we trained a convolutional recurrent neural network comprised of convolution layers to extract optical features followed by long short-term memory (LSTM) layers to account for the temporal aspect. The network was trained on video sequences acquired with the sensor hardware. To this end, test specimen with various blood concentrations and bubble sizes were prepared (see Section 2.8). The network was trained fully supervised. The ground truth was generated by manually marking the images in the sequences where a bubble just entered the view or almost completely left the view. The remaining images were classified automatically. Figure 6 depicts the network architecture. For details on the network and how it was trained, we refer the interested reader to [19].

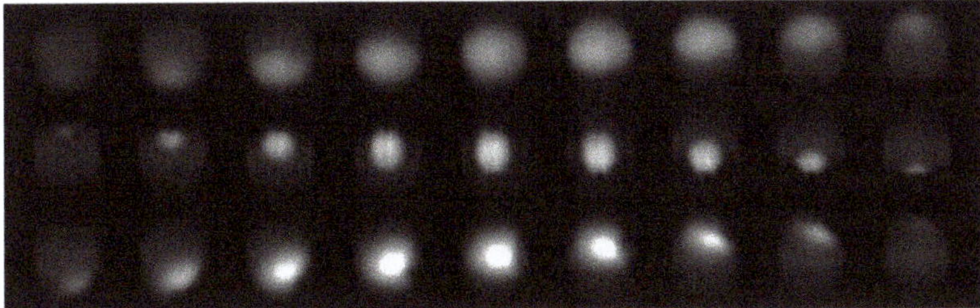

**Figure 5.** Examples of the appearance of bubbles in the sensor image. (**Upper row**): bubble passing the sensor close to the LED. (**Middle row**): bubble passing the sensor close to the camera. (**Lower row**): bubble passing near the center of the tube. For all images, the same specimen was used, so *Hb* concentration as well as bubble size are the same.

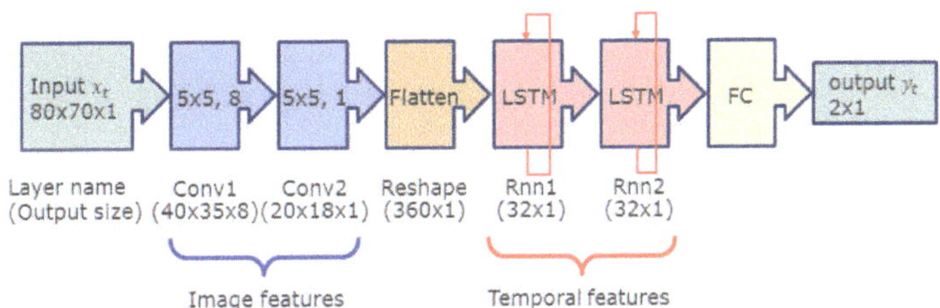


**Figure 6.** Network to detect bubbles in the optical path. The convolutional layers extract image features, such as bright and dark edges, bright spots, etc., while the LSTM layers account for the temporal components. The network is able to detect bubbles traversing in both directions and even accounts for those that remain static in the optical path for an extended time.

### 2.6. Sensor Signal Processing

Sensor signal processing, i.e., bubble detection, *Hb* value generation, and communication, was implemented on a Raspberry Pi 4B. The camera was connected via the USB 3 interface, which also provided power to the LED. Furthermore, the Raspberry Pi was used to communicate the data to the main system and to control additional sensors that are out of the scope of this paper. The operating system used was a completely stripped-down Ubuntu Linux with a custom compiled kernel.

The bubble detection network was run on the ARM-CPU exploiting the ONNX Runtime [20]. The camera provides images at a frame rate of 25 fps. All images are analyzed by the network. Sequences of 8 images are combined to guarantee stable outputs, even in the case that the network fails to detect a bubble or falsely detects a bubble in an image. The output frequency of the sensor was set to 10 measurements per second. Data were provided to the user side (C# DLL) by means of a double buffer. Overall, power consumption for the running system is below 700 mA.

To further support future integration of the sensor, we quantized the network to fixed point representation (Figure 7), using a quantization aware training (QAT) method. The resulting network was implemented on FPGA (Xilinx xc7z020-clg400-1). As shown in Table 1, the standard version of the network still leaves sufficient resources for additional hardware, such as a MIPI camera interface as well as the calibration tables to convert sensor readings into *Hb* values. Thus, the whole signal processing can be performed on a single SoC of a size less than $20 \times 20 \times 2$ mm. This not only reduces the overall sensor size, but also significantly lowers the overall power consumption and, in turn, enables mobile CBI monitoring.

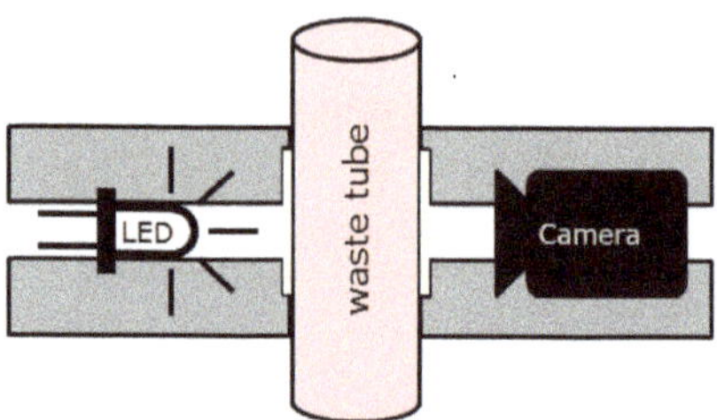


**Figure 7.** Schematic representation of the sensor. On the left side, an LED exerts light that is transmitted through the tube filled with irrigation waste fluid. The transmitted light is captured by a camera. The bright squares next to the tube represent thin glass plates that can be inserted into the optical pass to seal the LED and the camera and to enable effective cleaning and disinfection in clinical application.

**Table 1.** Listing of hardware resources used for bubble detection and MIPI interface on the FPGA. Two different implementations of the bubble detection network are given: one that complies with the current specifications regarding frame rates as well as a maximum performance version. Please note that the standard version leaves sufficient space to realize a MIPI interface to directly connect the camera.

| FPGA Utilization (c7z020-clg400-1) | | | |
|---|---|---|---|
| **Resource** | **Bubble Detection Standard** | **Bubble Detection Performance** | **MIPI Interface** |
| BRAM | 15% | 16% | 4% |
| DSP | 5% | 5% | 4% |
| FF | 20% | 26% | 0% |
| LUT | 73% | 99% | 6% |

### 2.7. Sensor Housing

A sensor used in clinical practice needs to be cleaned and disinfected. Neither the camera lens nor the LED are well suited for this process. Besides potential deterioration effects, camera lenses feature sharp edges where bacteria and other germs might survive. It is therefore mandatory to seal the optical components. A particularly suited material for the case at hand appears to be glass. We conducted a series of tests using glass cover slips for microscopic samples with a thickness of 0.15 mm $\pm$ 0.02 mm. Assuming circular openings for the camera lens ($\varnothing \approx 1.0$ cm) and the LED light channel ($\varnothing \approx 0.5$ cm), the cover slips appear to be sufficiently robust.

Since the positions of the glass plates, camera, and LED are rigid and only light frequencies in a very small band are used, we confirmed in tests that the influence of glass is a constant offset to the measurement. This was expected by altering Beer's Law in the following way:

$$I_t - \Delta I_t = (I_0 - \Delta I_0) \cdot 10^{-\varepsilon \Delta z c},$$

where $\Delta I_t$ and $\Delta I_0$ denote the attenuation of the exerted and captured intensity by the two glass slips.

The housing itself was 3D printed from black acrylonitrile butadiene styrene (ABS). Tests with white ABS showed significant light penetration in the sensor, which greatly affected the measurements. Although it would be possible to shield the measurement path by other means, using black material seemed to be the most direct approach.

In contrast to [13], we decided for a rigid housing instead of a clamp. While a clamp has the advantage that it can be most easily attached to a variety of tubes, it is much harder to know the true length of the measurement distance. Furthermore, the clamp pressure deforms the tube and thus introduces uncertainty. Lastly, it is much harder to avoid ambient illumination to impact on the measurement. An image of the sensor is given in Figure 8. Three versions of the sensor were assembled and provided for integration.

**Figure 8.** Image of the assembled sensor as used in the clinical validation study. Communication and power supply were realized using a USB 3 interface. A short tube segment was inserted to showcase sensor attachment to the CBI system.

*2.8. Preparation of Test Specimen*

Figure 9 depicts a set of test specimen used for sensor development. A complete set always covered a concentration range between 0% and 22% in 2% increments, i.e., 12 samples. Each specimen was manufactured in two different versions: one with saline irrigation fluid and one with an electrolyte-free irrigation fluid. The latter is mainly used during surgery, but opened bags are used up afterwards.

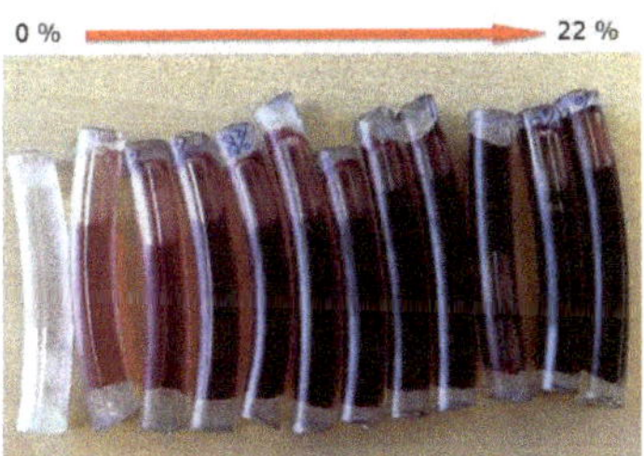


**Figure 9.** Test specimen samples. Tube segments were filled with diluted blood and sealed with hot glue. In order to test the effect of bubbles, a small fraction of air was left inside. This also prevented direct contact between the hot glue and test liquid. Samples were produced for blood concentration between 0% and 22% in increments of 2%. *Hb* concentration was measured using a hemoglobin analyzer.

To produce the test specimens, the drain tube of an irrigation system was cut into pieces approximately 6 cm long. One end of the tube segments was sealed with hot glue. Percentage indicators were written on each tube, using waterproof ink. This was done in a preparatory phase to avoid any delay when working with the blood.

The blood used for development was fresh porcine blood obtained immediately after slaughter. A total of 6 mL diluted Heparin (250 IU/mL) was added to 200 mL blood to prevent clotting (coagulation), and the mixture was carefully stirred to avoid sedimentation/separation. Since the proposed sensor actually measures the *Hb* concentration, the ground truth *Hb* value was measured using a hemoglobin analyzer (Measuring range: 0–25.5 g/dL; Imprecision (within run): CV < 1%; Calibrated against HiCN reference method.). Then, the blood was diluted with the respective irrigation fluid. The prepared liquid was poured into the tube segment, leaving approximately 0.5–1.0 cm of air, and the tube was sealed with hot glue. Leaving a small air bubble in the tube served two purposes: first, the presence of an air bubble allowed the effect of air bubbles to be studied accurately during scanning. Second, the air prevented possible adverse effects from the hot glue coming in direct contact with the liquid.

Samples prepared in this way provided valid measurements for one working day. Repeated measurements the following day showed significant deterioration, even when the samples were stored in the refrigerator.

In order to assess the effect of urine, additional test specimens were prepared. Human urine, collected immediately after waking up, was used for this purpose. With this urine, we were able to ensure a maximal content of salts and other metabolites. Tests on pure urine samples and samples composed of blood, urine and irrigation fluid revealed no significant effect on the measurements.

## 3. Results

In a planned self-experiment, the sensor was tested using blood from three human donors. Using human blood donations did not require ethics approval, according to the local ethics council. The full laboratory setup of the CBI monitoring system is shown in Figure 10 (left). The liquid flowed from the upper bag via a Foley catheter through a 3D printed bladder model into the urine bag. A pump controlled the rate of inflow. The *Hb* sensor was mounted to the outflow tube to detect the *Hb* level for data collection. To obtain five different blood concentrations, we prepared five bags filled with 500 mL saline solution and varied amounts of blood. As a reference measurement, the *Hb* concentration of each bag was measured via blood gas analysis (BGA). The average difference of the sensor-detected *Hb* level from the BGA-detected *Hb* level was 0.29 g/dL. The *Hb* concentration in human blood is 15 g/dL on average, so the measurement error with regard to the full measurement interval was estimated as ~10%. We note that the full measurement range of the sensor was tested in the laboratory trials. Since measurement accuracy drops significantly at higher concentrations, an increased mean deviation from the BGA was to be expected. Table 2 gives a rough estimate of the measurement accuracy for some blood concentrations. The highest blood concentration that can be quantified by the sensor is roughly 20%, i.e., 3 g/dL *Hb* concentration.

Additionally, we employed random *Hb* concentrations to model the rise and fall in *Hb* level to simulate a real-life complication scenario, presented as a graph of the *Hb* level over time as depicted in Figure 10 (right). To imitate acute bleeding, blood was injected into the 3D-printed bladder model. The blood supply was replaced with a saline solution to resemble stopped bleeding. In the experiments, the sensor time constants for the CBI system, with the sensor placed near the waste bag, were 33 s for rise and 119 s for fall. Accordingly, the sensor's response times were 2.7 min for rise and 9.9 min for fall. However, we would like to point out that these values strongly depend on temperature, irrigation flow, emissivity of the material and size, i.e., length and diameter of the respective irrigation tube, including the catheter. Furthermore, bubble detection was tested. The sensor reliably recognized random air bubbles or artificially generated air bubbles in the outflow tube.

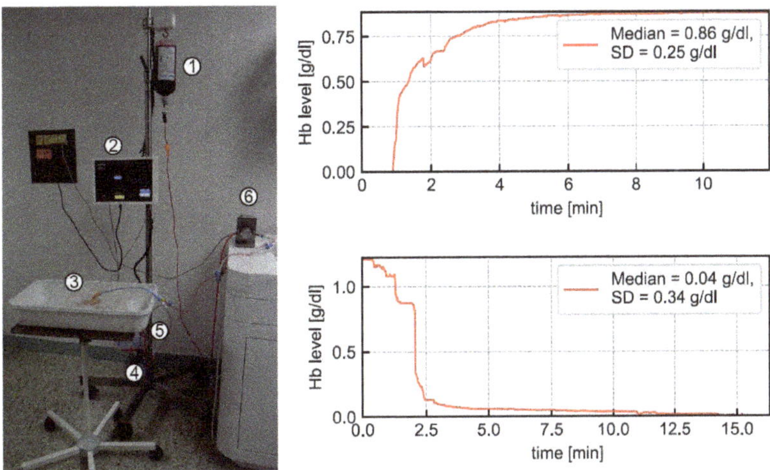

**Figure 10.** (**Left**): full system setup for clinical laboratory trials using blood from human donors: (1) blood donor, (2) display, (3) 3D bladder model, (4) *Hb* sensor, (5) urine bag, and (6) pump. The sensor was used in an extended setup, which is not in the scope of this paper. (**Right**): two exemplary measurement curves for increasing and decreasing *Hb* concentrations. Sensor time constants were estimated as 33 s for rise and 119 s for fall.

**Table 2.** Measurement errors with respect to the measured blood concentration. With increasing blood concentration the measurement accuracy drops due to the non-linearity of Beer's Law. We note that these values were acquired during development and were not part of the clinical study.

| Measurement Error of the Sensor | | | | | |
|---|---|---|---|---|---|
| Blood concentration | 4% | 5% | 15% | 19% | 21% |
| Absolute error | 0.11% | 0.17% | 0.73% | 1.5% | 2.5% |
| Relative error | 2.75% | 3.4% | 4.87% | 7.9% | 11.9% |

After receiving a positive ethical vote, a clinical trial involving 20 patients was conducted to test the sensor and other components of the CBI monitoring system. All patients were male (mean age: 73.1 years) and underwent TURB. CBI was administered as part of the standard treatment procedure. Further patient information was not collected to keep the data protection footprint low. For each patient, *Hb* levels were monitored for at least three hours, using the CBI monitoring system. As a reference measurement, we regularly checked the *Hb* level in the urine bag via BGA. We would like to point out that, unlike the *Hb* sensor, our BGA device did not allow to measure the *Hb* level continuously. The *Hb* level detected by BGA is the average of several minutes. Figure 11 depicts an example of the sensor-detected *Hb* level's development illustrated as a red line. As gray data points, ten BGA-detected *Hb* levels are presented. The median *Hb* values are at a comparable level. The mean deviation of the sensor-detected *Hb* level from that of the BGA-detected *Hb* level across all patients was −0.003 g/dL. Between minute 120 and 140, we noted a rise in *Hb* levels. The reason was an incorrectly configured inflow speed of saline solution. As the inflow speed was accelerated, the *Hb* level decreased to normal. Thus the sensor reliably measured *Hb* levels within a clinically acceptable deviation and with a high responsiveness across all experiments. In the patient study, the functionality of the *Hb* sensor's air bubble detection could also be positively confirmed.

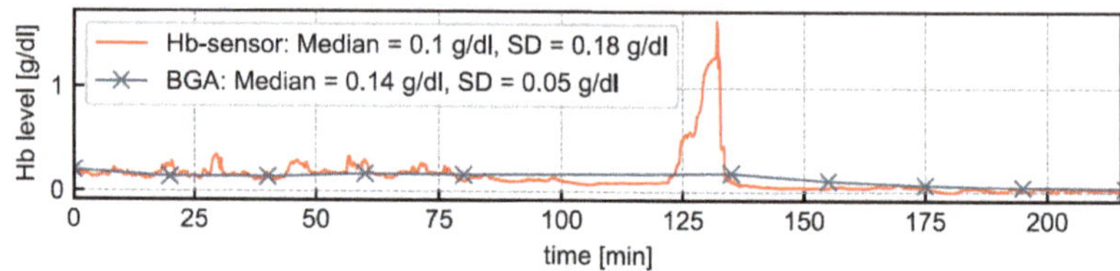



**Figure 11.** Measurement curve for a real patient. The blood gas analysis agrees well with the sensor readings. At approximately minute 125, the irrigation was inadvertently left off after the urine bag was emptied, resulting in a real-life example of the importance of the continuous monitoring of CBI.

## 4. Discussion

As shown in the results, we were able to design a sensor to measure the *Hb* concentration in the effluent of a CBI with decent accuracy ($\pm 0.003$ g/dL). While accuracy is highest for very small concentrations, it drops for higher concentrations, due to the exponential relation between concentration and absorption. However, for practical application this does not have negative implications. Already at 10% blood concentration, medical personnel should consider intervention. In case of a low irrigation flow, the flow should be increased to prevent cloth formation. In case of an already high irrigation flow, acute bleeding should be assumed. To increase measurement accuracy, we extended the measurement interval, using multiple exposure times. Further extension using even higher exposure times does not improve the situation, since total absorption occurs—given the fixed tube diameter.

We decided on a rigid sensor housing to achieve the most reliable measurements for system demonstration. This, of course, comes at the drawback that only a particular tube diameter is supported by the sensor. However, insets can be easily designed to account for various tube diameters. Vastly larger tube diameters decrease the measurement interval significantly since total absorption occurs earlier, while smaller tube diameters increase the measurement interval. In any case, an individual re-calibration of the sensor will be necessary.

The effect of salt on the measurement cannot be completely ignored. As can be found in the work of Pegau et al. [16], salt lowers the absorption of near infrared light by water and thus, can lead to an underestimation of the *Hb* concentration. As a consequence, it is definitely necessary to calibrate the sensor for the respective irrigation fluid. However, the amount of salt introduced by urine is negligible and does not have a significant effect on the measurement accuracy.

As stated in Section 2.8, we prepared the blood samples for the development from porcine blood. Although it is very similar to human blood, the calibration tables do not perfectly fit the practical application. A re-calibration with human blood was not conducted prior to the study since the developers had no access to human blood samples. A re-calibration at the hospital site was not possible since the procedure is technically involved and developers could not travel due to COVID-19 restrictions. Still, the achieved results during the clinical study were more than satisfying.

For system demonstration, we used a cheap camera without housing and other accessories. This camera was not designed for medical applications but rather for rapid prototyping. As such, it is relatively large ($40 \times 40 \times 38$ mm) and the main reason for the current overall sensor size of $44 \times 44 \times 85$ mm. Additionally, the constant current supply electronic is relatively large ($40 \times 23 \times 12$ mm). Using a miniaturized camera and an integrated electronic, the overall size of the sensor can easily be reduced to less than $15 \times 15 \times 30$ mm in total. Realizing a SoC version, the sensor including evaluation electronic could be realized in a package not larger than $25 \times 25 \times 30$ mm.

Additionally, an approach similar to that of Hageman et al. [11] can be extended into a non-invasive measurement solution, such as the one presented by Ding et al. [12]. However, such a solution is inferior to the absorption-based approach for several reasons. Firstly, in the absorption approach, the emitter and sensor can be placed in line with the test sample. This allows for a technically less-involved setup, because all geometries can be assumed to be planar and of a constant size, without significant sacrifice of accuracy. Any additional material, such as the tube or glass slides (c.f. Sections 2.5 and 2.7) only play a minor role when considering Beer's Law.

In contrast, a reflectivity-based system must consider a technically involved angled arrangement because the light emitter and sensor cannot be in the same position. More importantly, the tube has a significant, non-trivial impact on the reflectance characteristics. In particular, the coloring of the tube affects the perceived color, regardless of whether a color scale is used by personnel or a computerized evaluation is performed. Lastly, reflection spectra appear to be less specific than absorption spectra.

A completely different method for estimating blood concentration in the effluent could be developed based on the photoacoustic effect. Briefly, when a medium is irradiated with a sequence of light flashes, periodic heating (and cooling) occurs. The resulting alternation of volume expansion and contraction constitutes a source of sound [21]. Using laser sources with specific wavelengths and, for example, piezoelectric acoustic receivers, it is possible to measure $Hb$ concentration with high precision and specificity. In [22] a system for in vitro measurement of $Hb$ and $HbO_2$ is described. Integrating the laser emitter into a catheter, a similar system could even allow concentration measurements directly in the bladder, which would significantly reduce the response time of the sensor (cf. Section 3). On the other hand, a rather expensive optical fiber would have to be integrated into the catheter, which is a disposable product. In [23], an interesting work is presented that aims at miniaturizing a photoacoustic sensing system. However, we assume that current solutions are still relatively large, complex and expensive and often require much more energy, compared to the sensor presented here.

Lastly, the clinical study only evaluated sensor functionality, i.e., the quality of the $Hb$ measurement as well as bubble detection during clinical application. Further processing of sensor readings to, for example, generate alarms or treatment guidance, was neither performed nor part of the study. Sensor readings were acquired blindly and evaluated retrospectively.

## 5. Conclusions

Although further investigations are required as part of an approval study in accordance with the German as well as the international Medical Devices Act, the basic practical suitability of the sensor was demonstrated. Its non-invasive applicability allows a simple extension of the existing CBI systems and thus, a gapless monitoring of the irrigation procedure. Measuring the $Hb$ concentration using the absorption principle (Beer's Law) provides reliable data for the application at hand. In combination with a (mobile) communication system, monitoring data can be immediately transferred to nurse guard rooms or even mobile devices. This significantly reduces the burden on nursing staff to care for an entire ward, while still ensuring effective CDI for the benefit of patients

**Author Contributions:** Conceptualization, G.R. and A.M.; methodology, G.R.; software, X.T.; validation, X.T., G.R. and D.S.S. ; formal analysis, X.T. and L.K.; investigation, X.T.; resources, G.R. and A.M.; data curation, X.T. and L.K; writing—original draft preparation, G.R.; writing—review and editing, G.R., M.Y. and A.M.; visualization, X.T., G.R. and L.K.; supervision, G.R.; project administration, G.R.; funding acquisition, G.R. and A.M. All authors have read and agreed to the published version of the manuscript.

**Funding:** This research was partially funded by the German Ministry of Research and Education grant number BMBF VISIMON 16SV7861K.

**Institutional Review Board Statement:** Positive ethics vote was achieved from Ethics Committee of Albert-Ludwigs-Universität Freiburg, Freiburg im Breisgau, Germany, 20-1177 MPG §23b, 29 October 2020. The study was registered according to article 35 of the Declaration of Helsinki 2013 at Deutsches Register Klinischer Studien (DRKS) under ID DRKS00023647, 27 January 2021.

**Informed Consent Statement:** Informed consent was obtained from all subjects involved in the study.

**Data Availability Statement:** Not applicable.

**Acknowledgments:** The authors want to thank the VISIMON project partner DITABIS AG, Pforzheim, Germany for the integration of the sensor into a system fit for practical application in a patient study.

**Conflicts of Interest:** The authors declare no conflict of interest.

# References

1. Do, J.; Lee, S.W.; Jeh, S.U.; Hwa, J.S.; Hyun, J.S.; Choi, S.M. Overnight continuous saline irrigation after transurethral resection for non-muscle-invasive bladder cancer is helpful in prevention of early recurrence. *Can. Urol. Assoc. J.* **2018**, *12*, E480. [CrossRef] [PubMed]
2. Comez, Y.I. Laparoscopic Simple Prostatectomy. In *Prostatectomy*; Genadiev, T., Ed.; IntechOpen: Rijeka, Croatia, 2019; Chapter 5. [CrossRef]
3. Elzayat, E.; Habib, E.; Elhilali, M. Holmium Laser Enucleation of the Prostate in Patients on Anticoagulant Therapy or With Bleeding Disorders. *J. Urol.* **2006**, *175*, 1428–1432. [CrossRef]
4. Nojiri, Y.; Okamura, K.; Kinukawa, T.; Ozawa, H.; Saito, S.; Okumura, K.; Terai, A.; Takei, M. Continuous Bladder Irrigation Following Transurethral Resection of the Prostate (Turp). *Jpn. J. Urol.* **2007**, *98*, 770–775. [CrossRef] [PubMed]
5. Okorie, C.O. Is continuous bladder irrigation after prostate surgery still needed? *World J. Clin. Urol.* **2015**, *4*, 108–114. [CrossRef]
6. Kaplan, D.; Kohn, T.; Kieran, K.; Yates, J. Available online: https://www.auanet.org/education/auauniversity/for-medical-students/medical-students-curriculum/medical-student-curriculum/urologic-emergencies (accessed on 23 August 2021).
7. Vaidyanathan, S.; Singh, G.; Selmi, F.; Hughes, P.L.; Soni, B.M.; Oo, T. Complications and salvage options after laser lithotripsy for a vesical calculus in a tetraplegic patient: A case report. *Patient Saf. Surg.* **2015**, *9*, 1–8. [CrossRef]
8. Scholtes, S. Management of clot retention following urological surgery. *Nurs. Times* **2002**, *98*, 48–50. [PubMed]
9. Ng, C. Assessment and intervention knowledge of nurses in managing catheter patency in continuous bladder irrigation following TURP. *Urol. Nurs.* **2001**, *21*, 97–112. [PubMed]
10. Frank, T.; Stintzing, F.; Carle, R.; Bitsch, I.; Quaas, D.; Strass, G.; Bitsch, R.; Netzel, M. Urinary pharmacokinetics of betalains following consumption of red beet juice in healthy humans. *Pharmacol. Res.* **2005**, *52*, 290–297. [CrossRef]
11. Hageman, N.; Aronsen, T.; Tiselius, H.G. A simple device (Hemostick®) for the standardized description of macroscopic haematuria Our initial experience. *Scand. J. Urol. Nephrol.* **2006**, *40*, 149–154. [CrossRef] [PubMed]
12. Ding, A.; Cao, H.; Wang, L.; Chen, J.; Wang, J.; He, B. A novel automatic regulatory device for continuous bladder irrigation based on wireless sensor in patients after transurethral resection of the prostate. *Medicine* **2016**, *95*, e5721. [CrossRef] [PubMed]
13. Chan, Y.H.; Chen, K.W.; Wu, Q.; Chiong, E.; Ren, H. Pre-Clinical Proof-of-Concept Study of a Bladder Irrigation Feedback System for Gross Haematuria in a Lab Setup. *Multimodal Technol. Interact.* **2020**, *4*, 59. [CrossRef]
14. Timm, U.; Lewis, E.; Leen, G.; McGrath, D.; Kraitl, J.; Ewald, H. Non-invasive continuous online hemoglobin monitoring system. In Proceedings of the 2010 IEEE Sensors Applications Symposium (SAS), Limerick, Ireland, 23–25 February 2010; pp. 131–134. [CrossRef]
15. Zhang, Y.; Fan, N.; Zhang, L.; Hu, X.; Wang, L.; Wang, H.; Kaushik, D.; Rodriguez, R.; Wang, Z. Novel strategy to monitor fluid absorption and blood loss during urological endoscopic surgery. *Transl. Androl. Urol.* **2020**, *9*, 1192. [CrossRef] [PubMed]
16. Pegau, W.; Gray, D.; Zaneveld, J. Absorption and attenuation of visible and near-infrared light in water: Dependence on temperature and salinity. *Appl. Opt.* **1997**, *36*, 6035–6046. [CrossRef] [PubMed]
17. Palmer, K.; Williams, D. Optical properties of water in the near infrared. *Opt. Soc. Am.* **1974**, *64*, 1107–1110. [CrossRef]
18. Gonçalves, L.C.P.; de Souza Trassi, M.A.; Lopes, N.B.; Dörr, F.A.; dos Santos, M.T.; Baader, W.J.; Oliveira, V.X.; Bastos, E.L. A comparative study of the purification of betanin. *Food Chem.* **2012**, *131*, 231–238. [CrossRef]
19. Tan, X.; Reis, G.; Stricker, D. Convolutional Recurrent Neural Network for Bubble Detection in a Portable Continuous Bladder Irrigation Monitor. In *Artificial Intelligence in Medicine*; Springer: Cham, Switzerland, 2019; pp. 57–66. [CrossRef]
20. ONNX Runtime. Available online: https://www.onnxruntime.ai/ONNXRuntime (accessed on 23 August 2021).
21. Bell, A.G. On the production and reproduction of sound by light. *Am. J. Sci.* **1880**, *20*, 305–324. [CrossRef]
22. Lee, C.; Jeon, M.; Jeon, M.Y.; Kim, J.; Kim, C. In vitro photoacoustic measurement of hemoglobin oxygen saturation using a single pulsed broadband supercontinuum laser source. *Appl. Opt.* **2014**, *53*, 3884–3889. [CrossRef] [PubMed]
23. Stylogiannis, A.; Riobo, L.; Prade, L.; Glasl, S.; Klein, S.; Lucidi, G.; Fuchs, M.; Saur, D.; Ntziachristos, V. Low-cost single-point optoacoustic sensor for spectroscopic measurement of local vascular oxygenation. *Opt. Lett.* **2020**, *45*, 6579–6582. [CrossRef] [PubMed]

*Communication*

# A Cell's Viscoelasticity Measurement Method Based on the Spheroidization Process of Non-Spherical Shaped Cell

Yaowei Liu [1], Yujie Zhang [1], Maosheng Cui [2], Xiangfei Zhao [1], Mingzhu Sun [1] and Xin Zhao [1,*]

1   Institute of Robotics and Automatic Information System, Tianjin Key Laboratory of Intelligent Robotics, Nankai University, Tianjin 300071, China; liuyaowei@mail.nankai.edu.cn (Y.L.); zhangyujie1002@mail.nankai.edu.cn (Y.Z.); 1120170124@mail.nankai.edu.cn (X.Z.); sunmz@nankai.edu.cn (M.S.)
2   Institute of Animal Sciences, Tianjin 300112, China; tjsnykxyxmsyyjs@tj.gov.cn
*   Correspondence: zhaoxin@nankai.edu.cn

**Abstract:** The mechanical properties of biological cells, especially the elastic modulus and viscosity of cells, have been identified to reflect cell viability and cell states. The existing measuring techniques need additional equipment or operation condition. This paper presents a cell's viscoelasticity measurement method based on the spheroidization process of non-spherical shaped cell. The viscoelasticity of porcine fetal fibroblast was measured. Firstly, we introduced the process of recording the spheroidization process of porcine fetal fibroblast. Secondly, we built the viscoelastic model for simulating a cell's spheroidization process. Then, we simulated the spheroidization process of porcine fetal fibroblast and got the simulated spheroidization process. By identifying the parameters in the viscoelastic model, we got the elasticity (500 Pa) and viscosity (10 Pa·s) of porcine fetal fibroblast. The results showed that the magnitude of the elasticity and viscosity were in agreement with those measured by traditional method. To verify the accuracy of the proposed method, we imitated the spheroidization process with silicone oil, a kind of viscous and uniform liquid with determined viscosity. We did the silicone oil's spheroidization experiment and simulated this process. The simulation results also fitted the experimental results well.

**Keywords:** robotic cell manipulation; mechanical properties; elasticity measurement; viscosity measurement; cell mechanics

**Citation:** Liu, Y.; Zhang, Y.; Cui, M.; Zhao, X.; Sun, M.; Zhao, X. A Cell's Viscoelasticity Measurement Method Based on the Spheroidization Process of Non-Spherical Shaped Cell. *Sensors* **2021**, *21*, 5561. https://doi.org/10.3390/s21165561

Academic Editors: Vahid Abolghasemi, Hossein Anisi and Saideh Ferdowsi

Received: 26 July 2021
Accepted: 16 August 2021
Published: 18 August 2021

**Publisher's Note:** MDPI stays neutral with regard to jurisdictional claims in published maps and institutional affiliations.

## 1. Introduction

The mechanical properties of biological cells, especially the elastic modulus and viscosity of cells, can provide an important basis for the evaluation of cell viability and cell states and the judgment of biological activity [1–3], and is crucial for the understanding of cell structure and physiological function [4–6]. To measure cell viscoelasticity, scientists have developed methods such as atomic force microscopy (AFM) [7–12], magnetic tweezers technique [13,14], optical tweezers technique [15,16], microfluidic technique [17–19], and micropipette aspiration (MA) technique [20–24]. These methods are suitable for different situations. AFM technique detects the viscoelasticity of the cell by moving the cantilever probe in vertical direction and monitoring its bending displacement. Magnetic tweezers technique and optical tweezers technique apply a certain force to the magnetic beads or silicon beads adhered to cells through magnetic field or light field to deform cells and obtain the viscoelasticity of cells. Microfluidic technique obtains the viscoelasticity of cells by detecting the deformation of cells under different microchannels and different shear forces. MA technique obtains the viscoelasticity of cells by measuring the length of the cells aspirated into the micropipette under different pressures.

Among these techniques, MA technique has become widely used due to the reasons of no need to purchase or prepare additional equipment, lower measurement cost, and easier integration into existing commercial micro-operation systems [25]. However, the micropipette aspiration method has high requirements for the seal between the cell and the micropipette in the measurement process [26]. Slightly improper sealing will result in ineffective MA operations, which will have a great impact on the measurement results. Meanwhile, the measuring of results is highly dependent on the accuracy of the force sensor. The viscoelasticity differences of the same cells measured by the same research groups using the micropipette aspiration method will also be very large. For example, the elasticity of human chondrocytes measured by Jones et al. was $0.65 \pm 0.63$ kPa [27], wherein the standard deviation was as large as the measured value. As the shape of the cell might be non-spherical, it will be more difficult to seal the cell and micropipette. In order to eliminate the influence of sealing on the measurement results, it is necessary to design a cell viscoelastic measurement method based on the micropipette aspiration platform and with low requirements for sealing.

In this paper, we proposed a cell's viscoelasticity measurement method based on the spheroidization process of non-spherical shaped cell. The spheroidization process means the process of some deformable non-spherical objects turning into spherical shapes due to surface tension. We firstly introduced the method of recording the spheroidization process of porcine fetal fibroblast and recorded the fetal fibroblast's spheroidization process. Secondly, we built the viscoelastic model for simulating non-spherical shaped cell's spheroidization process based on the fact that the capsule-like porcine fetal fibroblast will finally become spherical (Figure 1). Then, we simulated the spheroidization process of porcine fetal fibroblast and got the simulated spheroidization process. By changing the parameters in the simulations, we got the elasticity and viscosity that best fitted the experiments. The magnitude of the elasticity and viscosity of fetal fibroblast was in agreement with those measured in other literatures. To verify the accuracy of this method, we imitated the spheroidization process with silicone oil, a kind of viscous and uniform liquid with determined viscosity. We did the silicone oil's spheroidization experiment and simulated this process. The simulation results fitted the experimental results well.

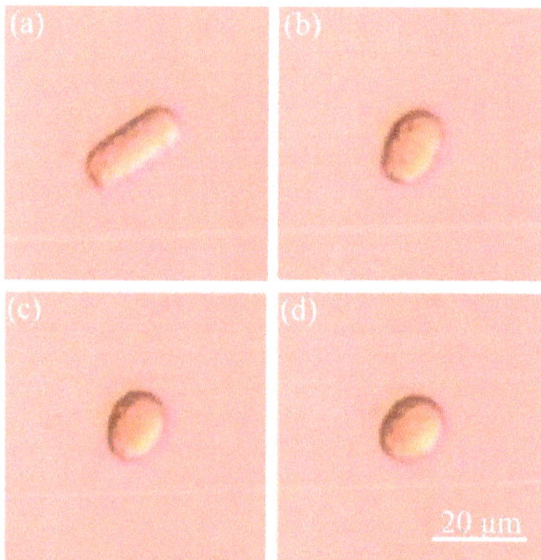


**Figure 1.** Spheroidization process of the capsule-like porcine fetal fibroblast. (**a–d**) From capsule-like cell to spherical-like cell.

## 2. Materials and Methods

### 2.1. System Setup

The spheroidization experiment of porcine fetal fibroblast was performed on the self-developed NK-MR601 micro-operation system [28–30] (Figure 2). The system consists a microscope (CK-40, Olympus, Tokyo, Japan); a CCD camera (W-V-460, Panasonic,

Osaka, Japan, frame rate: 20 frame/s); a motorized X-Y stage (travel range: 100 mm, repeatability: ±1 μm/s, maximum speed: 2 mm/s); two X-Y-Z manipulators (travel range: 50 mm, repeatability: ±1 μm/s, maximum speed: 1 mm/s); a self-developed micro-injector, providing negative pressure to aspirate the fetal fibroblast and positive pressure to eject the fetal fibroblast; a self-developed motion control box, controlling the micro-platform, micro-manipulators, and micro-injector through the host computer.

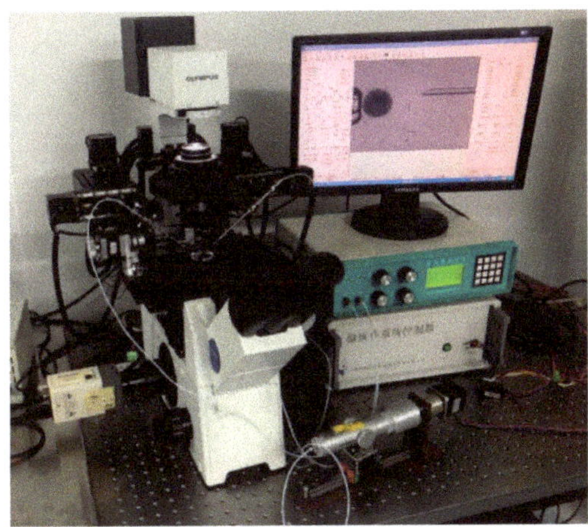

**Figure 2.** NK-MR601 micro-operation system.

The silicone oil spheroidization experiment was performed on NK-MR601 with the CCD replaced by a highspeed camera (C110, Miro, Wayne, NJ, USA, frame rate: 1000 frame/s).

The micropipettes used in the spheroidization experiments of porcine fetal fibroblast and silicone oil were made from borosilicate glass tubes with an outer diameter of 1 mm and an inner diameter of 0.8 mm. The micropipette used in the fetal fibroblast spheroidization experiments were pulled by the puller (MODEL P-97, Sutter Instrument, Novato, CA, USA), and fractured by the microforge (MF-900, NARISHIGE, Tokyo, Japan) with an inner diameter of 10 μm. The micropipette used in the silicone oil spheroidization experiments was pulled and fractured by hand by Yaowei Liu, with an outer diameter of 200 μm.

### 2.2. Preparation and Spheroidization of Porcine Fetal Fibroblast

We obtained the porcine fetal fibroblast from a sow at day 35 of pregnancy. After removal of head, internal organs and limbs, the remaining parts were cut into pieces at approximately 1 mm$^3$. We smeared the pieces evenly in a 35 mm dish and cultured in Dulbecco's modified Eagle's medium (DMEM), containing 15% fetal calf serum (FCS), 0.1 mM non-essential amino acids (NEAA), 6 μL/mL Gentamycin and 0.05 mM L-glutamine. Cells were cultured in a 37 °C humidified incubator containing 5% $CO_2$. Cells were trypsinized and cryo-preserved for use when cells grown to ~90% confluence.

The spheroidization experiments of porcine fetal fibroblast were carried out in Medium 199 (Sigma). Figure 3 shows the typical images of the porcine fetal fibroblast spheroidization process:

(1) Give negative pressure in the micropipette to aspirate the cell into the micropipette;
(2) Give positive pressure in the micropipette to eject the capsule-like porcine fetal fibroblast out of the micropipette;
(3) Record the length and the width of the non-spherical shaped cell;
(4) The end of the spheroidization process. The pressure was adjusted by hand. The cells were placed near the tip of micropipette initially and aspirated into the micropipette for more than 10 s. The images were captured with 50 frames per second

and measured with 2 frames per minute. The initial ratio was determined by the inner diameter of micropipette and the cell volume in the experiment. The method of detecting the size of capsule-like fetal fibroblast is described in Appendix A.

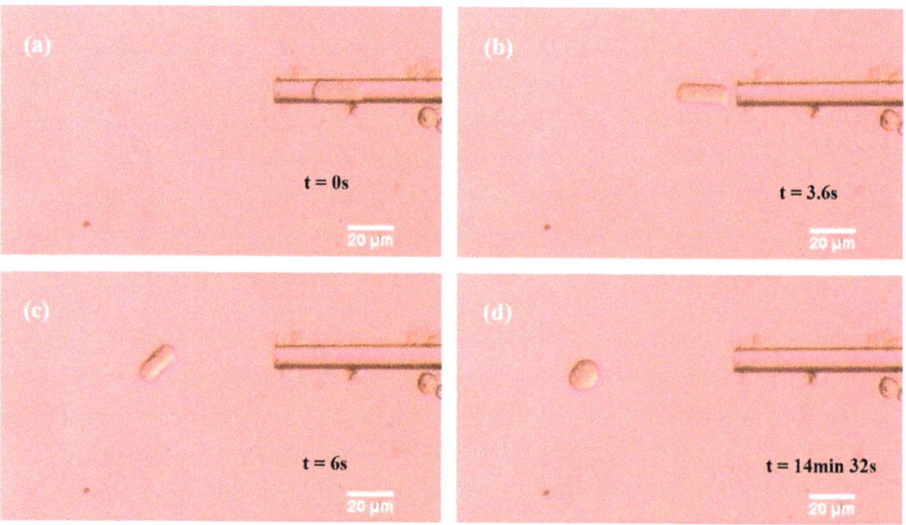


**Figure 3.** Typical images in the spheroidization process of porcine fetal fibroblast. (**a–d**) From porcine fetal fibroblast just coming out of the micropipette to porcine fetal fibroblast becoming a sphere.

### 2.3. Spheroidization of Silicone Oil

Figure 4 shows the method of recording the silicone oil spheroidization process:

(1) Drop culture medium M199 (Sigma) into a petri dish (Corning, 430165 35 mm × 10 mm). Overlay M199 drop with silicone oil (Sigma-Aldrich, St. Louis, MO, USA). The pink liquid in Figure 4 represents M199 and the blue liquid represents the silicone oil.

(2) Move the micropipette tip into the silicone oil drop. Give negative pressure in the micropipette to aspirate some silicone oil into the micropipette.

(3) Move the micropipette tip into M199 solution. Provide positive pressure in the micropipette to eject silicone into M199 solution. Record the silicone oil spheroidization process with a high-speed camera.

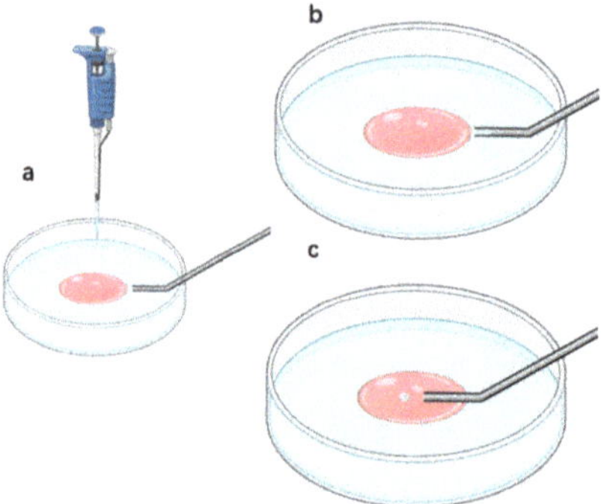


**Figure 4.** (**a**) Drop M199 into a petri dish and overlay M199 with silicone oil; (**b**) aspirate silicone oil into the micropipette; (**c**) move the micropipette tip into M199 and eject silicone oil.

### 2.4. Viscoelastic Model

We use a viscoelastic model to study the spheroidization process. The cell is modeled as homogeneous viscoelastic liquid, which is surrounded by infinitesimal thin cortical layer. We use the Jeffrey's viscoelastic fluid model (Equation (2)) because it is independent of the frame of reference and the motion as a whole in space [31]. Besides, it has only 2 additional parameters, while being able to imitate the viscoelastic behavior. More complex models (e.g., heterogeneous liquid) are hard to modify the parameters to obtain reliable results. In the simulation, the cortical layer is realized by surface tension. We made the following assumptions:

(1) The inner material of fibroblast is homogeneous and isotropic. Based on this assumption we can get global cell properties.
(2) The fibroblast is incompressible. It is for the ease of simulation.
(3) The influence of gravity and pressure variance because of different depth is negligible. It is reasonable by comparing the gravity and pressure variance with hydrostatic pressure (about $1/10^6$ in micron scale).

$$\overset{\triangledown}{T} = \frac{\partial T}{\partial T} + (v \cdot \nabla)T - \nabla v \cdot T - T \cdot (\nabla v)^T \tag{1}$$

$$\lambda \overset{\triangledown}{T} + T = 2\eta E \tag{2}$$

$$\rho \frac{Dv}{Dt} = \nabla \cdot (-pI + K + T) \tag{3}$$

$$K = 2\mu E \tag{4}$$

where $\overset{\triangledown}{T}$ denotes the upper convection derivative [32] of $T$ defined by Equation (1). $T$ is the viscoelastic stress tensor that changes with time according to Equation (2). $v$ is the velocity field. $\lambda$ is the characteristic time. $\eta$ is the viscosity in the viscoelastic term. $E$ is the strain-rate tensor. $\rho$ is the density of porcine fetal fibroblast and is assumed to be constant in the following simulation. $p$ is the pressure. $Dv/Dt$ is the material derivative of $v$. $I$ is the unit tensor. $K$ is the shear stress tensor which can be obtained from Equation (4). $\mu$ is another viscosity in the pure viscous term. Equation (3) is the Navier-Stokes equation.

An analogy of the model in one dimension can be illustrated as Figure 5. The total stress tensor (right terms in the bracket of Equation (3)) is composed of hydrostatic pressure $-pI$, viscous stress $K$ and viscoelastic stress $T$. In this figure, $E$ is the stiffness coefficient of spring, $\eta$ and $\mu$ are the viscosities of two dashpots, and $\lambda = \eta/E$. The bottom line represents a Maxwell model, for which the relationship of strain rate $e$ and stress $\sigma$ is $\lambda \dot{\sigma} + \sigma = \eta e$ in 1D case. By replacing the time derivative with upper convected derivative and extend the equation to 3D tensor form, we get Equation (2). The usage of upper convected derivative for continuum materials was argued by Oldroyd in [31].

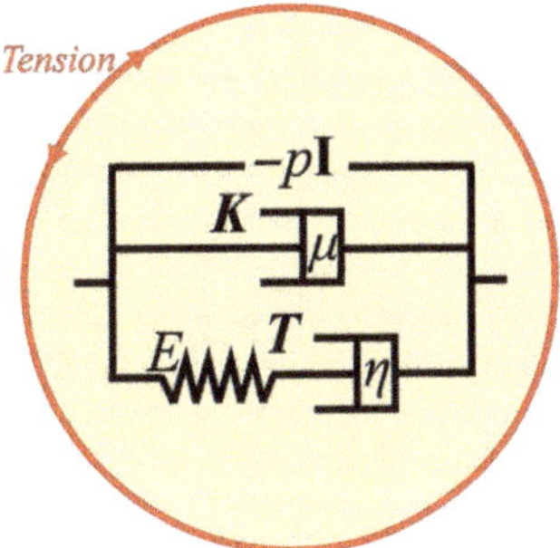


**Figure 5.** Illustration of viscoelastic model.

### 2.5. Simulation of the Spheroidization Process

We used Ansys Student Fluent software to simulate the spheroidization processes. Two-dimensional axisymmetric was adopted for efficiency.

For the simulation of fibroblast, we set the capsule-like porcine fetal fibroblast as a cylinder in the middle and two hemispheres at both ends, as shown in Figure 6 (only 1/4 part was used by applying the axisymmetric and symmetric condition). In this paper, the length was set as 20.8 μm, the width was set as 8.6 μm. The computation domain was a rectangle of $16 \times 10$ μm$^2$ which was divided into $0.1 \times 0.1$ μm$^2$ structural quadrilateral grids. The pressure variance in the scale of several micrometers is of the order of 0.01 Pa, which is far less than the barometric pressure. Besides, it is balanced by the gravity, so we neglected both pressure variance and the gravity. We used volume of fluid (VOF) model to introduce the surface tension. Laminar flow was adopted because of low Reynolds number. We set the four boundaries as axisymmetric, symmetric and pressure outlet, respectively. The densities were set as 1080 kg/m$^3$ [25] for fibroblast and 998.2 kg/m$^3$ (the density of water at 20 °C) for surrounding liquid. The surface tension coefficient $T$ was set as 10 μN/m [33]. The viscosity $\eta$ and elasticity $E$ ($E = \eta/\lambda$) were introduced with user-defined scalars (UDS, see Appendix B). Ansys Student Fluent software solves the momentum Equation (3) without viscoelastic stress term $T$ by default. We used user-defined scalars (UDS) to insert $T$ into the equation (see Appendix B for more details). To study the influence of viscosity and elasticity, we firstly set $\lambda = 1$ s, and changed viscosity $\eta$ as 10 Pa·s, 20 Pa·s, 50 Pa·s, 100 Pa·s, 200 Pa·s, and 500 Pa·s. Secondly, we set $\eta = 500$ Pa·s, and changed $\lambda$ as 500 s, 100 s, 20 s, 1 s, 0 s, 1 s and 0.02 s [33]. The timestep was 3 s. Based on the results obtained when the parameters selected in a wide range, we made more compact selections and compared the results with experimental data. The one that fitted best was viewed as measurement result.

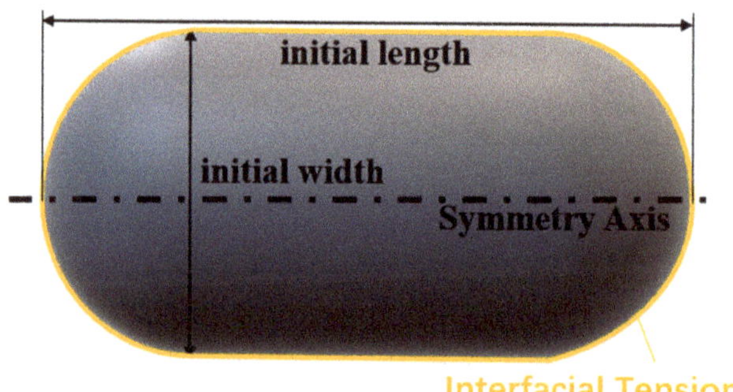


**Figure 6.** Capsule-like porcine fetal fibroblast.

For the simulation of silicone oil, the initial shape was set as axisymmetric while the contour being obtained by image processing procedure (see Appendix A). The computation domain was a rectangle of $500 \times 200$ μm$^2$ which was divided into $2 \times 2$ μm$^2$ structural quadrilateral grids. As is shown in Figure 7, one boundary is symmetric axis and others are pressure outlet. Using the contour obtained by image processing (Appendix A), a user-defined function sets the corresponding region as silicone oil (secondary phase), while the remainder as culture medium (primary phase). The viscosity $\mu$ was inserted by setting the material property of silicone oil in the software. The viscoelastic stress term $T$ was removed because it was considered as pure viscous liquid. Volume of fluid model and laminar flow was adopted. Then we run the simulation with 1 ms timestep. Please see the simulation procedure details in Appendix B.

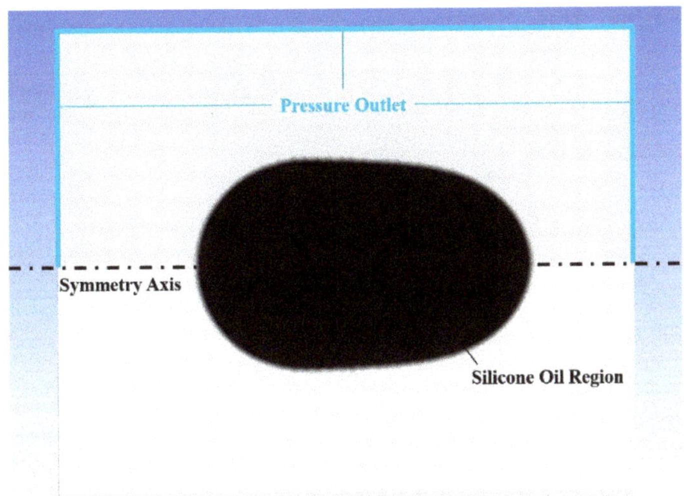


**Figure 7.** Geometric shape and boundary settings.

## 3. Results

### 3.1. Spheroidization Result of Porcine Fetal Fibroblast and Its Simulation

The porcine fetal fibroblast was used in the experiments.

The typical images in the spheroidization process of porcine fetal fibroblast have been shown in Figures 1 and 2 (Video S1). The length and the width changing process was shown in Figure 8. The whole spheroidization process took 15 min to reach a 90% width–length ratio.

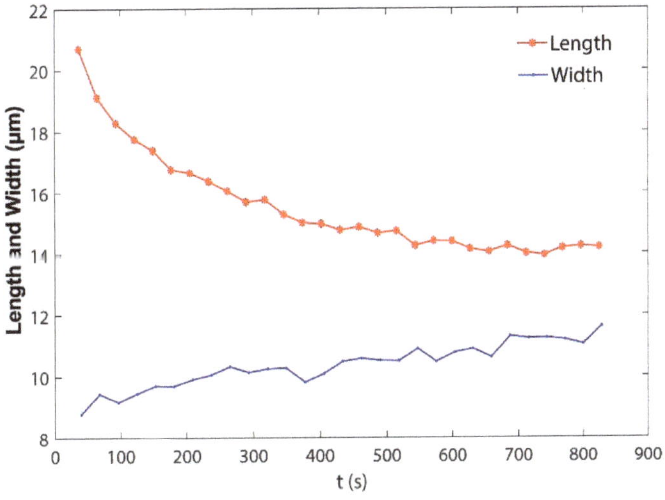


**Figure 8.** Variation of porcine fetal fibroblast length and width with time.

We got the simulated width–length ratio at the condition that $\lambda = 1$ and changed viscosity $\eta$, as shown in Figure 9. Video S2 shows the simulated spheroidization process of porcine fetal fibroblast. The results show that a larger $\eta$ will prevent the porcine fetal fibroblast from turning into a sphere, and the spheroidization time becomes longer.

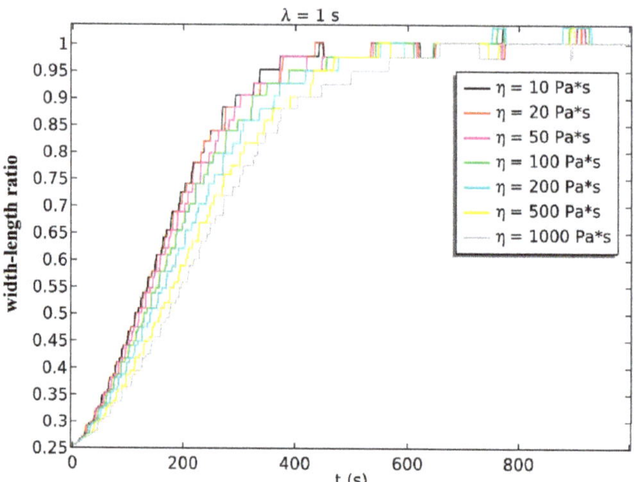

**Figure 9.** The simulated spheroidization process when $\lambda = 1$ and viscosity $\eta$ changed.

We got the simulated width–length ratio at the condition that $\eta = 500$ Pa·s and changed $\lambda$, as shown in Figure 10. The experimental results showed that the spheroidization process was more intense in the initial stage, but because the elasticity was smaller when $\lambda$ was larger, the small elasticity will bring a lag effect in the later stage of spheroidization, which would make the later stage of spheroidization slow down significantly.

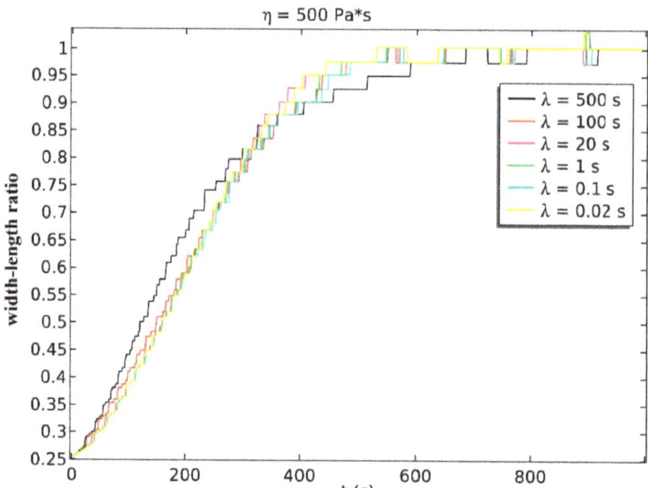

**Figure 10.** The simulated spheroidization process when $\eta = 500$ Pa·s and $\lambda$ changed.

Finally, by changing the values of $\eta$ and $\lambda$ in the simulation experiment, different curves of the spheroidization process of the simulated porcine fetal fibroblast were obtained. By comparing with the curves of the spheroidization process obtained in the real experiment, the elasticity and viscosity could be obtained. Figure 11 shows the length and width variation of porcine fetal fibroblast with time in the experiment and simulation. The viscosity $\eta$ obtained in this experiment is 10 Pa·s and the elasticity $E$ is 500 Pa. The magnitude of the results was in agreement with the measured results in [29].

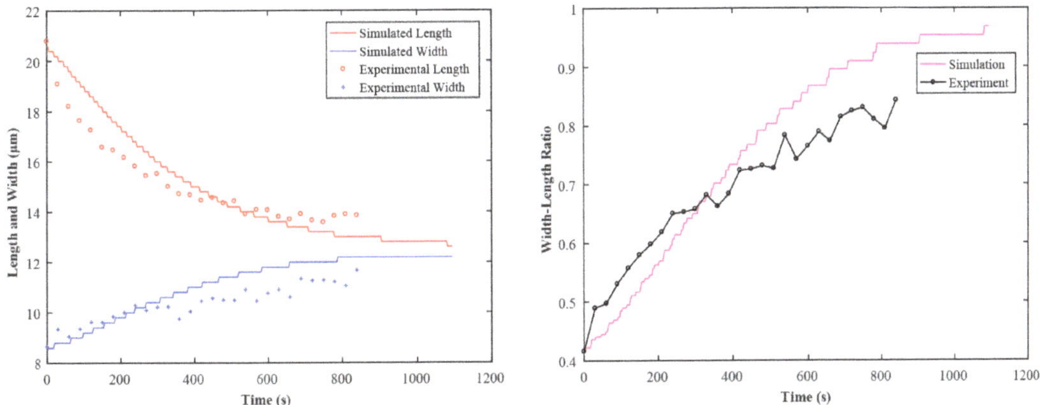

**Figure 11.** Variation of porcine fetal fibroblast length and width with time in the experiment and simulation.

### 3.2. Spheroidization Result of Silicone Oil and Its Simulation

Figure 12 shows the typical images in the spheroidization process of silicone oil. Video S3 shows this process of slowing down 100 times. Video S4 shows the simulated spheroidization process of silicone oil.

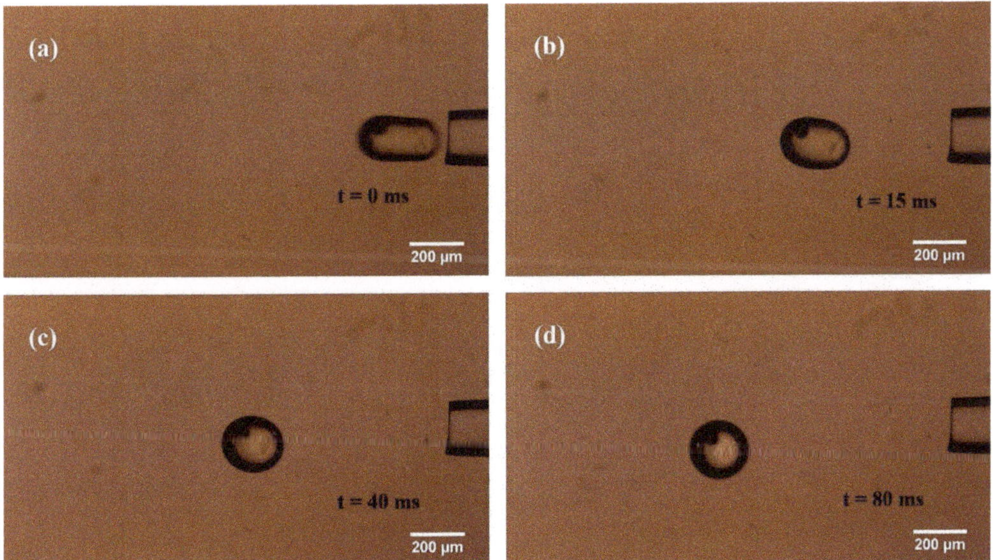


**Figure 12.** Typical images in the spheroidization process of silicone oil. (**a–d**) From silicone oil just coming out of the micropipette to silicone oil becoming a sphere.

We defined the time span from the release to 95% width–length ratio as spheroidization time, which is denoted as $t_s$. The $t_s$ was 80 ms in the experiment.

The viscosity and density of the silicone oil (Sigma-Aldrich) at 25 °C was 9.71 Pa·s and 0.971 g/mL. To simulate the spheroidization process of silicone oil, we also need to know the surface tension coefficient with culture medium of silicone oil. We measured the surface tension coefficient between silicone oil and culture medium by Du Noüy ring method [26], the coefficient was 0.024 N/m. The detail is shown in Appendix C. The method of detecting the contour of the silicone oil is shown in Appendix A.

Figure 13 shows the simulation results of silicone oil with different viscosities and surface tension coefficient. The R-square values of the fitted curves are 0.99 and 0.98. The results revealed that spheroidization time increases linearly as viscosity and the reciprocal of surface tension coefficient increases. The results showed that the spheroidization time $t_s$ was 100 ms in the simulation, which was similar to the real experiment (80 ms).

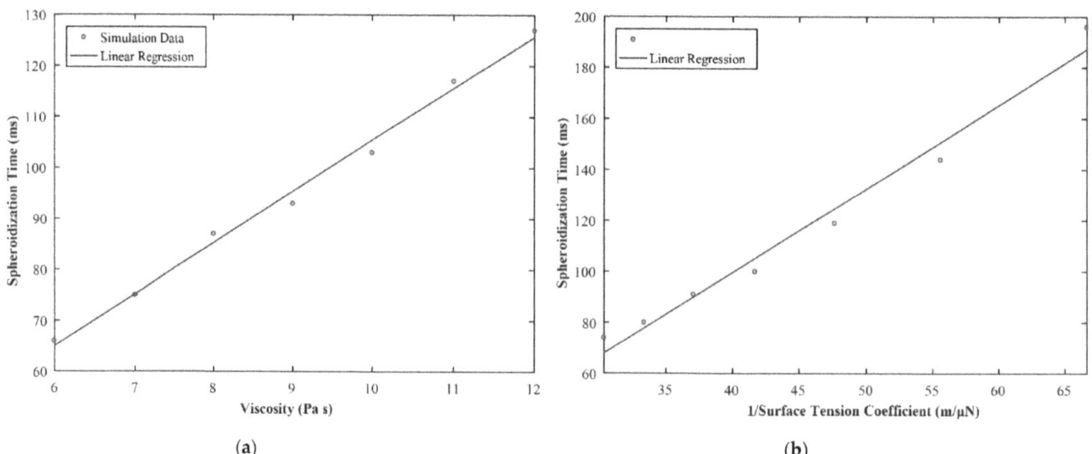



**Figure 13.** Simulation results: (**a**) spheroidization time increases linearly as viscosity increases. (**b**) Spheroidization time increases linearly as the surface tension coefficient increases.

## 4. Discussion

We should know that the results measured in this paper were based on the bulk measurements, by which the cells were assumed as isotropic, homogeneous. However, in reality cells are very heterogeneous and contain organelles. We also need to measure the local force and dissipative gradients, as well as map them across the cell surface [34–36]. Considering the measuring efficiency, only two parameters are necessary to describe the cellular mechanics, so the bulk measurement is more appropriate.

We used the cells in the suspension state instead of adherent state in this paper. Because our method needs to aspirate to the whole cell into the micropipette, and it is difficult to aspirate the adherent cells into the micropipette because of the adhesion. Since the whole suspension cell was sucked into the micropipette, the cell spheroidization process was only related to the shape of the micropipette. The seal between the cell and the micropipette will not affect the spheroidization recording results, which can avoid the influence of seal in the micropipette aspiration method.

We performed the cell experiments three times and the simulations 12 times per cell. As the cells were collected from one batch, the experimental curves were very similar. The parameters in the simulation were set not very accurate (just integers), so the results of these three cells were the same. The measurement results may be significantly different among different cell types or different cell batches. Our future work will focus on measuring the viscoelasticity of different cell types and improving the simulation accuracy by adjusting the parameters more accurately.

We could see that the simulation results shown in Figure 10 do not overlap with the experiment results shown in Figure 7 exactly. We supposed that there were three reasons:

(1) The whole spheroidization process would take a tremendously long time in a in vitro environment, which would influence the viscoelasticity of the cell a lot. We only recorded the spheroidization process when the cells reached a 90% width–length

ratio, which took about 15 min. Meanwhile, the simulation process recorded the whole spheroidization process. So, the experiment results and the simulation results could not overlap exactly.

(2) As mentioned above, the parameters in the simulation were not set very accurately (just integers), so the simulation results could not exactly fit the experiment results.

(3) The initial velocity was set to zero. This may have caused an initial acceleration stage from zero, while the non-zero initial shrink in velocity was found from the experiment. The problem was handled by running several steps in advance.

We know that if the viscous term of a viscoelastic body is increased, then it takes longer to get back to its original shape. Our results showed that a larger $\eta$ will prevent the porcine fetal fibroblast from turning into a sphere, which could further verify the validity of our simulation results.

## 5. Conclusions

This paper presents a cell's viscoelasticity measurement method based on the spheroidization process of a non-spherical shaped cell. We firstly introduced the process of recording the spheroidization process of porcine fetal fibroblast. We secondly built the viscoelastic model for simulating a cell's spheroidization process. We simulated the spheroidization process of porcine fetal fibroblast and got the simulated spheroidization process. Then we got the elasticity (500 Pa) and viscosity (10 Pa·s) of porcine fetal fibroblast by identifying the parameters in the viscoelastic model. The results showed that the magnitude of the elasticity and viscosity were in agreement with those measured by a traditional method. To verify the accuracy of the proposed method, we imitated the spheroidization process with silicone oil, a kind of viscous and uniform liquid with determined viscosity. We did the silicone oil's spheroidization experiment and simulated this process. The simulation results also fitted the experimental results well.

**Supplementary Materials:** The following are available online at https://www.mdpi.com/article/10.3390/s21165561/s1, Video S1: The spheroidization process of porcine fetal fibroblast (0 to 15 s: slowing down 5 times; 16 to 23 s speeding up 100 times). Video S2: The simulated spheroidization process of porcine fetal fibroblast. Video S3: The spheroidization process of silicone oil. Video S4: The simulated spheroidization process of silicone oil.

**Author Contributions:** X.Z. (Xin Zhao) and Y.L. conceived the idea for the study. X.Z. (Xin Zhao), Y.L., Y.Z. and M.S. designed the experiments. Y.L., Y.Z., M.C. and X.Z. (Xiangfei Zhao) performed the experiments. X.Z. (Xin Zhao) supervised the project. Y.L. and Y.Z. wrote the manuscript. Y.L., Y.Z. and M.C. contribute equally to this paper. All authors have read and agreed to the published version of the manuscript.

**Funding:** This research was jointly supported by the National Key R&D Program of China (2018YFB1304905, 2020YFB1313101), the National Natural Science Foundation of China (U1813210, 62003174, 62003173), and the China Postdoctoral Science Foundation (2020M680865).

**Institutional Review Board Statement:** All the procedures were approved by the Animal Care and Use Committee of Tianjin Animal Science and Veterinary Research Institute, and were performed in accordance with the NIH Guide for the Care and Use of Laboratory Animals (No. 8023, revised in 1996).

**Informed Consent Statement:** Not applicable.

**Data Availability Statement:** The raw data supporting the conclusions of this article will be made available by the authors, without undue reservation.

**Conflicts of Interest:** The authors declare no conflict of interest.

## Appendix A. Size Measurements of Porcine Fetal Fibroblast and Silicone Oil

In the initialization step of the simulation, we chose the image where the porcine fetal fibroblast and silicone oil were just out of the micropipette, then extracted its contour. The flow chart of image processing is shown in Figure A1. After the processes, we got a list of contour points where x axis was the axisymmetric axis. As the contour was symmetric, we used the upper half of it to conduct an axisymmetric simulation, which saved time and computational resources, and was more convenient to show the results. For the consecutive images, length and width were measured automatically for silicone oil and manually for fibroblast because of the low contrast ratio.

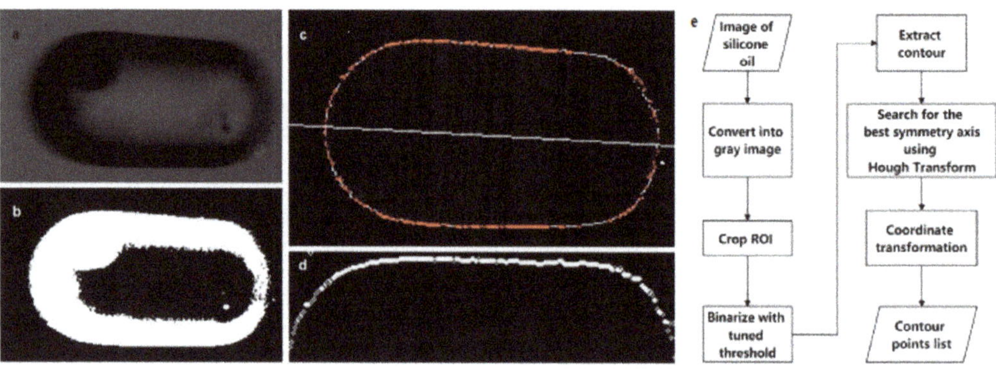

**Figure A1.** Image processing: (**a**) ROI of grayscale image; (**b**) binarized image; (**c**) axisymmetric contour; (**d**) upper side of contour; (**e**) flow chart of image processing.

### Appendix B. Simulation Procedure

We used Ansys Student Fluent to simulate the spheroidization process. The Fluent software provides the volume of fluid (VOF) module for multiphase simulation such that we can simulate the surface tension between cells and culture medium. Because the velocities in such processes are small (of the order of 1 nm/s), the Reynolds number is small, so the laminar model was adopted. For the cell simulation, we used user-defined scalars (UDS) to introduce the viscoelastic stress term. Details of the simulation are described below.

**Geometry:** rectangle region of $16 \times 10 \ \mu m^2$ for fibroblast and $500 \times 200 \ \mu m^2$ for silicone oil;

**Mesh:** structural quadrilateral grids of $0.1 \times 0.1 \ \mu m^2$ for fibroblast and $2 \times 2 \ \mu m^2$ for silicone oil;

**Boundary conditions:**
left and top—pressure outlet with 0 Pascal gauge pressure;
right—symmetric for fibroblast and pressure outlet for silicone oil;
down—axisymmetric;

**Fluid filed settings:**
General: 2D axisymmetric;
Models panel: multiphase—VOF (Phase Interactions—Surface Tension), Viscous (Laminar);

**Materials:** The density and viscosity are set in this stage. In this paper, the values are: $1080 \ kg/m^3$ of density and 500 Pa s of viscosity for fibroblast, $9.71 \ kg/m^3$ of density and 9.71 Pa s of viscosity for silicone oil, $998.2 \ kg/m^3$ (the density of water at 20 °C) of density and $1.003 \times 10^{-3}$ Pa s for surrounding liquid.

**Initialization and UDS equations:**
Ansys Student Fluent provides user-defined functions (UDF) for customizing fluid simulation. In this paper, a C code file utilizing predefined macros sets the initial fibroblast and silicone oil area, as well as introduces equations about viscoelastic stress $T$. To add $T$ into the momentum equation, four scalars were defined and added to the momentum equation by DEFINE_SOURCE macro. The usage of UDF can be found in official ANSYS Fluent UDF Manual. The initial velocities were set as zero.

For an arbitrary user-defined scalar $\phi$, Fluent solves the Equation (A1).

$$\frac{\partial \rho \phi}{\partial t} + \frac{\partial}{\partial x_i}\left(\rho u_i \phi - \Gamma_k \frac{\partial \phi}{\partial x_i}\right) = S \tag{A1}$$

where $\rho$ is the density, $u$ is the velocity, $\Gamma$ is the diffusion coefficient, $S$ is the source term. The terms in the equation represent unsteady term, convective flux, diffusion and source from left to right. Expanding Equation (2) and using Einstein summation convention, we get Equation (A2):

$$\lambda \frac{\partial \tau_{ij}}{\partial t} + \lambda u_k \frac{\partial \tau_{ij}}{\partial x_k} - \lambda \tau_{ik}\frac{\partial u_j}{\partial x_k} - \lambda \tau_{kj}\frac{\partial u_i}{\partial x_k} + \tau_{ij} = \mu\left(\frac{\partial u_i}{\partial x_j} + \frac{\partial u_j}{\partial x_i}\right) \tag{A2}$$

The first term was added with DEFINE_UDS_UNSTEADY macro. The second term was added with DEFINE_UDS_FLUX macro. Other terms were added with DEFINE_SOURCE macro. It should be noticed that for 2D axisymmetric setup in this paper, the gradients along normal direction of the geometry plane were all zero. Therefore, only four scalars $\tau_{11}, \tau_{12}, \tau_{21}, \tau_{22}$ were defined. Besides, a factor $2\pi$ should be multiplied for volume and area because of axisymmetric condition according to the manual.

**Solver**: 1 ms of timestep for silicone oil and 3 s for fibroblast. For each time step, iterate at most 100 steps with 0.001 as convergence absolute criteria.

**Notes:**

1. The quadrilateral grids are suggested for the meshing as it is more stable than triangular grids in the simulation. We suppose that it was because triangular grids cause larger curvature, which makes surface tension change dramatically within a local area.

2. The mesh size should not be too small, not only for the computation efficiency, but also for the convergence. When reducing the mesh size down to a certain scale, the simulation gets hard to converge. It is also supposed to be the result of large local surface tension.

### Appendix C. Surface Tension Coefficient Measurement of Silicone Oil

As is illustrated in Figure A2, when the interface between silicone oil and culture medium was flat and the platinum ring was on the interface, we have

$$F_0 + B = G \tag{A3}$$

where $F_0$ is the external force in this case, $B$ is the buoyancy, $G$ is the total weight of platinum ring and its frame.

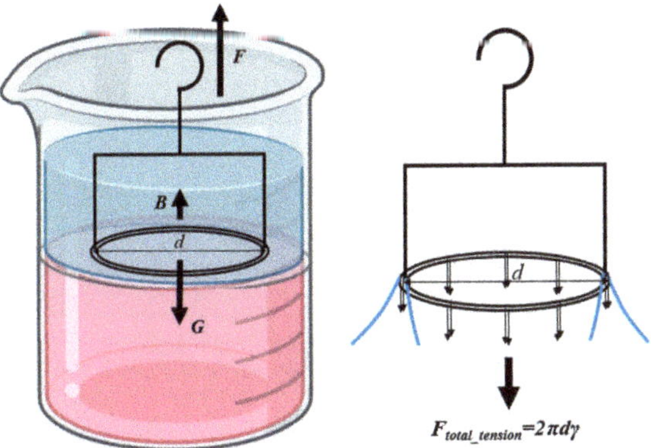


**Figure A2.** Du Noüy ring method to measure surface tension coefficient.

When we pull the platinum ring up, it stuck the interface, causing a circle of interface protruding (Figure A2), and an extreme thin film formed near the ring. Inner and outer interface both contributed to the surface tension, i.e.,:

$$F_{total\_tension} = 2\pi d\gamma \tag{A4}$$

where $\gamma$ was the surface tension coefficient.

The pull is a quasi-static process, we can get

$$F_m + B = G + F_{total\_tension} \tag{A5}$$

where $F_m$ was the maximum external force in the process. The change of $B$ was negligible compared with change of external force, so we used the same notation $B$. Combining (A3), (A4) and (A5), we get:

$$\gamma = \frac{F_m - F_0}{2\pi d} = \frac{\Delta F}{2\pi d} \tag{A6}$$

In this experiment, the diameter $d$ was 14 mm, and we got that $\Delta F$ was 2.11 mN in average, so $\gamma$ was 0.024 N/m.

## References


1. Liu, X.; Shi, J.; Zong, Z.; Wan, K.T.; Sun, Y. Elastic and viscoelastic characterization of mouse oocytes using micropipette indentation. *Ann. Biomed. Eng.* **2012**, *40*, 2122–2130. [CrossRef] [PubMed]
2. Murayama, Y.; Mizuno, J.; Kamakura, H.; Fueta, Y.; Nakamura, H.; Akaishi, K.; Anzai, K.; Watanabe, A.; Inui, H.; Omata, S. Mouse zona pellucida dynamically changes its elasticity during oocyte maturation, fertilization and early embryo development. *Hum. Cell* **2006**, *19*, 119–125. [CrossRef]
3. Jia, Z.; Feng, Z.; Wang, L.; Li, H.; Wang, H.; Xu, D.; Zhao, X.; Feng, D.; Feng, X. Resveratrol reverses the adverse effects of a diet-induced obese murine model on oocyte quality and zona pellucida softening. *Food Funct.* **2018**, *9*, 2623–2633. [CrossRef] [PubMed]
4. Pathak, A.; Kumar, S. Biophysical regulation of tumor cell invasion: Moving beyond matrix stiffness. *Integr. Biol.* **2011**, *3*, 267–278. [CrossRef] [PubMed]
5. Fletcher, D.A.; Mullins, R.D. Cell mechanics and the cytoskeleton. *Nature* **2010**, *463*, 485–492. [CrossRef]
6. Darling, E.M.; Topel, M.; Zauscher, S.; Vail, T.P.; Guilak, F. Viscoelastic properties of human mesenchymally-derived stem cells and primary osteoblasts, chondrocytes, and adipocytes. *J. Biomech.* **2008**, *41*, 454–464. [CrossRef]
7. Alcaraz, J.; Buscemi, L.; Grabulosa, M.; Trepat, X.; Fabry, B.; Farre, R.; Navajas, D. Microrheology of human lung epithelial cells measured by atomic force microscopy. *Biophys. J.* **2003**, *84*, 2071–2079. [CrossRef]
8. Efremov, Y.M.; Wang, W.H.; Hardy, S.D.; Geahlen, R.L.; Raman, A. Measuring nanoscale viscoelastic parameters of cells directly from AFM force-displacement curves. *Sci. Rep.* **2017**, *7*, 1–14. [CrossRef]
9. Garcia, P.D.; Garcia, R. Determination of the viscoelastic properties of a single cell cultured on a rigid support by force microscopy. *Nanoscale* **2018**, *10*, 19799–19809. [CrossRef]
10. Benaglia, S.; Amo, C.A.; Garcia, R. Fast, quantitative and high resolution mapping of viscoelastic properties with bimodal AFM. *Nanoscale* **2019**, *11*, 15289–15297. [CrossRef]
11. Parvini, C.H.; Saadi, M.A.S.R.; Solares, S.D. Extracting viscoelastic material parameters using an atomic force microscope and static force spectroscopy. *Beilstein J. Nanotechnol.* **2020**, *11*, 922–937. [CrossRef] [PubMed]
12. Parvini, C.H.; Cartagena-Rivera, A.X.; Solares, S.D. Viscoelastic Parameterization of Human Skin Cells to Characterize Material Behavior at Multiple Timescales. *bioRxiv* **2021**. [CrossRef]
13. Laurent, V.M.; Henon, S.; Planus, E.; Fodil, R.; Balland, M.; Isabey, D.; Gallet, F. Assessment of mechanical properties of adherent living cells by bead micromanipulation: Comparison of magnetic twisting cytometry vs optical tweezers. *J. Biomech. Eng.* **2002**, *124*, 408–421. [CrossRef] [PubMed]
14. Bausch, A.R.; Moller, W.; Sackmann, E. Measurement of local viscoelasticity and forces in living cells by magnetic tweezers. *Biophys. J.* **1999**, *76*, 573–579. [CrossRef]
15. Guo, H.L.; Liu, C.X.; Duan, J.F.; Jiang, Y.Q.; Han, X.H.; Li, Z.H.; Cheng, B.Y.; Zhang, D.Z. Mechanical properties of breast cancer cell membrane studied with optical tweezers. *Chin. Phys. Lett.* **2004**, *21*, 2543–2546.
16. Li, Y.J.; Wen, C.; Xie, H.M.; Ye, A.P.; Yin, Y.J. Mechanical property analysis of stored red blood cell using optical tweezers. *Colloids Surf. B* **2009**, *70*, 169–173. [CrossRef] [PubMed]
17. Rosenbluth, M.J.; Lam, W.A.; Fletcher, D.A. Analyzing cell mechanics in hematologic diseases with microfluidic biophysical flow cytometry. *Lab Chip* **2008**, *8*, 1062–1070. [CrossRef] [PubMed]
18. Shevkoplyas, S.S.; Yoshida, T.; Munn, L.L.; Bitensky, M.W. Biomimetic autoseparation of leukocytes from whole blood in a microfluidic device. *Anal. Chem.* **2005**, *77*, 933–937. [CrossRef]

19. Bow, H.; Pivkin, I.V.; Diez-Silva, M.; Goldfless, S.J.; Dao, M.; Niles, J.C.; Suresh, S.; Han, J. A microfabricated deformability-based flow cytometer with application to malaria. *Lab Chip* **2011**, *11*, 1065–1073. [CrossRef]
20. Evans, E.; Yeung, A. Apparent viscosity and cortical tension of blood granulocytes determined by micropipet aspiration. *Biophys. J.* **1989**, *56*, 151–160. [CrossRef]
21. Mohammadalipour, A.; Choi, Y.E.; Benencia, F.; Burdick, M.M.; Tees, D.F.J. Investigation of mechanical properties of breast cancer cells using micropipette aspiration technique. *FASEB J.* **2012**, *26*, 905–909. [CrossRef]
22. Sohail, T.; Tang, T.; Nadler, B. Micropipette aspiration of an inflated fluid-filled spherical membrane. *Z. Angew. Math. Phys.* **2012**, *63*, 737–757. [CrossRef]
23. Kamat, N.P.; Lee, M.H.; Lee, D.; Hammer, D.A. Micropipette aspiration of double emulsion-templated polymersomes. *Soft Matter* **2011**, *7*, 9863–9866. [CrossRef]
24. Hochmuth, R.M. Micropipette aspiration of living cells. *J. Biomech.* **2000**, *33*, 15–22. [CrossRef]
25. Liu, Y.; Cui, M.; Huang, J.; Sun, M.; Zhao, X.; Zhao, Q. Robotic Micropipette Aspiration for Multiple Cells. *Micromachines* **2019**, *10*, 348. [CrossRef]
26. Zhao, Q.; Wu, M.; Cui, M.; Qin, Y.; Yu, J.; Sun, M.; Zhao, X.; Feng, X. A novel pneumatic micropipette aspiration method using a balance pressure model. *Rev. Sci. Instrum.* **2013**, *84*, 123703. [CrossRef]
27. Jones, W.R.; Ting-Beall, H.P.; Lee, G.M.; Kelley, S.S.; Hochmuth, R.M.; Guilak, F. Alterations in the Young's modulus and volumetric properties of chondrocytes isolated from normal and osteoarthritic human cartilage. *J. Biomech.* **1999**, *32*, 119–127. [CrossRef]
28. Liu, Y.; Chen, D.; Cui, M.; Sun, M.; Huang, J.; Zhao, X. Evaluation of the deformability of the cell's zona pellucida based on the subpixel cell contour detection algorithm. In Proceedings of the 35th Chinese Control Conference (CCC), Chengdu, China, 29 August 2016; p. 9109.
29. Liu, Y.W.; Cui, M.S.; Sun, Y.M.; Feng, Z.Y.; Bai, Y.X.; Sun, M.Z.; Zhao, Q.L.; Zhao, X. Oocyte orientation selection method based on the minimum strain position in the penetration process. *J. Appl. Phys.* **2019**, *125*, 154701. [CrossRef]
30. Liu, Y.W.; Wang, X.F.; Zhao, Q.L.; Zhao, X.; Sun, M.Z. Robotic Batch Somatic Cell Nuclear Transfer Based on Microfluidic Groove. *IEEE Trans. Autom. Sci. Eng.* **2020**, *17*, 2097–2106. [CrossRef]
31. Oldroyd, J.G. *On the Formulation of Rheological Equations of State*; Royal Society: London, UK, 1950; p. 524.
32. Olsson, F.; Yström, J. Some properties of the Upper Convected Maxwell model for viscoelastic fluid flow. *J. Non-Newtonian Fluid Mech.* **1993**, *48*, 125–145. [CrossRef]
33. Lim, C.T.; Zhou, E.H.; Quek, S.T. Mechanical models for living cells - A review. *J. Biomech.* **2006**, *39*, 195–216. [CrossRef]
34. Benoit, M.; Gaub, H.E. Measuring cell adhesion forces with the atomic force microscope at the molecular level. *Cells Tissues Organs* **2002**, *172*, 174–189. [CrossRef]
35. Darling, E.M.; Zauscher, S.; Block, J.A.; Guilak, F. A thin-layer model for viscoelastic, stress-relaxation testing of cells using atomic force microscopy: Do cell properties reflect metastatic potential? *Biophys. J.* **2007**, *92*, 1784–1791. [CrossRef] [PubMed]
36. Cartagena, A.; Raman, A. Local Viscoelastic Properties of Live Cells Investigated Using Dynamic and Quasi-Static Atomic Force Microscopy Methods. *Biophys. J.* **2014**, *106*, 1033–1043. [CrossRef] [PubMed]

*Article*

# DMAS Beamforming with Complementary Subset Transmit for Ultrasound Coherence-Based Power Doppler Detection in Multi-Angle Plane-Wave Imaging

Che-Chou Shen * and Yen-Chen Chu

Department of Electrical Engineering, National Taiwan University of Science and Technology, Taipei 106335, Taiwan; m10807306@mail.ntust.edu.tw
* Correspondence: choushen@mail.ntust.edu.tw; Tel.: +886-2-27301229; Fax: +886-2-27376699

**Abstract:** Conventional ultrasonic coherent plane-wave (PW) compounding corresponds to Delay-and-Sum (DAS) beamforming of low-resolution images from distinct PW transmit angles. Nonetheless, the trade-off between the level of clutter artifacts and the number of PW transmit angle may compromise the image quality in ultrafast acquisition. Delay-Multiply-and-Sum (DMAS) beamforming in the dimension of PW transmit angle is capable of suppressing clutter interference and is readily compatible with the conventional method. In DMAS, a tunable $p$ value is used to modulate the signal coherence estimated from the low-resolution images to produce the final high-resolution output and does not require huge memory allocation to record all the received channel data in multi-angle PW imaging. In this study, DMAS beamforming is used to construct a novel coherence-based power Doppler detection together with the complementary subset transmit (CST) technique to further reduce the noise level. For $p = 2.0$ as an example, simulation results indicate that the DMAS beamforming alone can improve the Doppler SNR by 8.2 dB compared to DAS counterpart. Another 6-dB increase in Doppler SNR can be further obtained when the CST technique is combined with DMAS beamforming with sufficient ensemble averaging. The CST technique can also be performed with DAS beamforming, though the improvement in Doppler SNR and CNR is relatively minor. Experimental results also agree with the simulations. Nonetheless, since the DMAS beamforming involves multiplicative operation, clutter filtering in the ensemble direction has to be performed on the low-resolution images before DMAS to remove the stationary tissue without coupling from the flow signal.

**Keywords:** delay-and-sum (DAS); delay-multiply-and-sum (DMAS); signal coherence; power doppler detection; plane-wave (PW) imaging; complementary subset transmit (CST); coherent plane-wave compounding (CPWC)

**Citation:** Shen, C.-C.; Chu, Y.-C. DMAS Beamforming with Complementary Subset Transmit for Ultrasound Coherence-Based Power Doppler Detection in Multi-Angle Plane-Wave Imaging. *Sensors* **2021**, *21*, 4856. https://doi.org/10.3390/s21144856

Academic Editors: Vahid Abolghasemi, Hossein Anisi, Saideh Ferdowsi and Manuel Graña

Received: 2 June 2021
Accepted: 13 July 2021
Published: 16 July 2021

**Publisher's Note:** MDPI stays neutral with regard to jurisdictional claims in published maps and institutional affiliations.

## 1. Background

Delay-and-Sum (DAS) beamforming is routinely adopted to produce image output in medical ultrasound imaging by compensating the time delay of the received echo according to the geometric path of propagation before coherent summation [1]. However, it suffers from intrinsic limitations such as insufficient image resolution and noticeable off-axis clutter. For plane-wave (PW) imaging which depends on the unfocused transmit wave to illuminate a wide field-of-view [2], these limitations are notably evident. In single-angle PW imaging, the received backscattered echoes from one PW transmit event are processed using DAS beamforming in the direction of receiving channel to generate the corresponding low-resolution image at frame rate on the order of kHz. Therefore, PW imaging is also referred as ultrafast imaging. The image quality of PW imaging can be improved by coherent plane wave compounding (CPWC) in which synthetic transmit focusing is achieved from multi-angle PW transmit [3,4]. Specifically, low-resolution images are firstly acquired from several PW transmit angles and then coherently combined to achieve the final high-resolution CPWC image. Considering an imaging depth of 75 mm, the pulse-repetition-interval (PRI) for pulse-echo imaging will be 100 μs with the sound velocity of 1.5 mm/μs. This corresponds to a frame rate of 10 kHz for low-resolution PW imaging. Assuming that every 10 low-resolution images is coherently combined to improve the image quality, the resultant high-resolution CPWC images will be produced at a frame rate of 1 kHz. With this high frame rate, CPWC imaging has been utilized to detect the motion of imaged objects in transient elastography [4,5] and Doppler flow imaging [6–13]. However, it should be noted that the image quality in CPWC imaging relies on the number of low-resolution images involved in the compounding, and thus an inevitable trade-off between the image quality and the frame rate exists. In other words, with just a few PW transmit angles, the image quality in CPWC imaging is generally unsatisfactory due to the presence of high clutter artifacts.

Note that the aforementioned CPWC imaging for each image pixel is actually performed by coherently summing all the received channel data from all the PW transmit angles. These data can be represented as an echo matrix comprising two dimensions of the PW transmit angle and the receiving channel. Thus, the CPWC imaging is actually constructed using two-dimensional DAS beamforming. Note that the signal coherence of the two-dimensional echo matrix can be used to reject low-coherence clutters and thermal noises to further improve the multi-angle PW image quality. One particular example is Delay-Multiply-and-Sum (DMAS) beamforming. Originally, DMAS beamforming is developed to extract the signal coherence in the dimension of receiving channel by multiplying the received echoes between every possible channel pair after time compensation [14]. In order to improve the computational efficiency of the original DMAS beamforming, alternative high-order versions of DMAS beamforming have been recently proposed with flexibly tunable image quality in [15,16]. Take the BB-DMAS [16] as an example, where a rational $p$ value is used to represent the order of DMAS beamforming. Note that a higher image quality can be achieved by adopting a higher $p$ value to emphasize more spatial coherence in DMAS beamforming. The implementation of BB-DMAS beamforming involves the magnitude scaling of time-delayed channel signal by $p$-th root and the subsequent $p$-th power after channel sum. DMAS beamforming has also been extended to multi-angle PW imaging by extracting the signal coherence of two-dimensional echo matrix. In [17], DMAS beamforming is applied in the dimension of receiving channel to exploit the spatial coherence of synthesized echoes in different channels. The synthesized channel data is produced by summing the echo matrix in the dimension of PW transmit angle for synthetic transmit focusing as in CPWC imaging [17]. On the contrary, the two-dimensional spatial coherence can be also derived directly from the entire echo matrix using echoes in both dimensions [18].

In this study, a novel coherence-based DMAS power Doppler detection together with complementary subset transmit (CST) is proposed for multi-angle PW imaging. Power Doppler provides essential information of the backscattered power of flow signal and generally has higher sensitivity to small vessels than color Doppler [19,20]. This is because these small vessels may not be detectable using velocity estimation in color Doppler due to noises. The proposed method firstly adopts DMAS beamforming in the dimension of PW transmit angle to suppress the background noise and clutter. Then, the CST technique is used to further reduce the noise level in power Doppler detection by correlation of two complementary DMAS signals. Unlike the coherent flow power Doppler (CFPD) method [21,22] that relies on short-lag spatial coherence [23] to extract the coherence of blood flow signal in the dimension of receiving channel, the proposed DMAS beamforming is based on the signal coherence among low-resolution images from distinct PW transmit angles. It is compatible with current CPWC imaging and does not require huge memory allocation to retain the entire channel data in the echo matrix. In other words, the delayed channel data is firstly summed to one low-resolution image pixel. Then, DMAS beamforming in the dimension of PW transmit angle can be performed by magnitude-scaling these low-resolution images from distinct PW transmit angles before restoring the signal dimensionality to produce the high-resolution image after coherent compounding. Section 2 introduces the basics of DMAS beamforming in the dimension of PW transmit angle for power Doppler detection and the subsequent implementation of CST technique. Simulation methods in this study are described in detail in Section 3, together with experimental setups. In Section 4, image quality of the proposed DMAS-based power Doppler detection is quantitatively presented. Section 5 concludes our results with discussions.

## 2. Theory

### 2.1. Basics of Power Doppler Detection

In Doppler ultrasound imaging, the motion of red blood cells in the vessel is detected by repetitive pulse transmissions to observe the temporal variations of backscattered signals. For each image pixel, the recorded signal corresponding to the $f$-th pulse transmission is generally referred to as the $f$-th Doppler ensemble where $f$ is the index of ensemble ($f = 1, 2, \ldots, F$). In other words, there are a total of $F$ ensembles available for velocity estimation in color Doppler and/or power estimation in power Doppler. Note that, in order to separate the blood flow signal from the stationary tissue signal and the thermal noises, a temporal clutter filter has to be applied to the Doppler ensembles to extract signal components with frequencies within a low-order threshold and a high-order threshold. In other words, the Doppler ensembles is band-pass filtered in the ensemble direction which is also called the slow-time direction. In this study, the band-pass clutter filtering is implemented using singular-value decomposition (SVD) [24,25] whose low-order and high-order thresholds are both adaptively determined.

Conventionally, the blood flow estimation can be achieved by autocorrelation of these Doppler ensembles as proposed in [26]. Specifically, the Doppler power is represented using the zero-lag autocorrelation as

$$\mathrm{PD} = \sum_{f=1}^{F} \left| y^f \right|^2 \tag{1}$$

where $y^f$ is the $f$-th Doppler ensemble after beamforming. Note that the Doppler power in Equation (1) is simply the summation of the squared magnitude of each Doppler ensemble.

In multi-angle PW imaging, the DAS beamforming (i.e., CPWC image) is the coherent summation of low-resolution image pixels from distinct PW transmit angles. Therefore, given the low-resolution image pixel as $x_m$ where $m$ is the index of PW transmit angle ($m = 1, 2, \ldots, M$), the output of DAS beamforming is represented as

$$y_{\mathrm{DAS}} = \sum_{m=1}^{M} x_m \tag{2}$$

Note that the summation in Equation (2) is to produce the high-resolution CPWC image. After substituting Equation (2) into Equation (1), the conventional power Doppler detection of DAS beamforming in CPWC imaging is calculated as

$$\mathrm{PD}_{\mathrm{DAS}} = \sum_{f=1}^{F} \left( \left| \sum_{m=1}^{M} x_m \right|^2 \right) \tag{3}$$

In other words, the power of the high-resolution image is summed among ensembles to provide the final power Doppler estimation of DAS beamforming.

### 2.2. Power Doppler Detection of DMAS Beamforming

The DMAS beamforming in this study is implemented using baseband data to eliminate the need for oversampling of radio-frequency waveform [16]. Specifically, when the baseband data for low-resolution image pixel from the $m$-th PW transmit angle (i.e., $x_m$ in Equation (2)) is represented as $x_m = a_m e^{j\phi_m}$, DMAS beamforming in the dimension of

PW transmit angle is performed by first maintaining the phase of low-resolution pixel but adopting the $p$-th root to scale the pixel magnitude as $\hat{x}_m = \sqrt[p]{a_m}e^{j\varphi_m}$. Then, the $p$-th power of the summation of magnitude-scaled low-resolution image pixels from all the available PW transmit angles is performed to produce the final high-resolution image pixel. In other words, DMAS beamforming in the dimension of PW transmit angle can be defined as

$$y_{\text{DMAS}} = \left( \sum_{m=1}^{M} \hat{x}_m \right)^p = \left( \mathbf{w}^H \hat{\mathbf{x}} \right)^p \tag{4}$$

where

$$\mathbf{w} = \begin{bmatrix} 1 & 1 & 1 & \cdots & 1 \end{bmatrix}^T = \mathbf{1}$$

$$\hat{\mathbf{x}} = \begin{bmatrix} \hat{x}_1 & \hat{x}_2 & \hat{x}_3 & \cdots & \hat{x}_M \end{bmatrix}^T$$

Here, the symbol $H$ represents Hermitian transpose and the real-valued weighting vector $\mathbf{w}$ is actually a unity vector $\mathbf{1}$ to equally emphasize the contribution from all the available PW transmit angles. For power Doppler detection, the magnitude of DMAS image is averaged among consecutive ensembles before the calculation of image power. In other words, the power Doppler of DMAS beamforming in this study is formulated as

$$\text{PD}_{\text{DMAS}} = \left( \sum_{f=1}^{F} |y_{\text{DMAS}}| \right)^2 \tag{5}$$

Though the proposed DMAS beamforming is also applicable to B-mode imaging, it should be noted that the DMAS beamforming in this study is calculated from low-resolution images after SVD clutter filtering in order to remove both stationary tissue and noises for power Doppler estimation.

*2.3. Power Doppler Detection of DMAS Beamforming with CST (DMAS-CST)*

DMAS beamforming with CST technique depends on DMAS signals from two subsets of PW transmit angles. The idea of complementary subset is similar to that in [27–29] but is defined in the dimension of PW transmit angle instead of receiving channel. Specifically, DMAS beamforming is performed using the available PW transmit angles in each transmit subset and the beamforming output is denoted as $y_{\text{DMAS1}}$ and $y_{\text{DMAS2}}$, respectively, for subset 1 and subset 2:

$$y_{\text{DMAS1}} = \left( \mathbf{w}_1^H \hat{\mathbf{x}} \right)^p$$
$$y_{\text{DMAS2}} = \left( \mathbf{w}_2^H \hat{\mathbf{x}} \right)^p$$

where the weighting vector $\mathbf{w}_1$ for subset 1 is related to the weighting vector $\mathbf{w}_2$ for subset 2 by $\mathbf{w}_1 = \mathbf{1} - \mathbf{w}_2$ to ensure the complementary property. Note that, when the total number of available PW transmit angle is odd-valued, the subset 1 and 2 can share one specific PW transmit angle to equalize the number of PW transmit angle in each subset. For example, when there are totally seven PW angles in the transmit sequence, $\mathbf{w}_1$ and $\mathbf{w}_2$ can be respectively defined as [1 1 1 0.5 0 0 0] and [0 0 0 0.5 1 1 1] so that each subset comprises half of the total PW transmit angles. For even number of PW transmit angle, on the other hand, any PW transmit angle should belong to either one of the two complementary subsets. The two DMAS signals are then correlated to reduce the noise level and a square root of the correlation is performed to restore the dimensionality of DMAS signal. In other words, the DMAS-CST beamforming can be formulated as

$$y_{\text{DMAS-CST}} = \sqrt{y_{\text{DMAS1}}\, y_{\text{DMAS2}}^*} = \left( \mathbf{w}_1^H \mathbf{R}\, \mathbf{w}_2 \right)^{\frac{p}{2}} \tag{6}$$

where the symbol * is for complex conjugate and $\mathbf{R} = \hat{\mathbf{x}}\hat{\mathbf{x}}^H$ is the autocorrelation matrix of magnitude-scaled low-resolution image pixels from different PW transmit angles (i.e., $\hat{x}$ in Equation (4)). Note that the power Doppler detection of DMAS-CST beamforming is also implemented by ensemble averaging of $y_{\mathrm{DMAS-CST}}$ before power estimation as

$$PD_{\mathrm{DMAS-CST}} = \left( \sum_{f=1}^{F} y_{\mathrm{DMAS-CST}} \right) \left( \sum_{f=1}^{F} y_{\mathrm{DMAS-CST}} \right)^* \tag{7}$$

The signal flowchart of DMAS-CST beamforming for power Doppler estimation is schematically represented in Figure 1. It should be noted that, when both $\mathbf{w}_1$ and $\mathbf{w}_2$ are replaced with the unity vector 1, power Doppler of DMAS-CST beamforming in Equation (7) will degenerate to that of DMAS beamforming in Equation (5). In other words, the original DMAS beamforming can be understood as a special case of DMAS-CST beamforming. Moreover, it is expected that the achievable noise reduction in DMAS-CST beamforming should rely on the number of ensembles. Note that the complementary weighting vectors $\mathbf{w}_1$ and $\mathbf{w}_2$ can effectively eliminate the uncorrelated noise only when the noise component in the autocorrelation matrix $\mathbf{R}$ is diagonal. However, this statistically demands sufficient ensemble averaging for the noise component to converge to $\sigma_{\mathrm{N}}^2 \mathbf{I}$ where $\mathbf{I}$ is the identity matrix and $\sigma_{\mathrm{N}}^2$ is the noise variance of magnitude-scaled low-resolution image.

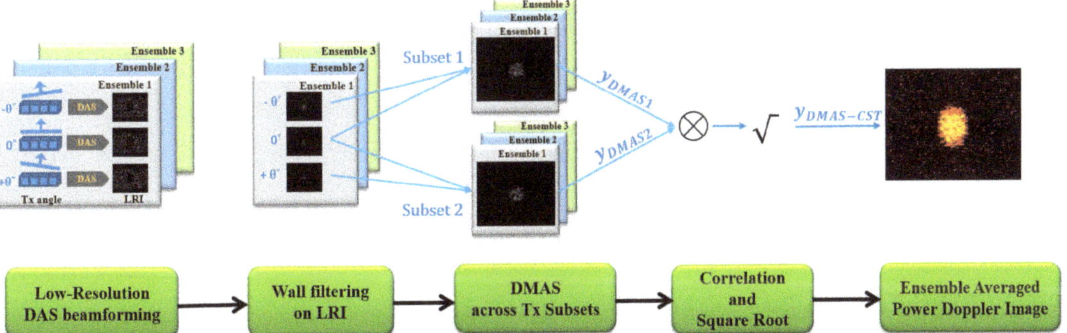


**Figure 1.** Schematic diagram of DMAS-CST beamforming for power Doppler detection in multi-angle PW imaging. Note that the wall filtering is performed on low-resolution images of each PW transmit angle before DMAS beamforming.

## 3. Methods

### 3.1. Simulation Setup

The Field II program [30,31] has been used for all simulations. The simulation schematic is shown in Figure 2. A flow channel with a radius of 2 mm is embedded in the speckle-generating tissue phantom to simulate the blood vessel with an inclined angle of 45°. The scatterers inside the flow channel are assumed to move according to a parabolic velocity distribution (i.e., laminar flow) with the peak velocity at the center of 15 mm/s. The scatterer density is set to contain about 10 scatterers per resolution cell and the scattering magnitude of the tissue is assumed to be 60 dB higher than that of the blood flow in the flow channel. White Gaussian noises are included into the simulated channel waveforms before beamforming to achieve a channel signal-to-noise ratio of 0 dB for the blood flow signal. A 128-elements linear array transducer was used for both transmission and reception in the simulations. The transmit frequency is set to be 5 MHz. A total of 7 plane waves evenly spanning an azimuthal angular range of −7.5° to +7.5° are sequentially transmitted with a pulse-repetition-frequency (PRF) of 3.9 kHz to produce low-resolution images of size $375 \times 128$ from distinct PW angles. Therefore, the compounded high-resolution imaging has a frame rate of approximately 556 Hz. Note that the low-resolution images are constructed using baseband DAS beamforming in the direction of receiving channels as conventional CPWC imaging. The PW transmit sequence is repeated 15 times to provide an ensemble number of 15. Other detailed parameters are shown in Table 1. For each PW transmit angle, the corresponding low-resolution images from different ensembles are clustered in the ensemble direction to form a three-dimensional matrix of $375 \times 128 \times 15$ for SVD clutter filtering to eliminate high-frequency noise and stationary tissue. In the simulation, the low-order and high-order thresholds of SVD clutter filter are set to 2 and 10, respectively. Finally, the filtered low-resolution images of different PW transmit angles were compounded using either DAS or DMAS processing to produce the high-resolution power Doppler image with ensemble averaging.

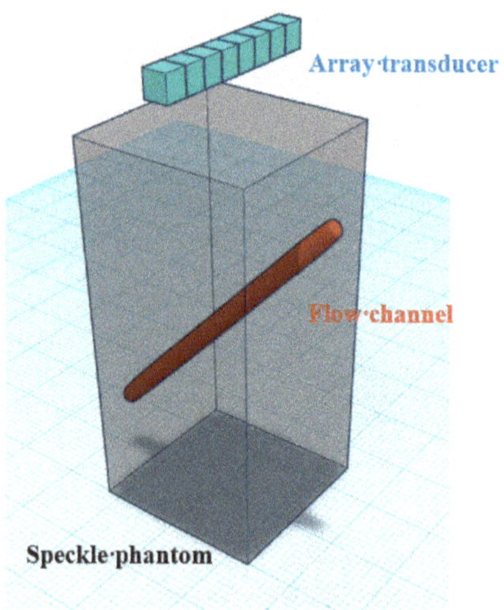


**Figure 2.** Schematic diagram of the simulated flow channel in the speckle phantom. Note that the intersection of the image plane of the array transducer and the cylindrical vessel will be an ellipse.

**Table 1.** The imaging parameter of Field II simulations.

| Imaging System | |
| --- | --- |
| Transducer | Linear Array |
| Pitch | 0.3 mm |
| Number of elements | 128 |
| Elevation focus | 30 mm |
| Sampling frequency | 20 MHz |
| Image size in pixels | 375 (axial) $\times$ 128 (lateral) |
| **Transmit Pulse** | |
| Center frequency | 5.0 MHz |
| Excitation | 3 cycles |
| PW transmit angle | $7 (-7.5°{\sim}+7.5°)$ |
| Ensemble | 15 |
| PRF | 3.9 kHz |
| **Phantom** | |
| Speed of Sound | 1540 m/s |
| Scattering magnitude | 60 dB (tissue clutter) |
| | 0 dB (blood flow) |

### 3.2. In Vivo Experimental Setup

The animal data was provided by the S-Sharp Corporation (New Taipei, Taiwan). The data was collected from a 6-month-old female New Zealand white rabbit. The rabbit was anesthetized with an intramuscular injection of Zoletil® 50 according to its body weight and placed on a warming pad to maintain its body temperature at 37 °C. The L154BH linear array with a center frequency of 6.4 MHz was used in the experiment. The experimental PW transmit sequence comprises 6 PW angles which equidistantly increasing from −5° to +5° with a PRF of 4 kHz. The corresponding low-resolution images were beamformed by Prodigy ultrasonic imaging system (S-sharp, New Taipei City, Taiwan) and then processed offline using Matlab (The MathWorks, Natick, MA, USA) for SVD clutter filtering. Therefore, the effective high-resolution frame rate is about 667 Hz. A total of 64 ensembles were acquired and each ensemble is compounded from low-resolution images from 6 angles. The detailed imaging parameters are shown in Table 2. In the in vivo experiment, the low-order and high-order thresholds of SVD clutter filter are set to be 10 and 40, respectively. These thresholds are determined using the descendent magnitude of each singular value. One typical example in Figure 3 demonstrates that the low-order threshold is regarded as the turning point at which the curve of singular value has a slope of −1 to indicate the beginning of the flattened curve. On the other hand, the high-order threshold corresponds to where the curve of singular value is about to decrease linearly. This is because the high-order singular value of white Gaussian noise should follow a linear distribution under the logarithm scale [32]. Similar to the simulations, the SVD clutter filter is also individually applied to the low-resolution images of each PW transmit angle. Then, these filtered low-resolution images are coherently compounded using either DAS or DMAS beamforming to produce the final high-resolution power Doppler image.

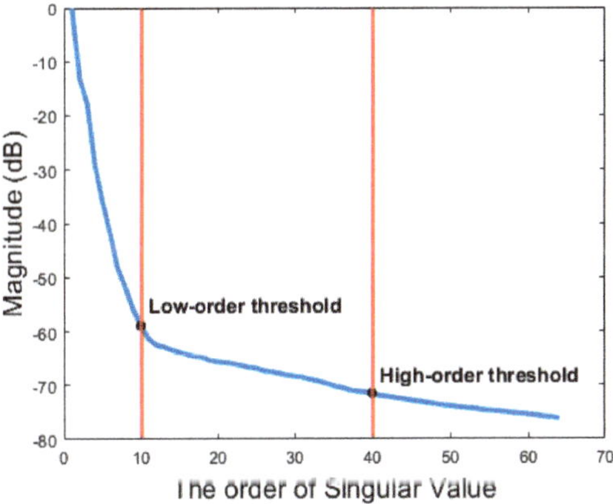


**Figure 3.** The curve of singular value of the in vivo experimental data and the corresponding low-order and high-order threshold for SVD clutter filtering.

**Table 2.** The imaging parameter of in vivo experiment.

| Prodigy Imaging System | |
| --- | --- |
| Transducer | L154BH |
| Pitch | 0.3 mm |
| Number of elements | 128 |
| Elevation focus | 20 mm |
| Sampling frequency | 25.6 MHz |
| Image size in pixels | 520 (axial) × 128 (lateral) |

**Table 2.** *Cont.*

| Transmit Pulse | |
| --- | --- |
| Center frequency | 6.4 MHz |
| Excitation | 5 cycles |
| PW transmit angle | 6 ($-5°{\sim}+5°$) |
| Ensemble | 64 |
| PRF | 4 kHz |

*3.3. Quantitative Analysis*

In order to quantitatively compare the image quality among different beamforming methods in power Doppler imaging, two region-of-interests (ROIs) are defined in the power Doppler image to respectively represent the blood flow area and the background area. The calculation of Doppler signal-to-noise ratio (SNR) and contrast-to-noise ratio (CNR) are defined as follows [33]:

$$SNR = 10 \cdot \log_{10}\left( \frac{\overline{M}_{blood}}{\overline{M}_{background}} \right)$$
$$CNR = 10 \cdot \log_{10}\left( \frac{|\overline{M}_{blood} - \overline{M}_{background}|}{\sigma_{background}} \right)$$

where $\overline{M}_{blood}$ and $\overline{M}_{background}$ are the mean power of blood flow and background signals, respectively, and $\sigma_{background}$ represent the standard deviation of background signals. For each power Doppler image in this study, its leftmost upper panel shows the corresponding ROIs for the blood flow region (blue box) and the background region (white box).

## 4. Results

*4.1. Simulations*

Power Doppler images of the simulated flow phantom in DMAS beamforming and DMAS-CST beamforming are respectively provided in the upper and lower panels of Figure 4. Seven PW transmit angles are used for coherent compounding (i.e., [$-7.5°$ $-5°$ $-2.5°$ $0°$ $+2.5°$ $+5°$ $+7.5°$]). The flow velocity in the simulation is 15 mm/s and the ensemble number for averaging is 15. The power Doppler images from left to right correspond to different $p$ values of 1.5, 2.0 and 2.5 in both DMAS and DMAS-CST beamforming while DAS beamforming is also provided as a reference in the leftmost panels. Note that the power level of Doppler image is represented using the brightness as shown in the color bar. A brighter image pixel means that the Doppler power in this spatial location is higher than that in other pixels. The background region (i.e., the white box) does not enclose any flow vessel and thus its power only comes from the random noises. Since the background region of DMAS image appears to be darker than that of DAS image, visual observations indicate that the power Doppler images with DMAS beamforming alone generally have a lower noise level in the background than that with DAS beamforming. Specifically, the Doppler SNR increases from 17.1 dB in DAS to 21.8 dB, 25.3 dB, and 28.4 dB in DMAS, respectively, with the $p$ value of 1.5, 2.0, and 2.5. Take the $p$ value of 2.0 in DMAS beamforming as an example, the improvement in Doppler SNR is 8.2 dB compared to the DAS counterpart. On the other hand, when DMAS beamforming is performed together with the CST technique, it is also apparent in Figure 4 that the background noise can be further suppressed to a lower level in DMAS-CST beamforming. The corresponding Doppler SNR improves by another 6.4 dB, 6.0 dB, and 4.5 dB, respectively, for the $p$ value of 1.5, 2.0, and 2.5. Note that the improvement in Doppler SNR due to the CST technique appears to decrease with the $p$ value in DMAS beamforming. This observation will be discussed later. Note that DAS beamforming with CST technique can also improve the Doppler SNR but only by 3.4 dB.

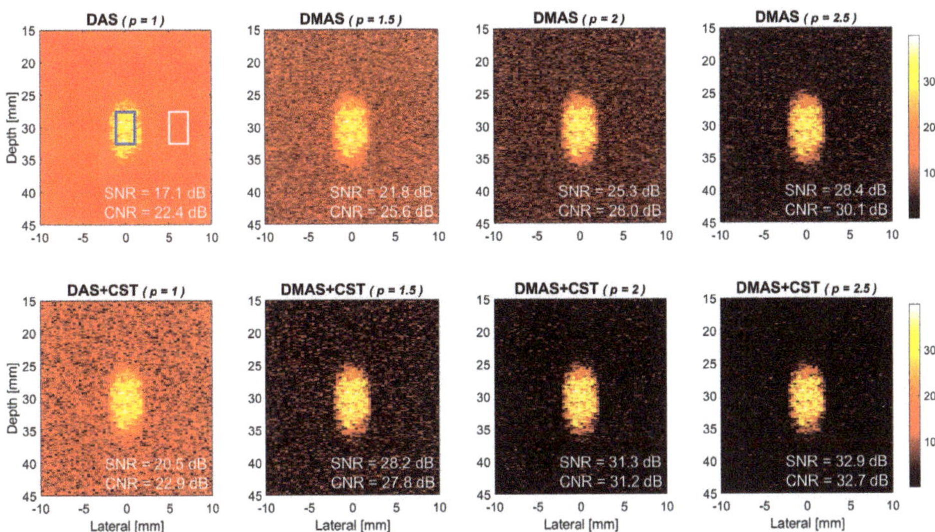

**Figure 4.** Simulated power Doppler images of flow phantom for DAS beamforming, DMAS beamforming with the $p$ value of 1.5, 2.0 and 2.5, respectively from left to right. Seven PW transmit angles are used (i.e., [$-7.5°$ $-5°$ $-2.5°$ $0°$ $+2.5°$ $+5°$ $+7.5°$]). Upper panels: without the CST technique. Lower panels: with the CST technique.

The Doppler SNR and CNR are also provided as a function of ensemble number ranging from 1 to 15 in Figures 5 and 6. In Figure 5, it should be noted that the Doppler SNR without the CST technique generally remains unchanged with the ensemble number in both DAS and DMAS beamforming. This is because the incoherent summation of power Doppler ensembles only helps to smooth the noise variation in the background but is not able to suppress the noise level. With the CST technique, on the contrary, the achievable Doppler SNR in both DAS-CST and DMAS-CST beamforming appears to consistently increase with the ensemble number. This is as expected since the CST technique depends on a diagonal autocorrelation matrix among PW transmit angles (i.e., $\mathbf{R}$ in Equation (6)) to remove the uncorrelated random noises. Nonetheless, for random noises, it takes sufficient realizations (i.e., sufficient power Doppler ensembles) for the autocorrelation matrix to converge to the diagonal form. This is why the SNR improvement due to the CST technique would increase with the ensemble number. Take DMAS beamforming with $p$ value of 2.0 as an example, the CST technique improves the Doppler SNR by 2.9 dB and 6.0 dB, respectively, when the ensemble number is 6 and 15. Similarly, the Doppler SNR in DAS beamforming also improves by 1.7 dB and 3.4 dB due to the CST technique, respectively for the ensemble number of 6 and 15. On the contrary, the Doppler CNR appears to increase with the ensemble number no matter whether the CST technique is performed or not. This comes from the reduction of noise variation in the process of ensemble averaging. For DAS beamforming, however, the Doppler CNR without and with the CST technique almost overlap with each other. In other words, the CST technique barely improves the Doppler CNR in DAS beamforming. For DMAS beamforming, on the other hand, the Doppler CNR markedly increases due to the CST technique for all $p$ values considered here. Take the $p$ value of 2.0 as an example, the CST technique improves the Doppler CNR in DMAS beamforming by 0.9 dB and 3.2 dB, respectively when the ensemble number is 6 and 15. Nonetheless, it should be noted that the CST technique could adversely lead to the decrease of Doppler SNR and CNR when the ensemble number is small (e.g., smaller than three in the simulation as shown in Figures 5 and 6).

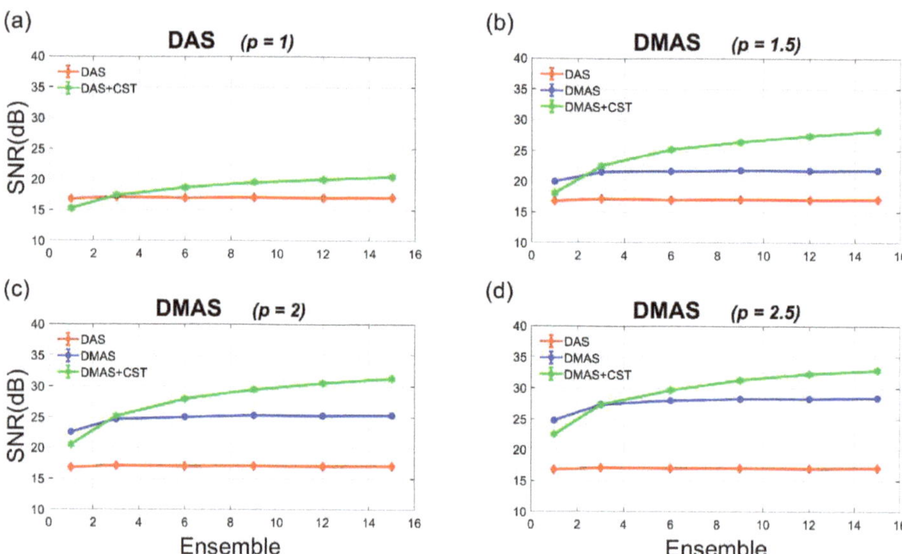

**Figure 5.** Quantitative analysis of Doppler SNR in the simulations as a function of ensemble number for (**a**) DAS beamforming and (**b–d**) DMAS beamforming with the *p* value of 1.5, 2.0, and 2.5, respectively.

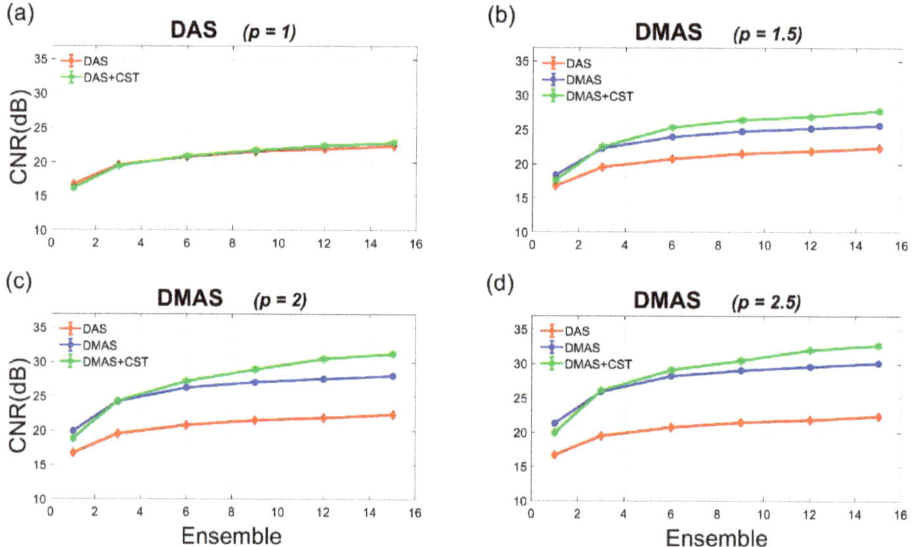

**Figure 6.** Quantitative analysis of Doppler CNR in the simulations as a function of ensemble number for (**a**) DAS beamforming and (**b–d**) DMAS beamforming with the *p* value of 1.5, 2.0, and 2.5, respectively.

Since the DMAS-based power Doppler detection in this study relies on the signal coherence among low-resolution images from different PW transmit angles, the effect of the number of PW transmit angle should be considered. Figure 7 shows that the simulated power Doppler images when the number of PW transmit angle is reduced to 5 (i.e., [−5° −2.5° 0° +2.5° +5°]). All other imaging parameters remain the same as those in Figure 4. Compared to its 7-angle counterpart in Figure 4, it should be noted that the 5-angle power

Doppler images in Figure 7 exhibit higher background noise level and the resultant Doppler SNR and CNR also decrease relative to those in Figure 4. For example, the Doppler SNR in DMAS beamforming without CST technique decreases from 21.8 dB, 25.3 dB, and 28.4 dB in Figure 4 to 19.8 dB, 22.6 dB, and 25.1 dB in Figure 7, respectively, with the $p$ value of 1.5, 2.0, and 2.5. In other words, when a smaller number of PW transmit angle is used to construct the final high-resolution power Doppler image for a higher frame rate, the Doppler SNR and CNR may decrease in DMAS beamforming. This implies that the coherence-based suppression of random noises performs better when more realizations of noise are available from distinct PW transmit angles. In DMAS-CST beamforming, on the other hand, the Doppler SNR also decreases from 28.2 dB, 31.3 dB, and 32.9 dB in Figure 4 to 26.2 dB, 28.3 dB, and 29.4 dB in Figure 7. Note that the Doppler SNR in DAS beamforming also decreases from 17.1 in Figure 4 to 15.7 in Figure 7. However, it should be taken into considerations that the ensemble number of power Doppler image is fixed to 15 for both 5-angle and 7-angle transmit sequences in this comparison. In practical applications, DMAS-CST beamforming may be expected to suffer less from a smaller number of PW transmit angle and will be discussed later.

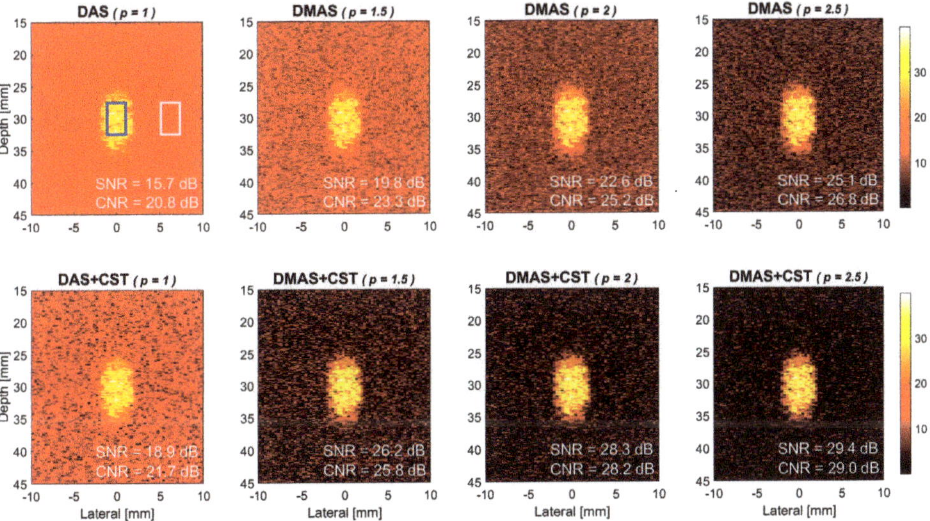

**Figure 7.** Simulated power Doppler images of flow phantom for DAS beamforming, DMAS beamforming with the $p$ value of 1.5, 2.0 and 2.5, respectively from left to right. Five PW transmit angles are used (i.e., [$-5°$ $-2.5°$ $0°$ $+2.5°$ $+5°$]). Upper panels: without the CST technique. Lower panels: with the CST technique.

It has been clearly indicated in the theory section that the DMAS-based power Doppler in this study is performed by averaging the magnitude of Doppler signal among different ensembles and then the power of the averaged Doppler signal is estimated as in Equation (5). This is different from the conventional approach in which the power of Doppler signal is averaged among ensembles as in Equation (1). The reason for performing ensemble averaging of signal magnitude instead of signal power can be justified by the power Doppler images as shown in Figure 8. In the upper panels, the power Doppler images are constructed using ensemble averaging of signal magnitude for both DAS beamforming and DMAS beamforming with the $p$ value of 1.5, 2.0, and 2.5, respectively, from left to right. In the lower panels, however, the power Doppler images are constructed using ensemble averaging of signal power as in the conventional approach. Therefore, some of the panels in Figure 8 are just duplicates of those in Figure 4. It is apparent that the power Doppler images in the upper panels consistently have a lower noise level than their counterpart in the lower panels, especially for DMAS beamforming. Specifically, the Doppler SNR in DMAS beamforming improves by 1.3 dB, 2.0 dB, and 2.8 dB, respectively for $p$ values of 1.5, 2.0, and 2.5 when the ensemble averaging is switched from power to magnitude. Note that the Doppler SNR in DAS beamforming also improves by 0.5 dB but this minor change in noise level is not visually detectable in the corresponding power Doppler images.

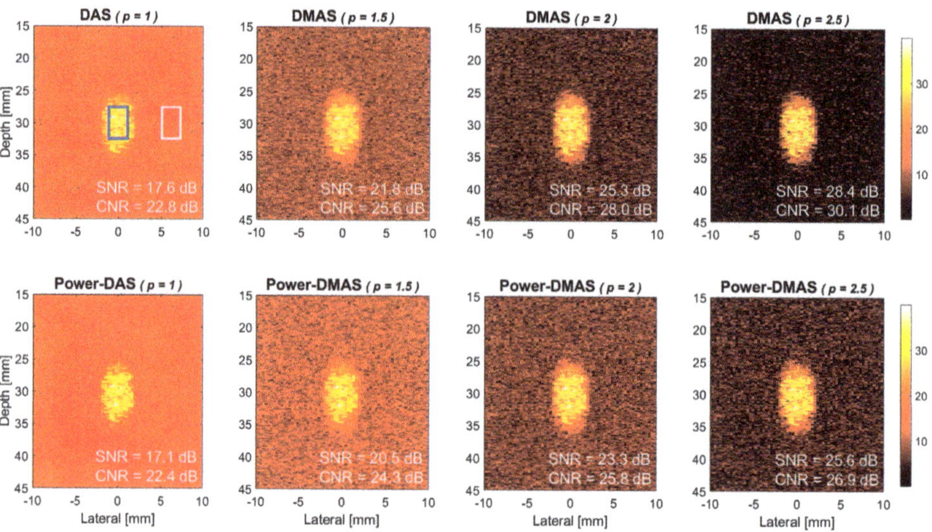

**Figure 8.** Simulated power Doppler images of flow phantom for DAS beamforming, DMAS beamforming with the $p$ value of 1.5, 2.0, and 2.5, respectively, from left to right. Seven PW transmit angles are used (i.e., $[-7.5° \; -5° \; -2.5° \; 0° \; +2.5° \; +5° \; +7.5°]$). Upper panels: ensemble averaging of signal magnitude. Lower panels: ensemble averaging of signal power.

### 4.2. Experiments

Experimentally acquired power Doppler images of the rabbit's kidney are provided in Figure 9 for DAS and DMAS beamforming without and with CST technique. Visual observations also demonstrate that the power Doppler images with DMAS beamforming generally has a lower noise level in the background than that with DAS beamforming. These observations on experimental images are in agreement with those on simulations. Specifically, the experimental Doppler SNR without CST technique increases from 25.6 dB in DAS to 28.5 dB, 30.9 dB, and 33.0 dB in DMAS, respectively, with the $p$ value of 1.5, 2.0, and 2.5. Take the $p$ value of 2.0 in DMAS beamforming as an example, the improvement in experimental Doppler SNR is 5.3 dB compared to the DAS counterpart. On the other hand, when DMAS beamforming is performed together with CST technique, it is also apparent in Figure 9 that the background noise can be further suppressed to a lower level in DMAS-CST beamforming. Specifically, the experimental Doppler SNR in DMAS-CST beamforming improves by 6.5 dB, 5.9 dB, and 5.0 dB, respectively, with the $p$ value of 1.5, 2.0, and 2.5, compared to those in DMAS beamforming alone. The efficacy of CTS technique on alleviating uncorrelated noises is also consistent between the experimental and the simulation results. Besides, though the CST technique in the experiments does help to further boost the Doppler SNR, the achievable improvement also decreases with the $p$ value in DMAS beamforming.

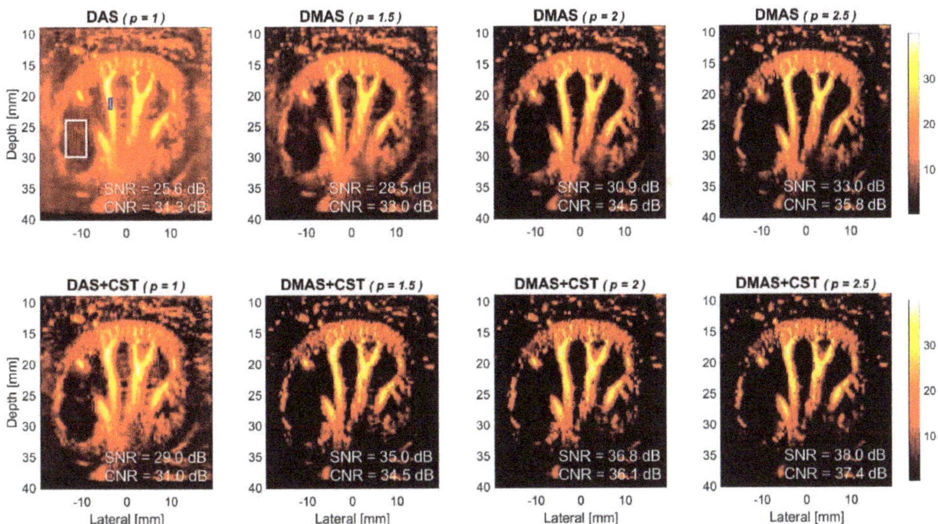

**Figure 9.** Experimental power Doppler images of rabbit's kidney for DAS beamforming, DMAS beamforming with the *p* value of 1.5, 2.0, and 2.5, respectively from left to right. Six PW transmit angles are used (i.e., [−5° −3° −1° +1° +3° +5°]). Upper panels: without the CST technique. Lower panels: with the CST technique.

The experimental Doppler SNR and CNR are quantitatively provided as a function of ensemble number ranging from 1 to 64 in Figures 10 and 11. It should be noted that the experimental Doppler SNR without the CST technique in Figure 10 generally remains unchanged with the ensemble number for both DAS and DMAS beamforming. This phenomenon agrees with the simulation results in Figure 5. In contrast, when the CST technique is performed together with either DAS or DMAS beamforming, the experimental Doppler SNR increases with the ensemble number. This is also consistent with that in the simulations because sufficient power Doppler ensembles would allow the autocorrelation matrix to be diagonal for the CST technique to remove uncorrelated noises. With the *p* value of 2.0, the experimental Doppler SNR improves from 30 dB, 30.3 dB, and 30.9 dB in DMAS beamforming alone to 31.6 dB, 34.5 dB, and 36.8 dB in DMAS-CST beamforming, respectively, when the ensemble number increases from 16, 32, and 64. When the ensemble number is small, however, it should be noted that the CST technique may adversely compromise both Doppler SNR and Doppler CNR. For example, with the *p* value of 2.0, the Doppler SNR with only one ensemble actually decreases from 26.8 dB in DMAS beamforming to 24.3 dB in DMAS-CST beamforming. This observation also agrees with that in simulations. Actually, Figure 10 shows that the CST technique demands an ensemble number larger than eight in the experiments to provided improvement in Doppler SNR for both DAS and DMAS beamforming.

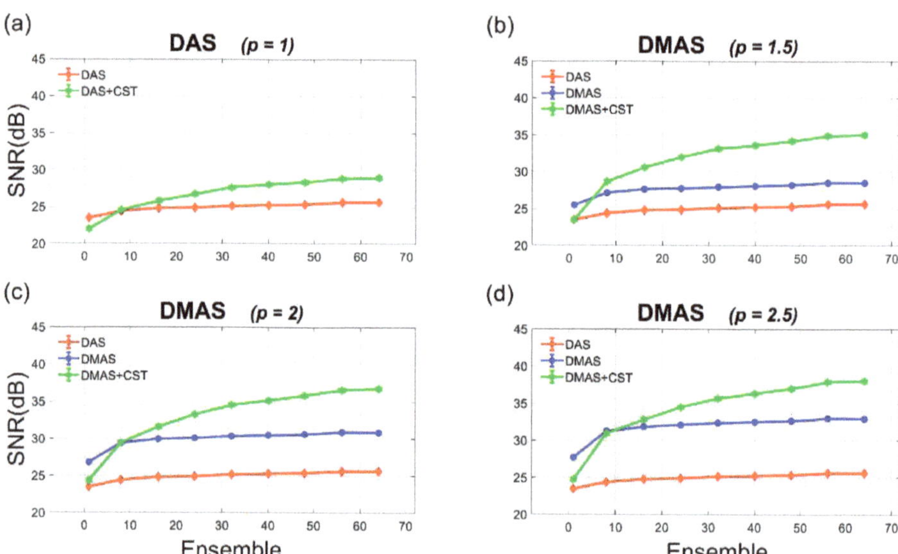

**Figure 10.** Quantitative analysis of Doppler SNR in the experiments as a function of ensemble number for (**a**) DAS beamforming and (**b**–**d**) DMAS beamforming with the $p$ value of 1.5, 2.0, and 2.5, respectively.

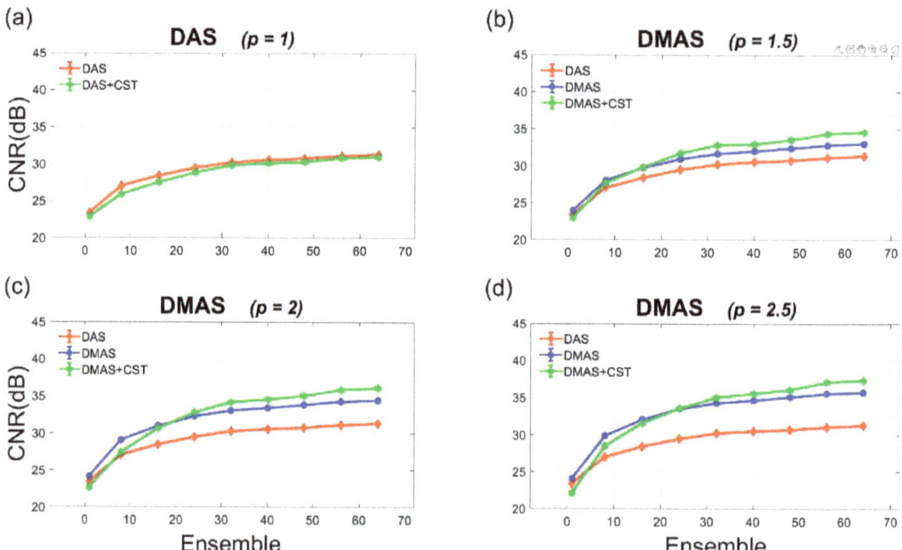

**Figure 11.** Quantitative analysis of Doppler CNR in the experiments as a function of ensemble number for (**a**) DAS beamforming and (**b**–**d**) DMAS beamforming with the $p$ value of 1.5, 2.0, and 2.5, respectively.

In contrast, the Doppler CNR in Figure 11 increases with the ensemble number due to the reduced variation of noise no matter whether the CST technique is performed or not for both DAS and DMAS beamforming. For DAS beamforming, however, the CST technique appears to barely improve the Doppler CNR for all number of ensembles. For DMAS beamforming, on the other hand, the CST technique with sufficient ensembles could provide noticeable improvement in Doppler CNR due to the suppressed noise background. For example, the CST technique improves the Doppler CNR in DMAS beamforming by

1.2 dB and 1.6 dB, respectively, with the ensemble number of 32 and 64 when the $p$ value is 2.0.

## 5. Discussion and Conclusions

In this study, DMAS beamforming of low-resolution images from distinct PW transmit angles is used to construct a novel coherence-based power Doppler detection in multi-angle PW imaging. Moreover, the CST technique is also developed to further reduce the noise level in power Doppler detection by correlation of two DMAS signals from complementary subset transmit. Since the proposed method is based on signal coherence of the echo matrix in the dimension of PW transmit angle, it can be readily applied to boost the performance of conventional CPWC imaging by replacing DAS beamforming of low-resolution images with DMAS beamforming. Note that the proposed DMAS beamforming of low-resolution images does not require the raw channel data and thus is relatively free from huge memory allocation to record the entire echo matrix in multi-angle PW imaging. Specifically, DMAS beamforming in the dimension of PW transmit angle is performed by first maintaining the phase of low-resolution pixel but adopting the $p$-th root to scale the pixel magnitude. After the summation of magnitude-scaled low-resolution image pixels from available PW transmit angles, the $p$-th power is performed to produce the final high-resolution image pixel. Here, the $p$ value represents the degree of signal coherence considered in DMAS beamforming and thus a higher $p$ value generally produces higher-quality images. For the implementation of CST technique, complementary transmit subsets can be defined from the available PW transmit angles to produce the corresponding signals in DMAS beamforming. Then, the two DMAS signals can be correlated to reduce the noise level in the final power Doppler imaging. Note that the CST technique is also applicable to DAS beamforming by correlating two complementary DAS signals from conventional CPWC imaging.

It should be emphasized that both the DMAS beamforming and the CST technique improve the quality of power Doppler image by including the signal coherence into the image output. They are intrinsically different from a simple nonlinear mapping of the pixel value which would darken any low-intensity pixel regardless of whether the pixel belongs to noise or blood flow and thus degrade the image contrast. In order to validate this, a weaker flow is simulated as shown in Figure 12 by reducing the peak velocity to only 5 mm/s while all other simulation parameters and signal processing remain unchanged to those in Figure 4. Note that the weaker flow intensity is demonstrated both by the lower brightness of the flow region and the corresponding Doppler SNR in Figure 12 than its counterpart in Figure 4 for each panel. With the same noise level in the simulation of both Figures 4 and 12, their difference in Doppler SNR actually represents the image contrast of power Doppler detection between the stronger and the weaker flow signals. Take the DMAS-CST beamforming in the lower panels as an example, the Doppler SNR decreases from 28.2 dB, 31.3 dB, and 32.9 dB in Figure 4 to 23.6 dB, 26.5 dB, and 27.8 dB in Figure 12, respectively, with the $p$ value of 1.5, 2.0, and 2.5. Therefore, the image contrast of DMAS-CST beamforming between Figures 4 and 12 is respectively 4.6 dB, 4.8 dB, and 5.1 dB. Compared to the DAS reference whose image contrast is 4.3 dB (i.e., 17.1–12.8), DMAS-CST beamforming exhibits no significant change in image contrast with marked suppression in background noises. In other words, the proposed DMAS-CST beamforming can preserve the image contrast of conventional DAS beamforming while improving the Doppler SNR significantly.

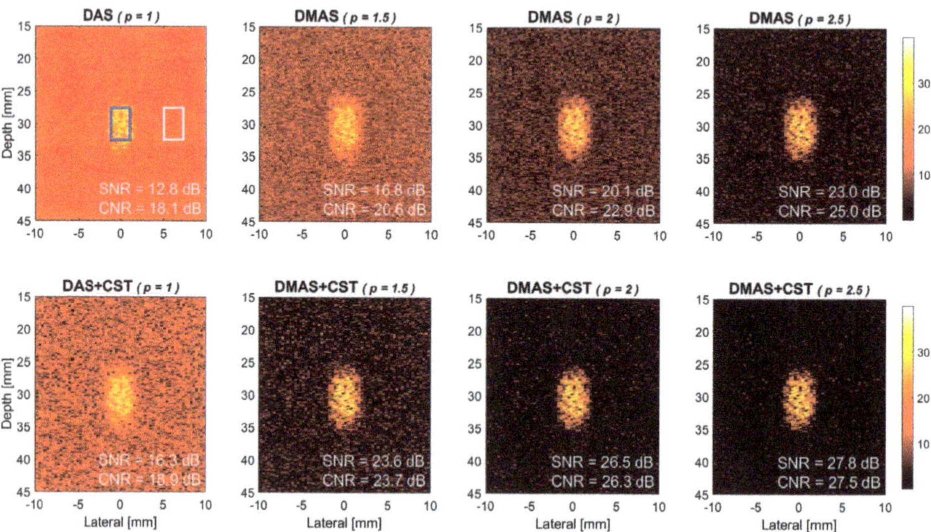

**Figure 12.** Simulated power Doppler images of flow phantom for DAS beamforming, DMAS beamforming with the $p$ value of 1.5, 2.0, and 2.5, respectively from left to right. Upper panels: without the CST technique. Lower panels: with the CST technique. All simulation parameters and signal processing remain the same as those in Figure 4 except that the peak flow velocity is reduced from 15 mm/s to 5 mm/s to produce a weaker flow after clutter filtering. Note that the brightness of each panel is normalized to that of its counterpart in Figure 4.

Moreover, in the proposed DMAS and DMAS-CST beamforming, the Doppler power is estimated using a square-of-sum approach as defined in Equations (5) and (7). Compared to the conventional sum-of-square power estimation in Equations (1) and (3), the square-of-sum approach additionally includes the cross-correlation among ensembles into the estimated Doppler power. Therefore, the square-of-sum power can be understood as the sum-of-square power compensated by the cross-correlation among ensembles. Note that, with a sufficient number of ensembles, the cross-correlation term will statistically approach zero for uncorrelated thermal noises but will remain large for true flow signal. This is exactly why the averaging of signal magnitude before taking power (i.e., square of sum) in upper panels of Figure 8 always provides higher Doppler SNR and CNR than the averaging of signal power (i.e., sum of square) in the corresponding lower panels. This observation is also similar to that reported in [34] which uses larger lag of autocorrelation to represent the Doppler power.

Both simulations and experiments have been performed to validate the DMAS-based power Doppler imaging. Results indicate that, since the random noises have a low coherence among low-resolution images, the proposed DMAS beamforming is capable of producing a lower background noise level than the DAS counterpart and thus the achievable Doppler SNR increases with the $p$ value in DMAS beamforming. Besides, when the CST technique is integrated with DMAS beamforming, the corresponding Doppler SNR further improves by another 6.4 dB, 6.0 dB, and 4.5 dB in the simulations for DMAS-CST beamforming with the $p$ value of 1.5, 2.0, and 2.5, respectively. Note that the improvement in Doppler SNR due to the CST technique decreases with the $p$ value. Our experimental results also confirm the decrease of achievable improvement in Doppler SNR with the $p$ value in DMAS-CST beamforming. This is because the DMAS beamforming without CST technique already helps to suppress not only the low-coherence image clutter but also the uncorrelated random noises. Consequently, when the random noises have been largely suppressed by adopting a higher $p$ value in DMAS beamforming, there will be fewer residual noises left for the CST technique to remove. This is probably why the efficacy of CST technique on Doppler SNR appears to degrade with the increasing $p$ value in DMAS-CST beamforming. On the other hand, the CST technique in DAS beamforming barely leads to any improvement in Doppler CNR, as demonstrated by the overlap of Doppler CNR without and with CST technique in both Figures 6a and 11a. This observation is consistent with that reported in [28] even though their complementary subsets are defined in the receiving aperture while ours are defined in the PW transmit angle. Nonetheless, it should be noted that the CST technique in DMAS beamforming does provide a marked improvement in Doppler CNR. Moreover, in order to remove the random noises effectively, the CST technique demands sufficient ensembles to ensure the diagonal autocorrelation matrix of noises from distinct PW transmit angles. Consequently, the CST technique may adversely degrade the quality of power Doppler detection when the number of ensembles is small. As the two complementary weighting vectors $\mathbf{w}_1$ and $\mathbf{w}_2$ are selected to respectively correspond to the negative and the positive PW transmit angles in this study, it can be generalized to any complementary pair. Theoretically, the complementary pair with interleaved PW transmit angles should be preferred to minimize the angle difference between the two subsets. This is because, when the imaged features have a certain orientation, a large difference in PW transmit angle between the two subsets could make the imaged features more visible in one transmit subset than the other. In this case, these particular features will be relatively suppressed by the correlation of the two complementary DMAS signals as compared to other features without obvious orientations.

Our results also indicate that the performance of DMAS-based power Doppler imaging would improve with the number of PW transmit angles. This is because the image clutter and noises can be better distinguished from the true flow signal by comparing among the low-resolution image pixels from more PW transmit angles. Nonetheless, it should be noted that the aforementioned observation is based on the same number of ensemble for two PW transmit sequences with different number of PW transmit angle. In practical applications, the number of PW transmit angle is actually related to the achievable number of ensemble for averaging. For example, with a temporal window of 1 s for Doppler detection and a PRI of 100 μs, the number of high-resolution ensembles will be 2000 and 2500, respectively for a PW transmit sequence with 5 angles and 4 angles. For DMAS beamforming without CST technique, since the corresponding Doppler SNR in both simulations and experiments generally remains unchanged with the number of ensemble, the 5-angle PW transmit sequence will be preferred due to its larger number of PW angles for better coherence estimation in DMAS beamforming. For DMAS-CST beamforming, on the contrary, the corresponding Doppler SNR noticeably increase with the number of ensemble and thus the advantage of the 5-angle PW transmit sequence could be compromised by its smaller number of ensemble compared to that of the 4-angle PW transmit sequence. In other words, DMAS-CST beamforming may suffer less from the smaller number of PW transmit angle due to the corresponding increase in the number of ensemble.

One major limitation of DMAS-based power Doppler imaging may be its computational efficiency. Since the proposed DMAS beamforming involves multiplicative operation of the low-resolution images from distinct PW transmit angles, the low-resolution images have to be firstly grouped according to its PW transmit angle and then each group is individually band-pass filtered in the direction of ensemble using SVD to remove the stationary tissue before DMAS beamforming. Otherwise, if the band-pass filtering is performed after DMAS beamforming, the multiplicative coupling between the blood flow and stationary tissue will be no longer removable. Consequently, the band-pass filtering has to be repetitively performed by $M$ times where $M$ is the total number of PW transmit angle for DMAS beamforming. For DAS beamforming (i.e., CPWC imaging), on the other hand, its linear operation allows the band-pass clutter filter to be implemented in the final high-resolution images to ease the computational burden. Note that, however, the computational complexity in clutter filtering increases not only for the proposed DMAS beamforming but also for any nonlinear beamforming such as CFPD in [21]. In this case, a simpler filter such as Finite Impulse Response may be preferred instead of the SVD filter in this study for real-time implementation of DMAS-based power Doppler imaging.

**Author Contributions:** Conceptualization, C.-C.S.; Data curation, Y.-C.C.; Formal analysis, Y.-C.C.; Funding acquisition, C.-C.S.; Project administration, C.-C.S.; Supervision, C.-C.S.; Writing—original draft, Y.-C.C.; Writing—review & editing, C.-C.S. Both authors have read and agreed to the published version of the manuscript.

**Funding:** This work was supported by the Ministry of Science and Technology of Taiwan under Grant No. 108-2221-E-011-072-MY3 and 108-2221-E-011-071-MY3.

**Institutional Review Board Statement:** S-Sharp Corporation acquires the experimental data with the approval of the Institutional Animal Care and Use Committee in National Taiwan University Hospital.

**Data Availability Statement:** Restrictions apply to the availability of these data. Experimental data was obtained from S-Sharp Corporation (New Taipei, Taiwan) and are available with the permission of S-Sharp Corporation.

**Acknowledgments:** Support from S-Sharp Corporation (New Taipei, Taiwan) is highly acknowledged for providing the experimental data.

**Conflicts of Interest:** The authors declare no conflict of interest.

# References

1. So, H.; Chen, J.; Yiu, B.; Yu, A. Medical Ultrasound Imaging: To GPU or Not to GPU? *IEEE Micro* **2011**, *31*, 54–65. [CrossRef]
2. Sandrin, L.; Catheline, S.; Tanter, M.; Hennequin, X.; Fink, M. Time-resolved pulsed elastography with ultrafast ultrasonic imaging. *Ultrason. Imaging* **1999**, *21*, 259–272. [CrossRef]
3. Tanter, M.; Fink, M. Ultrafast imaging in biomedical ultrasound. *IEEE Trans. Ultrason. Ferroelectr. Freq. Control* **2014**, *61*, 102–119. [CrossRef]
4. Montaldo, G.; Tanter, M.; Bercoff, J.; Benech, N.; Fink, M. Coherent plane-wave compounding for very high frame rate ultrasonography and transient elastography. *IEEE Trans. Ultrason. Ferroelectr. Freq. Control* **2009**, *56*, 489–506. [CrossRef]
5. Zhang, Y.; Li, H.; Lee, W.N. Imaging heart dynamics with ultrafast cascaded-wave ultrasound. *IEEE Trans. Ultrason. Ferroelectr. Freq. Control* **2019**, *66*, 1465–1479. [CrossRef]
6. Bercoff, J.; Montaldo, G.; Loupas, T.; Savery, D.; Mézière, F.; Fink, M.; Tanter, M. Ultrafast compound Doppler imaging: Providing full blood flow characterization. *IEEE Trans. Ultrason. Ferroelectr. Freq. Control* **2011**, *58*, 134–147. [CrossRef]
7. Osmanski, B.F.; Pernot, M.; Montaldo, G.; Bel, A.; Messas, E.; Tanter, M. Ultrafast Doppler imaging of blood flow dynamics in the myocardium. *IEEE Trans. Med. Imaging* **2012**, *31*, 1661–1668. [CrossRef]
8. Ekroll, I.K.; Swillens, A.; Segers, P.; Dahl, T.; Torp, H.; Lovstakken, L. Simultaneous quantification of flow and tissue velocities based on multi-angle plane wave imaging. *IEEE Trans. Ultrason. Ferroelectr. Freq. Control* **2013**, *60*, 727–738. [CrossRef]
9. Mace, E.; Montaldo, G.; Osmanski, B.F.; Cohen, I.; Fink, M.; Tanter, M. Functional ultrasound imaging of the brain: Theory and basic principles. *IEEE Trans. Ultrason. Ferroelectr. Freq. Control* **2013**, *60*, 492–506. [CrossRef]
10. Ricci, S.; Bassi, L.; Tortoli, P. Real-time vector velocity assessment through multigate Doppler and plane waves. *IEEE Trans. Ultrason. Ferroelectr. Freq. Control* **2014**, *61*, 314–324. [CrossRef] [PubMed]
11. Yiu, B.Y.S.; Yu, A.C.H. Least-squares multi-angle Doppler estimators for plane-wave vector flow imaging. *IEEE Trans. Ultrason. Ferroelectr. Freq. Control* **2016**, *63*, 1733–1744. [CrossRef]
12. Ekroll, I.K.; Voormolen, M.M.; Standal, O.K.-V.; Rau, J.M.; Lovstakken, L. Coherent compounding in Doppler imaging. *IEEE Trans. Ultrason. Ferroelectr. Freq. Control* **2015**, *62*, 1634–1643. [CrossRef]
13. Leow, C.H.; Bazigou, E.; Eckersley, R.J.; Yu, A.C.H.; Weinberg, P.D.; Tang, M.X. Flow velocity mapping using contrast enhanced high-frame-rate plane wave ultrasound and image tracking: Methods and initial in vitro and in vivo evaluation. *Ultrasound Med. Biol.* **2015**, *41*, 2913–2925. [CrossRef]
14. Matrone, G.; Savoia, A.S.; Caliano, G.; Magenes, G. The delay multiply and sum beamforming algorithm in ultrasound b-mode medical imaging. *IEEE Trans. Med. Imaging* **2015**, *34*, 940–949. [CrossRef]
15. Polichetti, M.; Varray, F.; Béra, J.C.; Cachard, C.; Nicolas, B. Nonlinear beamformer based on p-th root compression—Application to plane wave ultrasound imaging. *Appl. Sci.* **2018**, *8*, 599. [CrossRef]
16. Shen, C.C.; Hsieh, P.Y. Ultrasound baseband delay-multiply-and-sum (BB-DMAS) nonlinear beamforming. *Ultrasonics* **2019**, *96*, 165–174. [CrossRef]
17. Matrone, G.; Savoia, A.; Caliano, G.; Magenes, G. Ultrasound plane-wave imaging with delay multiply and sum beamforming and coherent compounding. In Proceedings of the 2016 International Conference of the IEEE Engineering in Medicine and Biology Society (EMBC), Orlando, FL, USA, 16–20 August 2016.
18. Shen, C.C.; Hsieh, P.Y. Two-dimensional spatial coherence for ultrasonic DMAS beamforming in multi-angle plane-wave imaging. *Appl. Sci.* **2019**, *9*, 3973. [CrossRef]
19. Rubin, J.M.; Bude, R.O.; Carson, P.L.; Bree, R.L.; Adler, R.S. Power Doppler US: A potentially useful alternative to mean frequency-based color Doppler US. *Radiology* **1994**, *190*, 853–856. [CrossRef] [PubMed]

20. Bude, R.O.; Rubin, J.M. Power Doppler sonography. *Radiology* **1996**, *200*, 21–23. [CrossRef]
21. Li, Y.L.; Dahl, J.J. Coherent flow power Doppler (CFPD): Flow detection using spatial coherence beamforming. *IEEE Trans. Ultrason. Ferroelectr. Freq. Control* **2015**, *62*, 1022–1035. [CrossRef] [PubMed]
22. Li, Y.L.; Hyun, D.; Abou-Elkacem, L.; Willmann, J.K.; Dahl, J.J. Visualization of small-diameter vessels by reduction of incoherent reverberation with coherent flow power Doppler. *IEEE Trans. Ultrason. Ferroelectr. Freq. Control* **2016**, *63*, 1878–1889. [CrossRef]
23. Lediju, M.A.; Trahey, G.E.; Byram, B.C.; Dahl, J.J. Short-lag spatial coherence of backscattered echoes: Imaging characteristics. *IEEE Trans. Ultrason. Ferroelectr. Freq. Control* **2011**, *58*, 1377–1388. [CrossRef]
24. Demené, C.; Deffieux, T.; Pernot, M.; Osmanski, B.F.; Biran, V.; Gennisson, J.L.; Sieu, L.A.; Bergel, A.; Franqui, S.; Correas, J.M.; et al. Spatiotemporal clutter filtering of ultrafast ultrasound data highly increases Doppler and fUltrasound sensitivity. *IEEE Trans. Med. Imaging* **2015**, *34*, 2271–2285. [CrossRef]
25. Provost, J.; Papadacci, C.; Demene, C.; Gennisson, J.L.; Tanter, M.; Pernot, M. 3-D ultrafast Doppler imaging applied to the noninvasive mapping of blood vessels in vivo. *IEEE Trans. Ultrason. Ferroelectr. Freq. Control* **2015**, *62*, 1467–1472. [CrossRef]
26. Loupas, T.; Peterson, R.B.; Gill, R.W. Experimental evaluation of velocity and power estimation for ultrasound blood flow imaging, by means of a two-dimensional autocorrelation approach. *IEEE Trans. Ultrason. Ferroelectr. Freq. Control* **1995**, *42*, 689–699. [CrossRef]
27. Seo, C.H.; Yen, J.T. Sidelobe suppression in ultrasound imaging using dual apodization with cross-correlation. *IEEE Trans. Ultrason. Ferroelectr. Freq. Control* **2008**, *55*, 2198–2210. [PubMed]
28. Stanziola, A.; Leow, C.H.; Bazigou, E.; Weinberg, P.D.; Tang, M.X. ASAP: Super-contrast vasculature imaging using coherence analysis and high frame-rate contrast enhanced ultrasound. *IEEE Trans. Med. Imaging* **2018**, *37*, 1847–1856. [CrossRef]
29. Leow, C.H.; Bush, N.L.; Stanziola, A.; Braga, M.; Shah, A.; Hernandez-Gil, J.; Long, N.J.; Aboagye, E.O.; Bamber, J.C.; Tang, M.X. 3-D microvascular imaging using high frame rate ultrasound and ASAP without contrast agents: Development and initial in vivo evaluation on nontumor and tumor models. *IEEE Trans. Ultrason. Ferroelectr. Freq. Control* **2019**, *66*, 939–948. [CrossRef] [PubMed]
30. Jensen, J.A. FIELD: A program for simulating ultrasound systems. *Med. Biol. Eng. Comput.* **1996**, *34*, 351–352.
31. Jensen, J.A.; Svendsen, N.B. Calculation of pressure fields from arbitrarily shaped, apodized, and excited ultrasound transducers. *IEEE Trans. Ultrason. Ferroelectr. Freq. Control* **1992**, *39*, 262–267. [CrossRef] [PubMed]
32. Song, P.; Manduca, A.; Trzasko, J.D.; Chen, S. Ultrasound small vessel imaging with block-wise adaptive local clutter filtering. *IEEE Trans. Med. Imaging* **2017**, *36*, 251–262. [CrossRef] [PubMed]
33. Chang, C.C.; Chen, P.Y.; Huang, H.; Huang, C.C. In vivo visualization of vasculature in adult zebrafish by using high-frequency ultrafast ultrasound imaging. *IEEE Trans. Biomed. Eng.* **2019**, *66*, 1742–1751. [CrossRef] [PubMed]
34. Tremblay-Darveau, C.; Bar-Zion, A.; Williams, R.; Sheeran, P.S.; Milot, L.; Loupas, T.; Adam, D.; Bruce, M.; Burns, P.N. Improved contrast-enhanced Power Doppler using a coherence-based estimator. *IEEE Trans. Med. Imag.* **2017**, *36*, 1901–1911. [CrossRef] [PubMed]

*Article*

# Learning U-Net Based Multi-Scale Features in Encoding-Decoding for MR Image Brain Tissue Segmentation

**Jiao-Song Long, Guang-Zhi Ma *, En-Min Song and Ren-Chao Jin**

School of Computer Science and Technology, Huazhong University of Science and Technology, Wuhan 430074, China; jiaosonglong92@163.com (J.-S.L.); esong@hust.edu.cn (E.-M.S.); jrc@hust.edu.cn (R.-C.J.)

* Correspondence: maguangzhi@hust.edu.cn; Tel.: +86-027-8779-2212

**Abstract:** Accurate brain tissue segmentation of MRI is vital to diagnosis aiding, treatment planning, and neurologic condition monitoring. As an excellent convolutional neural network (CNN), U-Net is widely used in MR image segmentation as it usually generates high-precision features. However, the performance of U-Net is considerably restricted due to the variable shapes of the segmented targets in MRI and the information loss of down-sampling and up-sampling operations. Therefore, we propose a novel network by introducing spatial and channel dimensions-based multi-scale feature information extractors into its encoding-decoding framework, which is helpful in extracting rich multi-scale features while highlighting the details of higher-level features in the encoding part, and recovering the corresponding localization to a higher resolution layer in the decoding part. Concretely, we propose two information extractors, multi-branch pooling, called MP, in the encoding part, and multi-branch dense prediction, called MDP, in the decoding part, to extract multi-scale features. Additionally, we designed a new multi-branch output structure with MDP in the decoding part to form more accurate edge-preserving predicting maps by integrating the dense adjacent prediction features at different scales. Finally, the proposed method is tested on datasets MRbrainS13, IBSR18, and ISeg2017. We find that the proposed network performs higher accuracy in segmenting MRI brain tissues and it is better than the leading method of 2018 at the segmentation of GM and CSF. Therefore, it can be a useful tool for diagnostic applications, such as brain MRI segmentation and diagnosing.

**Keywords:** magnetic resonance images; brain tissue segmentation; multi-scale feature learning; multi-branch pooling; multi-branch dense prediction; multi-branch output

**Citation:** Long, J.-S.; Ma, G.-Z.; Song, E.-M.; Jin, R.-C. Learning U-Net Based Multi-Scale Features in Encoding-Decoding for MR Image Brain Tissue Segmentation. *Sensors* **2021**, *21*, 3232. https://doi.org/10.3390/s21093232

Academic Editor: Ahmed Toaha Mobashsher

Received: 15 March 2021
Accepted: 28 April 2021
Published: 7 May 2021

**Publisher's Note:** MDPI stays neutral with regard to jurisdictional claims in published maps and institutional affiliations.

## 1. Introduction

The segmentation of brain tissues from magnetic resonance (MR) images is of primary importance for subsequent diagnosis, pathological analysis, prognosis assessment, and brain development monitoring [1]. MR images have different kinds of modalities, including T1, T1C, T2, PD, T1IR, and FLAIR, and each reflects particular characteristics of tissue regions in brain.

For example, both T2 and FLAIR sequences describe low signals in the white matter region and high signals in the gray matter region. T2 depicts marked high signals for the cerebrospinal fluid, where FLAIR shows low or no intensity signals [2,3]. Hence, we can aggregate these multiple modalities to capture richer information to improve brain tissue segmentation performance.

Generally, the goal of brain segmentation is to classify brain voxels as three major brain structures: gray matter (GM), white matter (WM), and cerebrospinal fluid (CSF). Traditional manual segmentation is time-consuming and tedious, and it is easy to produce bias due to the operator's subjective experience. Thus, the research on automatic brain tissue segmentation algorithm has been receiving extensive attention [4–7].

A few machine learning methods for automatic brain tissue segmentation have been proposed in literature, including methods based on hand-crafted features [7–10] and methods based on multi-atlas registration [11,12]. However, the performances of these methods are limited, owing to the fuzzy brain tissue edge [13], the multi-source noise, and the inhomogeneous intensity in brain MR images.

Recently, deep learning has been extensively applied in medical image segmentation; for example, segmenting local lesions such as tumors [14–16] and organs such as brain tissues [5,17,18]. By pooling features with different resolutions in the encoding path and recovering sharp object boundaries in the decoding path, the U-Net [19] can capture rich contextual information because of this encoding-decoding manner. The U-Net framework and its extensions have become the most common deep neural networks used in medical image segmentation.

However, it still faces challenges considering the complex anatomical structures and variable shapes of brain tissues. Five examples are shown in Figure 1, where the intensity of white matter is similar to the gray matter in the rugged edge (in the yellow box), hence, it is difficult to segment these brain tissues successfully because of the description of confused boundaries.

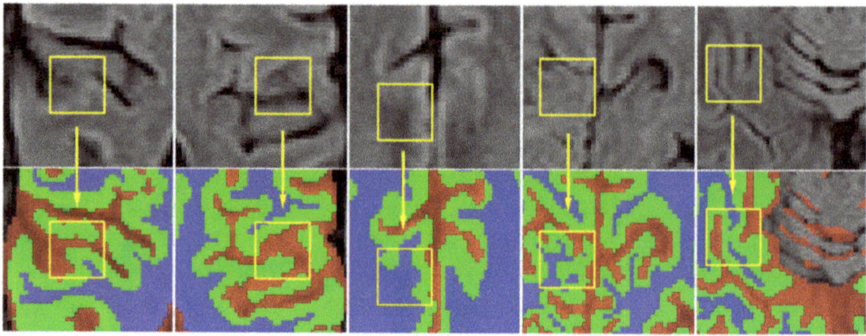

**Figure 1.** Illustration of complex anatomical structures and variable shapes in MRBrain2013S dataset. The first row lists the brain MR images in different areas. The second row shows the corresponding ground truth labels, where the colors denote different regions of the brain: red represents the cerebrospinal fluid (CSF), green the gray matter (GM), and blue the white matter (WM). Other tissues are represented with gray.

In terms of segmenting brain tissues accurately, we discovered that the problem of the U-Net-based models is the lack of multi-scale context information with a suitable receptive field. Unfortunately, the exploitation of multi-scale CNN features for semantic segmentation is a challenging task.

Conventionally, the multi-scale technique can be divided into two typical strategies: pooling at multiple scales and convoluting at multiple fields-of-views. For the former, [20] applies pooling operations with different grid scales. However, without a suitable number of grid scales, the detailed boundary information will be lost. For the latter, mainstream methods [21,22] adopt multiple rates of atrous convolution with a larger receptive field to harness multi-scale context information. However, although they can capture global information by multiple rates of atrous convolution, it is easy to encourage irrelevant redundant information [23] if without a suitable receptive field. In [21,22], extracting the multi-scale information is encoded in the last feature map; however, extracting multi-scale information in the previous feature layer is equally important, especially in medical image processing.

In addition, the above methods focus on extracting the multi-scale feature information on the spatial dimension. To learn better feature representation, the channel dimension-based multi-scale feature extracting is crucial; however, the related study is still lacking. Zhang et al. [24] suggest that a structure called "Densely Adjacent Prediction" might be used to encode spatial information into channels, and utilizes the adjacent channel information to predict results; however, it lacks the complementary multi-scale features [25].

To solve the aforementioned problems, https://orcid.org/0000-0001-7365-0053, (accessed on 29 April 2021) jointly obtain high-precision multi-scale CNN features. In this work, we propose to segment brain tissues with a novel Multi-scale Spatial and Channel Dimension U-Net (MSCD-UNet).

Our proposed architecture is based on UNet and influenced by the information extractors named multi-branch pooling (MP) and multi-branch dense prediction (MDP). To overcome the limitation of the 3D-UNet network, we propose a novel network by embedding the MP and MDP into 3D-UNet. The embedded network can capture more context cues while enhancing the details of multi-scale information by using the extractor MP in the encoding part and recovering the corresponding localization to a higher resolution layer by using the extractor MDP in the decoding part. Extensive experiments on three benchmarks with MRBrain2013, IBSR18, and ISeg2017 datasets demonstrate that our approach performs competitively against other state-of-the-art methods. The contributions of our paper are itemized in the following:

1. We have proposed a novel network by introducing spatial dimension and channel dimension-based multi-scale CNN feature information extractors into its encoding-decoding framework. In the encoding part, we propose the multi-branch pooling information extractor, called MP, to capture multi-scale spatial information for the information compensating. As pooling is easy to lose the useful spatial information when the feature map resolution is reduced, we propose the MP by using multiple max pooling with different kernel sizes in parallel to reduce the information missing and collect the neighborhood information with a suitable receptive field;

2. In the decoding part, we propose the multi-branch dense prediction, an information extractor, called MDP, to capture multi-scale channel information for the information compensating. During the decoding phase, after the maps resolution upsizing, the spatial information in these decompressed feature maps is fixed and the detailed information is represented more in channel dimension, so we consider that the prediction results at the adjacent position are related to the result of the center position. We divided the prediction result into multiple channel groups, and the multi-scale channel information of the center position can be created by averaging these groups for the purpose of information compensation. In addition, we designed a multi-branch output structure with MDP in the decoding part to form more accurate edge-preserving predicting maps by integrating the dense adjacent prediction features at different scales.

The two proposed ideas are first used in this paper. We carry out extensive experiments on three benchmarks (MRBrainS12, IBSR18, and ISeg2017) to evaluate our method. The results have proved the feasibility of our proposed method and the performance of improvement.

The remainder of the paper is structured as follows. The related work of brain tissue segmentation is described in Section 2. In Section 3, a detailed scheme of our solution is presented, including spatial-based multi-scale feature extractor in encoding, channel-based multi-scale feature extractor in decoding, multi-branch output structures, and MSCD-UNet. We perform MSCD-UNet experiments with MRBrain2013, IBSR18 and ISeg2017 datasets in Section 4, and discuss the results in Section 5. Finally, we conclude the paper with future work suggestions in Section 6.

## 2. Related Works

In this section, we briefly describe the related work of MRI brain tissue segmentation. Subsequently, we list the typical brain segmentation approaches in three categories: atlas-based registration, traditional machine learning-based, and deep learning-based. Atlas-based approaches are widely used in multi-modal circumstances [26,27]. These methods rely on registering several atlases to the target image, and then propagating the manual labels to this image. The label fusion strategy [28–30] is used to adjust the registered labels of different atlases to form the final segmentation. Because the accuracy of the registration processing is the key affecting the final segmentation result, it needs a large number of target templates to adapt the difference of brain anatomy, and these approaches are computationally expensive and perform poorly.

To address the above problems, many traditional methods based on machine learning are applied to segment brain tissues. For example, [31] adopted both intensity and spatial features to complete brain segmentation by using support vector machine. Tong et al. [17] used discriminative dictionary learning and sparse coding techniques to label brain tissues. Wang et al. [32] effectively integrated 3D Haar-like features from multi-source images together by utilizing the random forest technique to perform tissue segmentation. Zhang et al. [33] proposed a novel hidden Markov random field (HMRF) model which can encode spatial information through the mutual influences of neighboring sites to improve its accuracy and robustness. K. Mishro et al. [34] proposed a type-2 AWSFCM clustering algorithm to perform segmentation tasks. It assigned the problematic equidistant pixels to a single cluster by offering larger weight to pixel closing to the expected decision boundary. However, the main limitation of these traditional methods is that the intensity profiles of more detailed brain tissues overlap [16], and it is hard to distinguish between tissues in different brain regions.

Recently, deep learning methods based on CNN have become a powerful tool for segmenting brain tissues, which can overcome the drawback of atlas-based registration and traditional machine learning models. Zhang et al. [35] trained a CNN model for infant brain tissue segmentation by harnessing 2D single patches on axial plane slices of T1, T2, and FLAIR images. Moeskops et al. [36] introduced multiple patch sizes and multiple convolution kernel sizes into CNN to obtain multi-scale information to recognize the detailed information for brain tissue segmentation. Chung et al. [37] proposed to combine the dynamic random walker with the decay region of interest into CNN to acquire smooth segmentation of subcortical structures. However, these patch-based voxel classification methods still face troubles such as the limitation of local information and the complexity of boundaries surrounded by adjacent voxels.

Recently, fully CNN (FCNN) has been widely applied in brain segmentation to solve the above problems, as they predict the labels of voxels within the input patch simultaneously. Nie et al. [38] trained a shared network for each modality image, then fused their high-layer features in the final predicting layer. Xu et al. [39] regarded three serial slices as input of three channels to predict the middle slice by using the fully CNN. Chen et al. [40] proposed a model named VoxResNet to segment brain MR images, which can jointly encourage features of high-level context information and low-level image appearance to compensate the missing information at different levels. Dolz et al. [41] proposed Hyper-DenseNet, which can learn more complex combinations between modalities to expand the learning ability of all levels of abstraction and representation. Li et al. [42] captured and aggregated multi-scale features of brain tissues by using a multi-modality aggregation network named MMAN to accomplish brain segmentation with better accuracy. Chen et al. [43] presented a Dense-Res-Inception network to segment the cerebrospinal fluid, which is able to produce distinct features in terms of intensity, location, shape, and size. Lei et al. [44] proposed a dual aggregation network to adaptively aggregate different information of infant brain MRI modalities. Qamar et al. [18] proposed to combine dense connection, residual connection, and inception module to achieve excellent results. Yu et al. [45] developed a densely connected 3D-DenseVoxNet to preserve maximum information flow to ease the network training. Taoc et al. [46] presented a network very deep in architecture based on dense convolution network for volumetric brain segmentation. They used a model of bottleneck with compression to reduce the number of feature maps in each dense block, so as to reduce the number of learned parameters and result in computational efficiency. Dolz et al. [47] proposed a FCNN that adopts 3D spatial context of triplanar data and both global and local information for MRI brain segmentation. Sun et al. [48] proposed a volumetric feature recalibration (VFR) layer, which could richly capture the spatial contextual information, then leverage it for volumetric weighting between spatial layers.

An in-depth summarization of some of the related works in brain MRI segmentation along with techniques, advantages, and limitations is documented in Table 1.

**Table 1.** An overview of some related works on brain MRI segmentation problems.

| Paper | Technique | Advantage | Limitation |
|---|---|---|---|
| [27–31] | Atlas-based registration | Robustness to weak edges, strong adaptability. | Limited by the fuzzy brain tissue edge, multi-source noise, and inhomogeneous intensity. |
| [31] | SVM | Preserves information in the training images, and easy to implement. | Response time increase dramatically with dataset size. Slow training, memory intensive, and performance patient-specific learning. |
| [17] | Discriminative dictionary learning | | |
| [32] | Hidden Markov random field | | |
| [33] | Clustering algorithm | | |
| [35–37] | Patch-wise CNN | Fast, easy to implement, and low resource hungry. Capture discriminative features from a large input patch. | Sensitive to the patch size, lack of global information, difficult to converge small dataset. |
| [18,41,43–47] | FCNN with dense connection | Extract more reasonable and contextual information. | Large training time and storage space. High computational complexity. |
| [48] | FCNN with richer spatial information | Learn required weight for spatial feature extracting. | |

In this paper, we present a 3D U-Net-based architecture that includes multi-branch pooling and multi-branch dense prediction to capture the multi-scale features, which are the important factors that enable a FCNN to capture the complex contextual information and enlarge its limited receptive field.

### 3. Materials and Methods

Deep learning, one of the most effective methods in computer vision, is widely used. As illustrated in Figure 2, we designed a novel, fully convolutional neural network (FCNN) constructed by a 3D UNet with the proposed feature information extractors (MP and MDP). The proposed network is called Multi-scale MSCD-UNet. The details of the proposed approach are listed in the next subsection.

### 3.1. Model Overview

In Figure 2, the input slices were randomly cropped with the same center point from 3 modalities (T1, FLARI, T1_IR); thus, they have the corresponding position information. The concrete architecture of the MSCD-UNet consists of three main modules: MP, MDP, and multi-branch output. We exploit MSCD-UNet to capture the rich multi-scale semantic information in the encoding path by using multiple max pooling with different kernel sizes in parallel, and allow the detailed object boundary recovering in the decoding path by dividing the dense prediction maps into multiple groups. For each scale in the decoding path, we use a concatenation operation to connect these dense prediction maps for the information compensating. The multi-branch output module under a deeply supervised network component aims at largely discovering the learning ability of CNN from bottom to top layers, and producing more precise segmentation results by integrating the predicting maps of identical size at the last layer.

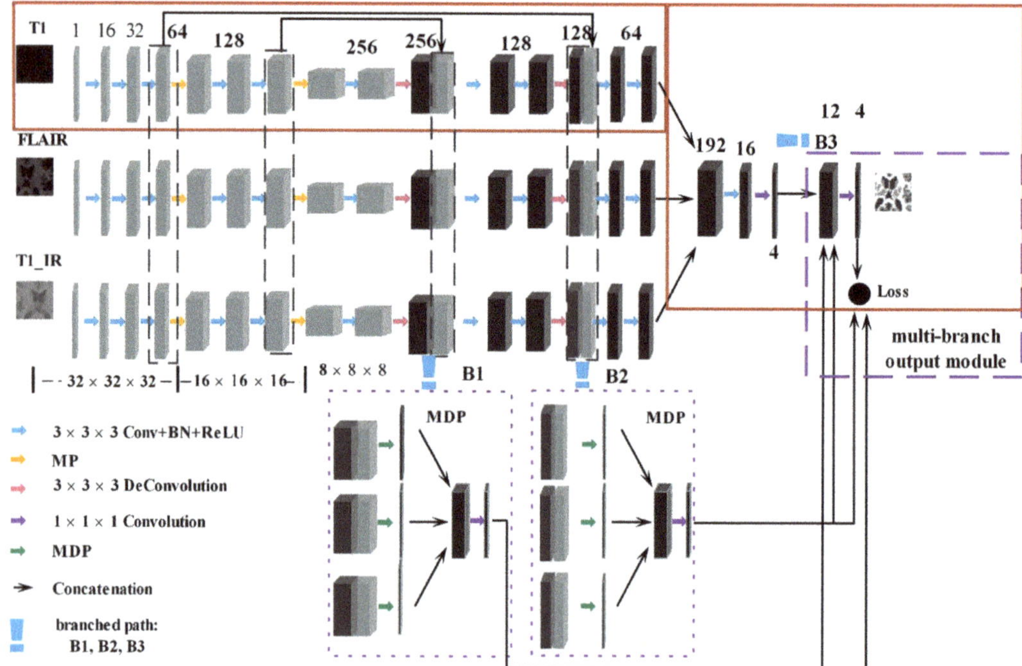

**Figure 2.** The architecture of proposed MSCD-UNet for brain segmentation consisting of MP, MDP, and the multi-branch output module. Three input samples with size 32 × 32 × 32 were randomly cropped with a same center point from 3 modalities (T1, T2-FLARI, T1_IR), they have the corresponding position information. Solid red box represents the subnetwork for T1 MR image.

### 3.2. Multi-Branch Pooling and Multi-Branch Dense Prediction

The information loss of down-sampling and up-sampling operations of an FCNN-based model is a common problem, which is mentioned as the weak ability of feature extracting in the encoding and decoding paths. In the encoding path, the repeated accumulation of pooling and convolution with strides at consecutive layers meaningfully reduces the spatial resolution of feature maps, then causing a loss of spatial information. In the decoding path, deconvolutional layers have been used to recover the corresponding localization for the higher resolution layer; it will result in great losses in channel dimension. In order to enhance the ability of feature extracting in spatial and channel dimensions, we propose to utilize a multi-scale spatial and channel dimensions-based network to capture higher semantic information during encoding and gradually recover the spatial information during decoding.

Multi-branch pooling (MP): pooling is employed to improve the invariants of the transformed image, the compact representations of semantic information, and the better robustness to noise and clutter [49]. The size of the feature map can be reduced by using different pooling scales, which will effectively ensure the validity of information and speed up the calculation. Empirically, max-pooling is widely used in the field of medical image processing; however, it is easy to lose the useful spatial contextual information when the feature map resolution is reduced. In order to reduce the loss of information, inspired by [20], they have adopted multiple rates of atrous convolution in parallel to harness multi-scale context information. However, although they can capture global information by multiple rates of atrous convolution, it is easy to encourage irrelevant redundant information without a suitable receptive field. Thus, we propose multi-branch pooling to collect the multi-scale spatial information during the encoding procedure, which in parallel consists of multiple max pooling with different kernel sizes. The parallel max-pooling separates the feature maps into different adjacent regions and produces pooled representations for the same location, while the neighborhood information with a suitable receptive field can be captured for the information compensating. After the MP operation, these parallel feature maps pooled with different kernels finally have identical size, and each time the feature map size is reduced by factor of two. In addition, we can see from Figure 3, the intensities of different brain tissues in different local regions of the brain are close to each other; thus, a lot of redundant information will be produced by using atrous convolution with a large receptive field. However, the proposed MP, as illustrated in Figure 4, can capture the multi-scale context information with a suitable receptive field.

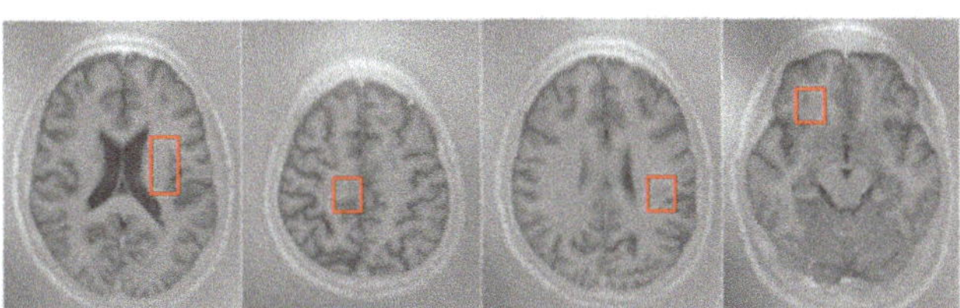

**Figure 3.** Example of the modality of T1_IR from patient no. 5 MRI. In this example, the intensities of different brain tissue in the different local brain regions are close to each other, like the examples in the red boxes.

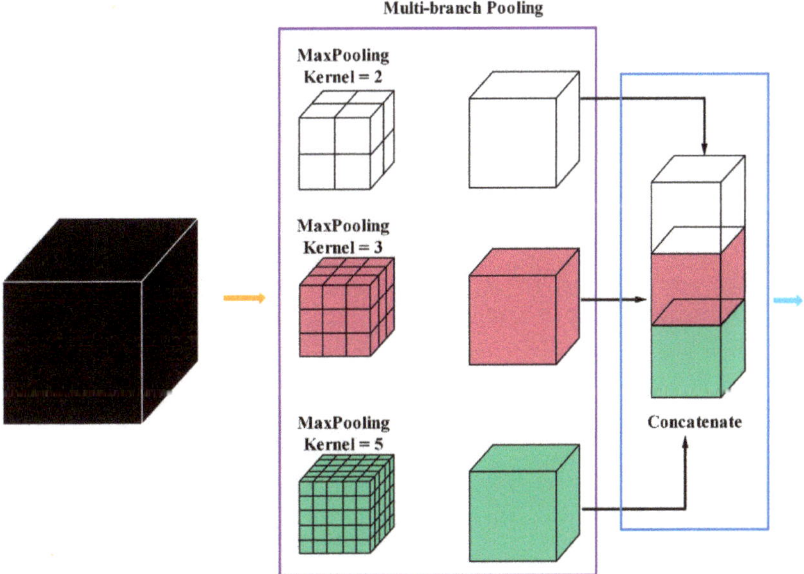


**Figure 4.** Encoding path with multi-branch max pooling.

Our proposed MP contains a three-branch structure with bin size $2 \times 2 \times 2$, $3 \times 3 \times 3$, and $5 \times 5 \times 5$ in first pooling stage, and a two-branch structure with bin size $2 \times 2 \times 2$ and $3 \times 3 \times 3$ in last pooling stage. The key idea of MP is to use suitable kernels, whose size is controlled by the parameter K. In order to gain the optimal combination of kernel size K, we enumerate different kernel sizes and validate the performance respectively; the results are detailed in Section 4.1. Additionally, we perform extensive experiments to compare the performance between the max pooling and the average pooling in Section 4.1.

Multi-branch dense prediction (MDP): as in the work of [19], the decoding module consists of a series of simple bilinear up-samplings by a consecutive factor of 2, which could be regarded as a naive decoding module. However, this naive decoding module may not fully recover the segmented object details. During the decoding phase, the compressed feature maps from the deepest encoding layer will be used to recover feature maps resolution by using deconvolution and up-sampling operation. After the maps resolution upsizing, the spatial information in these decompressed feature maps is fixed so the detailed information is represented more in channel dimension; thus, it implies we will be supposed to focus on the collection of complex information in channel dimension. Inspired by [24], considering that the predict results at the adjacent position are related to the result of the center point, they have divided the feature channels into one group in each up-sampling operation, where the number of feature channels has been fixed, resulting in a loss of information. In order to enhance the ability of feature extracting in channel, we design a channel-based multi-scale feature extractor (see Figure 5), named MDP, in which the feature channels are divided into multiple groups to free the fixed feature channels; the result of center point can be created by averaging these groups for the information compensating.

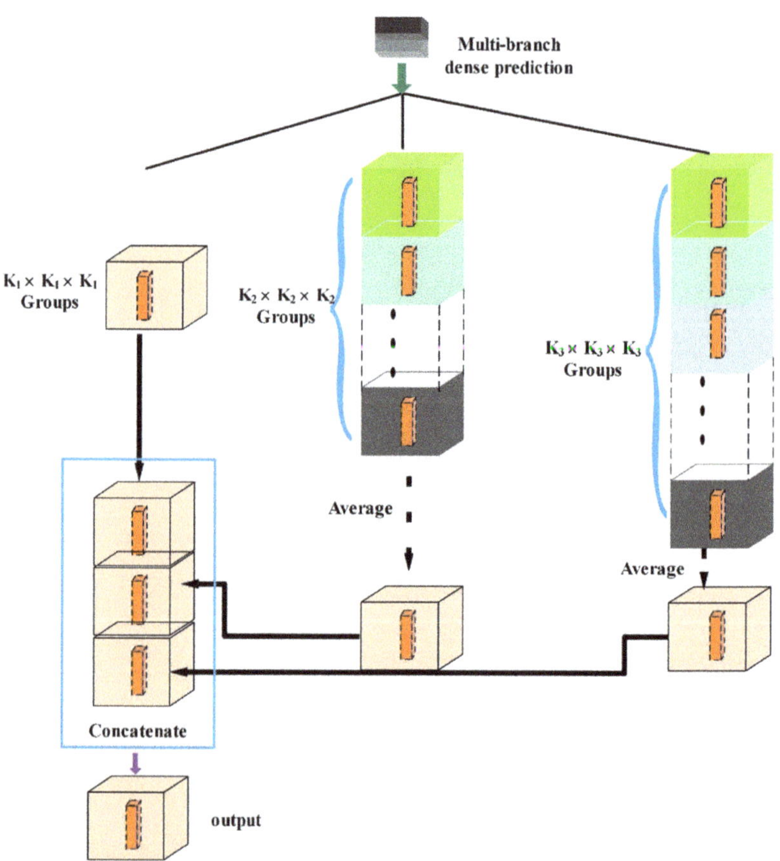


**Figure 5.** The components of multi-branch dense prediction (MDP).

For the decoding path, the feature point at the spatial location $(l, n, m)$ is responsible for its semantic information. In order to collect as much spatial information as possible into channels, this information extractor can be considered to predict results at the adjacent position, e.g., $(l - 1, n + 1, m + 1)$. When obtaining the final predicted results, results at the center position $(l, n, m)$ can be created by averaging the related scores. Concretely, supposing that the three window sizes are $k_1 \times k_1 \times k_1$, $k_2 \times k_2 \times k_2$, $k_3 \times k_3 \times k_3$, respectively, we divided the feature channels into three groups $k_1 \times k_1 \times k_1$, $k_2 \times k_2 \times k_2$, $k_3 \times k_3 \times k_3$, respectively. The outputs of MDP R are formed as follows:

$$R^{k1}_{l,m,n} = \frac{1}{k_1 \times k_1 \times k_1} \sum_{0 \leq r,s,t < k_1} y^{(r \times k_1 + s + t)}_{l+r-\lfloor \frac{k_1}{2} \rfloor, \ m+s-\lfloor \frac{k_1}{2} \rfloor, \ n+t-\lfloor \frac{k_1}{2} \rfloor}, \tag{1}$$

$$R^{k2}_{l,m,n} = \frac{1}{k_2 \times k_2 \times k_2} \sum_{0 \leq r,s,t < k_2} y^{(r \times k_2 + s + t)}_{l+r-\lfloor \frac{k_2}{2} \rfloor, \ m+s-\lfloor \frac{k_2}{2} \rfloor, \ n+t-\lfloor \frac{k_2}{2} \rfloor}, \tag{2}$$

$$R^{k3}_{l,m,n} = \frac{1}{k_3 \times k_3 \times k_3} \sum_{0 \leq r,s,t < k_3} y^{(r \times k_3 + s + t)}_{l+r-\lfloor \frac{k_3}{2} \rfloor, \ m+s-\lfloor \frac{k_3}{2} \rfloor, \ n+t-\lfloor \frac{k_3}{2} \rfloor}, \tag{3}$$

where $R_{l,n,m}$ represents the result at the position $(l, n, m)$ and $y^c_{l,n,m}$ is the feature map at position $(l, n, m)$ belonging to channel group c. The MDP scheme is illustrated in Figure 5.

We employed MDP as the output of our decoding module (see Figure 2). We set $k_1 = 1$, $k_2 = 3$, $k_3 = 4$ to conduct our experiments. In order to prove the validity of MDP, we tested the baseline model U-Net only with $k_1 = 1$ in the experimental section, and the results show that the MDP can improve the final performance. The results are detailed in Section 4.2.

### 3.3. Multi-Branch Output Modules and Loss Functions

The idea of multi-branch output modules is widely used in the deeply supervised network. In view of our proposed network, collecting multi-scale information in the decoding path can encourage more reliable and accurate predictions of the final results. Thus, we integrate multiple branch output in each scale after MDP operation (see Figure 2 for an illustration). Concretely, given a total $H$ branch output, each output will generate the prediction by an up-sampling operation with the associated weights. The multiple loss function of the whole network can be defined as a weighted sum of all of the branch output loss; its calculation formula is as follows:

$$Loss_{side}(W, w, gT) = \sum_{h=1}^{H} \beta_h l^h_{side}\left(W, w^h, gT\right), \tag{4}$$

where $\beta_h$ stands for the weight of the $h_{th}$ output loss function, $l^h_{side}$ is the cross-entropy loss function, and the count of the additional output $H$ is set to 3. $l_{side}$ is unfolded with the following formula:

$$l_{side}\left(W, w^h, gT\right) = - \sum_{i \in gT} \sum_c \omega_c gT_c log P\left(W, w^h\right), \tag{5}$$

where $gT$ is the label of ground truth, c denotes the $c_{th}$ classification label and $\omega_c$ is the associated weight, and $P(\cdot)$ indicates the output of network as the probabilistic prediction in the $c_{th}$ output way. Finally, a fusion layer can be applied to aggregate the prediction from each additional output by:

$$Loss_{fuse}(W, w, f) = \varnothing\left(gT, \sigma\left(\sum_{h=1}^{H} f_n A p^h_{side}\right)\right), \tag{6}$$

where $f_n$ represents the fusion weight, $Ap^h_{side}$ indicates the activation of the $h_{th}$ output way, $\sigma$ denotes the softmax activation function, and $\varnothing$ is the cross-entropy loss function. Finally, the final loss function of the network can be formed as:

$$Loss_{final} = Loss_{fuse}(W, w, f) + Loss_{side}(W, w). \tag{7}$$

### 3.4. Network Architecture

The U-Net [19] has been widely applied in medical image segmentation, which adequately combines the low-level high resolution and the high-level low resolution feature maps. Our proposed MSCD-UNet is similar to the 3D-UNet [50], but it can make up for the deficiency of information missed in U-Net by using MP and MDP to capture rich multi-scale context information.

The architecture of MSCD-UNet in this paper is shown in Figure 2. We follow the strategy in [48], where sub-volumes of $32 \times 32 \times 32$ are used as input for training. Instead of using the standard 3D U-Net with multi-channel inputs, we use a parallel feed forward network with different modalities and fuse their deep-high level features for voxel-wise prediction. The parallel feed forward network consists of three parts: input part, encoding part, and decoding part. The input part is divided into three parallel paths where the input data are T1, T2-FLAIR, and T1-IR, respectively. The encoding part includes two stages, each stage contains two $3 \times 3 \times 3$ convolution layers and each is followed by a batch normalization (BN) and a non-linear activation function (ReLU). At the end of each stage, the MP is attached to reduce resolution. The number of feature channels is doubled after each stage. Similarly, the decoding part also contains two stages, each stage consists of a deconvolution layer of $2 \times 2 \times 2$ followed by BN and ReLU. There are also two $3 \times 3 \times 3$ convolution layers each followed by BN and ReLU. Additionally, MDP is used to collect complex multi-scale channel information to recover the corresponding localization to higher resolution layer in each stage. Finally, a fusion layer can integrate the prediction result from each MDP output to produce more accurate edge-preserving segmentation results.

### 3.5. Dataset Introduction

Our proposed method is successful on the MRBrainS13 dataset of brain segmentation challenge. The method is evaluated in this section by three different datasets. MRBrainS13, IBSR18, and ISeg2017.

(1) MRBrainS13 is from the official website [51]. In the training dataset, it has five brain MR images, including 2 male subjects and 3 female subjects, and each subject is associated with 3 modality-channels (i.e., T1, T1_IR, FLAIR) and the manually marked labels of 4 classes, namely, gray matter (GM), white matter (WM), cerebrospinal fluid (CSF), and background, as shown in Figure 6. In the test dataset, it has 30 brain MR images. All the modality has been bias-corrected and the data of each subject is aligned. The voxel size is 0.958 mm $\times$ 0.958 mm $\times$ 3 mm for all modalities. Each modality of the MRI data is represented by a $240 \times 240 \times 48$ volume;

(2) IBSR18 is also used to evaluate our MSCD-UNet [52]. The IBSR18 training dataset contains 18 subjects, each subject in training data has a single T1-weighted modality. All volumes have a size of $256 \times 256 \times 128$ voxels, with voxel space ranging from 0.8 mm $\times$ 0.8 mm $\times$ 1.5 mm to 1.0 mm $\times$ 1.0 mm $\times$ 1.0 mm. A total of 4 anatomical brain structures are targeted for segmentation.

(3) ISeg2017 is also used to evaluate our MSCD-UNet [53]. ISeg2017 dataset has the combined modalities of T1w and T2w. MRT1 images are obtained with 144 sagittal slices utilizing the following parameters: flip angle = $7°$, *TR/TE* = 1900/4.38 ms, and resolution = $1 \times 1 \times 1$ mm$^3$. Likewise, MR-T2 images are obtained with 64 axial slices by using: flip angle = $150°$, *TR/TE* = 7380/119 ms, and resolution = $1.25 \times 1.25 \times 1.95$ mm$^3$. Ten infant subjects with manual labels were provided for training.

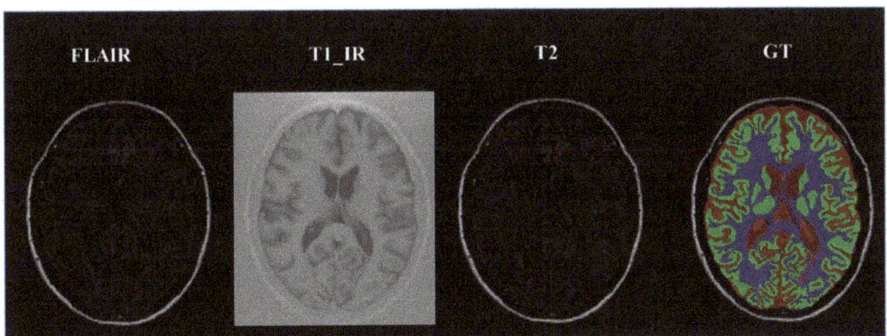


**Figure 6.** Example of MR images with different image modalities and the labels manually marked by experts; the first three images from left to right are FLAIR, T1, and T1_IR. The fourth image is the ground truth labels where the colors denote different regions of brain tissues: red represents cerebrospinal fluid (CSF), green the gray matter (GM), and blue the white matter (WM). Gray denotes the other tissues.

*3.6. Evaluation Metrics*

The following common segmentation indicators are employed to evaluate and compare our model with other state-of-the-art methods. The Dice Coefficient (DC), the 95th percentile of the Hausdorff Distance (HD), and the Absolute Volume Difference (AVD) are applied on MRBrainS13 to complete our experiments. For the IBSR18, DC is used for evaluation [54]. For the ISeg2017, DC and ASD is used for evaluation.

Dice coefficient (DC) is defined by the area overlap between the ground truth and segmentation prediction results as:

$$DC(G, P) = 2\frac{G \cap P}{G + P} \times 100\%, \tag{8}$$

where $G$ is the ground truth and $P$ represents the predicted segmentation result. $DC$ is a metric of area overlap between the predicted segmentation result $P$ and the ground truth $G$.

Because the conventional Hausdorff distance is very sensitive to the outliers, the $K_{th}$ ranked distance, i.e., $h_{95} = K^{th}_{p \in P} min_{g \in G} \| g - p \|$, is used as to suppress the outliers [52]; it is defined as:

$$HD(G, P) = max\{h_{95}(G, P), h_{95}(P, G)\}, \tag{9}$$

A smaller value $HD(G, P)$ represents a higher proximity between ground truth and segmentation result.

The absolute volume difference (AVD) is used to evaluate the difference between the predicted volume and the true volume as:

$$AVD(G, P) = \frac{|V_g - V_p|}{V_g} \times 100\%, \tag{10}$$

where $V_p$ is the volume of prediction and $V_g$ is the volume of truth. A lower value of AVD means the ground truth and prediction result are closer to each other.

The Average Surface Distance (ASD) is used to calculate for the predicted result $P$ and the corresponding ground truth $G$; it is defined as:

$$ASD(G, P) = \frac{1}{2}\left(\frac{\Sigma_{a \in G} min_{b \in p} d(a, b)}{\Sigma G} + \frac{\Sigma_{b \in p} min_{a \in g} d(b, a)}{\Sigma P}\right), \tag{11}$$

where $d(a, b) = \| a - b \|$ represents Euclidean distance between points $a$ and $b$.

### 3.7. Implementation Details

Tensorflow is used on the workstation with a NVIDIA GTX_1080Ti GPU in our experiments. In the pre-processing step for the MRBrainS13, IBSR18, and ISeg2017 datasets, MR images are normalized with the zero-mean method, which is calculated as follows: (1) each image is processed by subtracting a Gaussian smoothed image and applying a contrast-limited adaptive histogram equalization to enhance local contrast, (2) the resulting intensity value is subtracted by the mean intensity value and then divided by the standard deviation.

In the training phase, to avoid overfitting, data augmentation techniques (flipping, rotation, elastic stretching, shifting, zoom) are applied in the training procedure to get good performance. The network is trained for 18,000 iterations with ADAM optimizer and Xavier initialization, and the epoch is set as 1. The learning rate is set as 0.001, then being reduced by a factor after every 5000 iterations. Due to the limited capacity of GPU memory, for the input samples and the label samples, both of them with size $32 \times 32 \times 32$, are randomly cropped with a same center point from 4 modalities (T1, FLARI, T1_IR, the label image); thus, they have the corresponding position information. A total of around 72,000 sub-volume samples are extracted by random sampling to feed into the network. For the loss function, the weight of $hth$ output loss function $\beta_h$ is set as [1,1,1], the associated weight of the $cth$ class label $\omega_c$ is set as [1,1,2,2], and the fusion weight $f_n$ is set as [1,1,1].

In the test phase, the final prediction result is obtained by the majority voting strategy on the results of overlapping with a stride of 8.

## 4. Results

We performed an ablation study to investigate the efficacy of employing multi-branch pooling (MP), multi-branch dense prediction (MDP), and multi-branch output module by using five-fold cross-validation.

### 4.1. Ablation for Multi-Branch Pooling (MP)

In order to gain the optimal combination kernel sizes of MP, we enumerated different kernel sizes and test the performance on the MRBrain13 training dataset. We tried different kernel sizes K ranging from 2 to 7 to exploit the optimal combination in the two pooling stages. We named the combination of kernel in the first pooling stage "FP", and the combination of kernel in the second pooling stage "SP". In the case K = 7, which roughly equals to the feature map size ($8 \times 8$), the structure becomes "really global pooling". The results are presented in Table 2. From the results, we can find that the performance is better when the "FP" is the combination kernel size of 5, 3, 2, and "SP" is the combination kernel size of 3, 2. When the "FP" is 2 and "SP" is 2, it represents the standard 3D-UNet.

**Table 2.** Performances of the combination kernel sizes in the two pooling stages by 5-fold cross-validation in MRBrain13 training dataset (DC:%, HD:mm, AVD:%). The "FP" represents the first pooling stage, the "SP" represents the second pooling stage, and the "K" represents the combination of kernels.

| | K | | K | GM | | | WM | | | CSF | | |
|---|---|---|---|---|---|---|---|---|---|---|---|---|
| | | | | DC | HD | AVD | DC | HD | AVD | DC | HD | AVD |
| FP | 7,5,3,2 | SP | 7,5,3,2 | 82.47 | 2.10 | 7.99 | 83.59 | 3.61 | 7.71 | 78.22 | 3.30 | 8.61 |
| FP | 5,3,2 | SP | 5,3,2 | 83.12 | 1.94 | 7.79 | 85.40 | 2.89 | 7.52 | 81.45 | 2.58 | 8.59 |
| FP | 5,3,2 | SP | 3,2 | 86.08 | 1.71 | 6.76 | 89.02 | 1.76 | 6.71 | 84.15 | 2.24 | 7.82 |
| FP | 5,3,2 | SP | 2 | 84.50 | 1.75 | 7.01 | 86.04 | 2.75 | 7.17 | 83.23 | 2.44 | 8.02 |
| FP | 3,2 | SP | 3,2 | 85.98 | 1.90 | 7.22 | 88.90 | 2.32 | 6.59 | 84.63 | 2.18 | 8.13 |
| FP | 3,2 | SP | 2 | 82.25 | 2.03 | 8.07 | 84.34 | 3.48 | 7.42 | 83.01 | 2.98 | 8.66 |
| FP | 2 | SP | 2 | 85.94 | 1.85 | 7.09 | 88.83 | 2.39 | 6.82 | 83.79 | 2.31 | 8.30 |

In additional, in order to exploit the collecting ability of spatial information between max pooling and average pooling, each max pooling was replaced with average pooling in MP. The result of UNet_MP_Aver is shown for MP using average pooling in Table 3. It indicates that the UNet_MP_Max achieves higher performance over the UNet_MP_Aver. Comparing with average pooling, max pooling can effectively reduce the collection of redundant information.

**Table 3.** Performances of UNet, UNet_MP_Max, and UNet_MP_Aver by 5-fold cross-validation (DC:%, HD:mm, AVD:%).

| Tissue | GM | | | WM | | | CSF | | |
|---|---|---|---|---|---|---|---|---|---|
| **Evaluation Metric** | **DC** | **HD** | **AVD** | **DC** | **HD** | **AVD** | **DC** | **HD** | **AVD** |
| UNet | 85.94 | 1.85 | 7.09 | 88.83 | 2.39 | 6.82 | 83.79 | 2.31 | 8.30 |
| UNet_MP_Max | 86.08 | 1.71 | 6.76 | 89.02 | 1.76 | 6.71 | 84.15 | 2.24 | 7.82 |
| UNet_MP_Aver | 85.08 | 1.98 | 8.05 | 88.27 | 2.23 | 7.47 | 82.71 | 2.70 | 8.84 |

*4.2. Ablation for Multi-Branch Output with Multi-Branch Dense Prediction (MDP)*

As described in Section 3.2, we utilized MDP on the feature maps after using the concatenation layer. To analyze the performance of using MDP at each branch output, Table 4 provides the results of each branch output (B1, B2, B3) with MDP in each scale, in which B1-MDP is 1/4 scale of output, B2-MDP stands for 1/2 scale, and B3-MDP represents 1/1 scale. Additionally, B1, B2, and B3 respectively represent the branch output without MDP. According to the results (displayed in Table 4), it can be seen that the performance is improved by increasing the scale of feature maps and the results of Dice score on WM, GM, and CSF satisfy B1-MDP < B2-MDP < B3-MDP, and B1 < B2 < B3. The fusion of multi-branch output is the key prediction result in the proposed network because it controls the network prediction compensation and performance in different scales. When fusing the branch output prediction with B1-MDP + B2-MDP + B3, named as B4, the segmentation performance is obviously improved for the evaluation metrics on GM and CSF compared with those of two other fusions, B5 (B1 + B2 + B3) and B6 (B1-MDP + B2-MDP + B3 MDP).

**Table 4.** Performances of B1, B2, B3, B1-MDP, B2-MDP, B3-MDP, B4, B5, B6, and MSCD-UNet by 5-fold cross-validation (DC:%, HD:mm, AVD:%).

| Tissue | GM | | | WM | | | CSF | | |
|---|---|---|---|---|---|---|---|---|---|
| **Evaluation Metric** | **DC** | **HD** | **AVD** | **DC** | **HD** | **AVD** | **DC** | **HD** | **AVD** |
| B1 | 72.35 | 3.12 | 8.37 | 75.97 | 2.79 | 9.31 | 70.59 | 3.78 | 11.47 |
| B2 | 75.42 | 3.06 | 7.92 | 79.49 | 2.37 | 8.92 | 77.51 | 3.69 | 10.15 |
| D3 (UNet) | 85.94 | 1.85 | 7.09 | 88.83 | 2.39 | 6.82 | 83.79 | 2.31 | 8.30 |
| B1-MDP | 73.04 | 2.05 | 8.14 | 74.33 | 2.93 | 9.56 | 71.06 | 3.02 | 9.75 |
| B2-MDP | 76.08 | 2.19 | 7.66 | 76.02 | 2.53 | 8.74 | 77.15 | 3.24 | 9.82 |
| B3-MDP | 85.88 | 2.01 | 8.05 | 88.87 | 2.23 | 7.47 | 83.81 | 2.70 | 8.84 |
| B4 | 86.12 | 1.91 | 6.81 | 88.30 | 2.06 | 7.17 | 83.98 | 2.23 | 8.43 |
| B5 | 85.96 | 1.93 | 7.05 | 89.03 | 1.88 | 7.03 | 83.62 | 2.40 | 8.11 |
| B6 | 85.97 | 1.99 | 6.81 | 89.30 | 2.09 | 7.12 | 83.86 | 2.31 | 8.55 |
| MSCD-UNet | 86.41 | 1.52 | 5.76 | 89.18 | 2.13 | 7.21 | 84.29 | 2.16 | 7.73 |

Figure 7 provides a visual comparison of the segmentation results produced by the trained UNet and our MSCD-UNet on the MRBrainS13 dataset. It shows that, with MP and MDP, more accurate segmentation results can be generated. Specifically, additional details are preserved, including boundaries and edges.

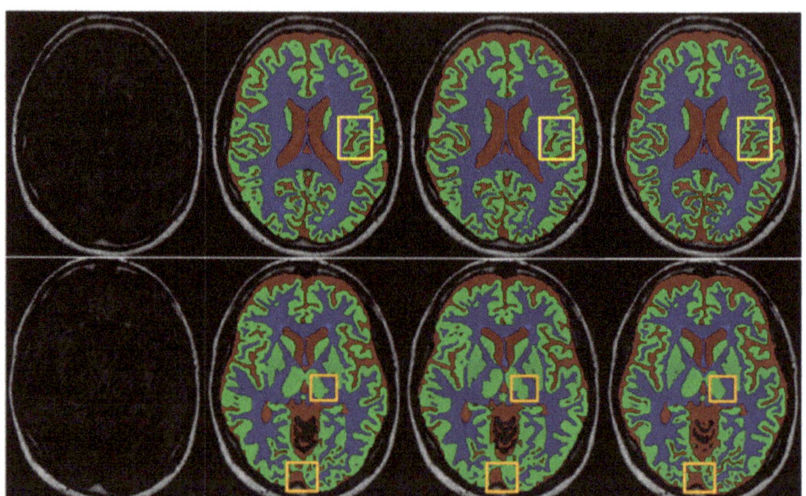

**Figure 7.** Segmentation results of the UNet and MSCD-UNet on the MR BrainS13 dataset. The rows show the segmentation results of different slices. From first column to last column: FLAIR, manual segmentation, segmentation result of UNet, segmentation result of MSCD-UNet. Center patch in solid yellow box of each segmentation result is highlighted. Each color denotes different brain tissue class, i.e., gray matter (blue), white matter (green), cerebrospinal fluid (red), and other tissues (gray).

Finally, it is observed that the result using MSCD-UNet (UNet_MP_Max + B4) is visually more accurate than those of other fusion strategies.

We also evaluate the MP and MDP on IBSR18 by five-fold cross-validation, where the IBSR18 consists of a larger single-modality T1-weighted MRI with more tissue labels. The evaluation is performed by using five-fold cross-validation on 18 subjects. However, the proposed MSCD-UNet has three channels as the input. Thus, a single subnetwork (e.g., subnetwork for T1 MR images presented in Figure 2) was reserved in MSCD-UNet while the remaining network structures were removed. The results are shown in Table 5. The Dices on GW, WM, and CSF are 85.39%, 89.08%, and 88.14% for UNet, respectively, and 89.82%, 91.18%, and 90.57% for MSCD-UNet, respectively. It reveals that, along with the using of MP and MDP, the performance of MSCD-UNet is obviously improved. Figure 8 provides a visual comparison of the segmentation results produced by the trained UNet and MSCD-UNet on the IBSR18 dataset.

**Table 5.** Cross-validation results of MRI brain segmentation using UNet and MSCD-UNet on IBSR18. (DC:%).

| Evaluation Metric | | DC | |
| --- | --- | --- | --- |
| Tissue | GM | WM | CSF |
| UNet | 85.39 | 89.08 | 88.14 |
| MSCD-UNet | 88.42 | 90.31 | 90.57 |

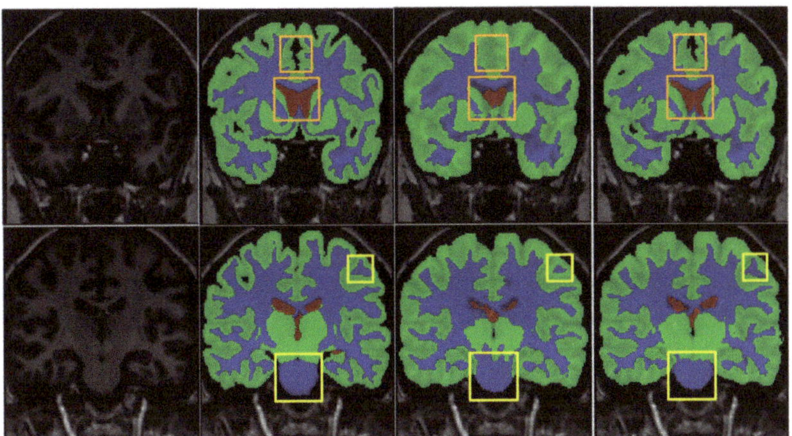

**Figure 8.** Segmentation results on the IBSR dataset using UNet and MSCD-UNet. The rows show the segmentation results of different slices. From first column to last column: T1, manual segmentation, segmentation result of UNet, segmentation result of MSCD-UNet. Center patch in solid yellow box of each segmentation result is highlighted. Each color denotes different brain tissues classes, i.e., gray matter (green), white matter (blue), cerebrospinal fluid (red), and other tissues (gray).

We have evaluated our proposed MSCD-UNet on ISeg2017, where the ISeg2017 consists of T1W, T2W, and label image. Like [44], the evaluation is performed by using nine subjects for training and one subject for validation. We evaluated our results on the ninth subject of the dataset. However, the proposed MSCD-UNet has three channels as the input. Thus, a subnetwork (e.g., subnetwork for T1, FLARI MR images presented in Figure 2) was reserved in MSCD-UNet while the remaining network structures were removed. The results are shown in Table 6. The Dices on GW, WM, and CSF are 91.36%, 89.91%, and 94.70% for UNet, respectively, and 92.17%, 90.47%, and 95.60% for MSCD-UNet, respectively. We can see that using the MP and MDP can yield improvements over the baseline of 3D-UNet.

**Table 6.** The validation results of MRI brain segmentation using UNet and MSCD-UNet on ISeg2017.

| Methoed | GM | | WM | | CSF | | Average |
|---|---|---|---|---|---|---|---|
| | DSC | ASD | DSC | ASD | DSC | ASD | DSC |
| UNet | 0.9136 | 0.354 | 0.8991 | 0.385 | 0.9470 | 0.135 | 0.9136 |
| Ours | 0.9217 | 0.322 | 0.9047 | 0.362 | 0.956 | 0.110 | 0.9274 |

### 4.3. Comparison with Existing State-of-the-Art Methods

We compare the results between our proposed MSCD-UNet and the state-of-the-art approaches on MRBrainS13 online test dataset. The segmentation of WM, GM, and CSF is evaluated by using the three metrics. A comparison listed in Table 7 indicates that the MSCD-UNet achieves better performance than many state-of-the-art methods [39–41,46,55,56]. The reason that our MSCD-UNet performs better is that our model can capture multi-scale information in spatial and channel dimensions by using MP and MDP to alleviate the lack of contextual information and the information loss during the encoding and decoding. Comparing with the similar U-Net architectures [42,48], Li et al. [42] have proposed a Dilated-Inception block to extract multi-scale features from brain MRI; however, it is easy to harness the irrelevant redundant information by using a larger dilation rate. In order to avoid harnessing the irrelevant redundant information, the proposed MP can capture multi-scale feature information with a suitable receptive field. From Table 7, we can see that our proposed architecture achieves better performance than [42]. Sun et al. [48] had the leading method in 2018; however, our proposed method obtained the best score on the GM and CSF, although [48] has a higher score on the CSF. Additionally, our architecture is parameter more efficient compared to [48], with 15 million learned parameters, less than [48], which has 20 million learned parameters. Our proposed multi-branch pooling (MP) and multi-branch dense prediction (MDP) can capture multi-scale feature information with a suitable receptive field, and it is sensitive to segment these brain tissues in edge because the intensity of tissues in edge vary greatly. Thus, our method achieves the best performance on the GM and CSF due to the greatly variation of intensity in the edge.

**Table 7.** A comparison with the state-of-the-art methods on MRBrainS2013 online test Dataset.

| Tissue | GM | | | WM | | | CSF | | |
|---|---|---|---|---|---|---|---|---|---|
| Evaluation Metric | DC | HD | AVD | DC | HD | AVD | DC | HD | AVD |
| MSCD-UNet | 86.69 | 1.23 | 5.65 | 89.73 | 1.75 | 6.21 | 85.15 | 1.66 | 5.70 |
| Sun [48] | 86.58 | 1.29 | 5.75 | 89.87 | 1.73 | 5.47 | 84.81 | 1.84 | 6.84 |
| Li [42] | 86.40 | 1.38 | 5.72 | 89.70 | 1.88 | 6.28 | 84.86 | 2.03 | 6.75 |
| Dolz [41] | 86.33 | 1.34 | 6.19 | 89.46 | 1.78 | 6.03 | 83.42 | 2.26 | 7.31 |
| Chen [40] | 86.15 | 1.44 | 6.60 | 89.46 | 1.93 | 6.05 | 87.25 | 2.19 | 7.68 |
| Bui [46] | 86.06 | 1.52 | 6.60 | 89.00 | 2.11 | 5.54 | 83.76 | 2.32 | 6.77 |
| Geraud [39] | 86.03 | 1.44 | 6.05 | 89.29 | 1.86 | 5.83 | 82.44 | 2.28 | 9.03 |
| Andermatt [55] | 85.40 | 1.54 | 6.09 | 88.98 | 2.02 | 7.69 | 84.13 | 2.17 | 7.44 |
| Stollenga [56] | 84.89 | 1.67 | 6.35 | 88.53 | 2.07 | 5.93 | 83.47 | 2.22 | 8.63 |

We also compare the results between our proposed MSCD-UNet and the state-of-the-art approaches on ISeg2017. The segmentation of WM, GM, and CSF is evaluated by using the three metrics. The results are shown in Table 8. The Dices on GW, WM, and CSF are 92.17%, 90.47%, and 95.60%, respectively, for our method. Compared to four other approaches [18,44–46], the performance has a higher average Dice score than [45,46]. Although the average Dice is lower than [18], the Dice on GM is higher; additionally, the optimal parameters are waiting to be found, and we will further exploit the potential of MP and MDP in future work.

**Table 8.** A comparison between proposed architecture and other 3D-based state-of-art methods in terms of DSC and ASD.

| Method | GM | | WM | | CSF | | Average |
|---|---|---|---|---|---|---|---|
| | DSC | ASD | DSC | ASD | DSC | ASD | DSC |
| Ours | 0.9217 | 0.322 | 0.9047 | 0.362 | 0.956 | 0.110 | 0.9274 |
| Lei [44] | 0.926 | 0.307 | 0.908 | 0.353 | 0.959 | 0.114 | 0.931 |
| Yu [45] | 0.8851 | - | 0.8546 | - | 0.9371 | - | 0.8922 |
| Qamar [18] | 0.9205 | - | 0.9050 | - | 0.958 | - | 0.9278 |
| Taoc [46] | 0.9157 | - | 0.9125 | - | 0.9469 | - | 0.9250 |

## 5. Discussion

In this paper, we proposed a Multi-scale Spatial and Channel Dimension-based U-Net for MRI brain segmentation. In our approach, an information extractor multi-branch pooling (MP) is used to capture spatial information in the encoding part, and an information extractor multi-branch dense prediction (MDP) is used to collect as much spatial information as possible into channels in the decoding part. As the intensity of white matter is similar to the gray matter in the rugged edge, enlarging the size of receptive field can improve the recognition performance. In our experiments, we validated that using multiple max pooling with different kernel sizes in parallel can dramatically improve the segmentation performance comparing to the standard 3D U-Net. For example, as shown in Table 2, the Dice coefficients of GM, WM, and CSF by using five-fold cross-validation are 85.94%, 88.83%, and 83.79, respectively, while using the MP can improve the Dice to 86.08%, 89.02%, and 84.15%, respectively. Integration of the multi-scale spatial information in the encoding part can further improve the segmentation accuracy.

Regarding the decoding part, this naive decoding module may not fully recover the segmented object details. During the decoding phase, the compressed feature maps from the deepest encoding layer will be used to recover feature map resolution by using deconvolution and up-sampling. After the maps resolution upsizing, the spatial information in these decompressed feature maps is fixed, so the detailed information is represented more in channel dimension. Hence, it is necessary to collect the complex information in channel dimension. To probe the influence of channel-based multi-scale feature extractor (MDP), we conducted the experiments with and without MDP. The evaluation performance results including DC, HD, and AVD can be seen in Table 4. From these results, we can see the performance of GM, WM, and CSF segmentation improved from 85.94% to 86.41%, 88.83% to 89.18%, and 83.79% to 84.29% on Dice, respectively.

However, our study has some limitations. Although our analysis shows that the MP and MDP with multi-branch output are effective in segmentation of GM, WM, and CSF, if the combination of different kernel sizes in MP and different groups in MDP are selected by a manual setting, which may be tedious and prone to errors if applied in some extreme cases. Nevertheless, this is evidence of the capability of MP and MDP in brain tissue segmentation tasks, indicating the need of further study on this issue to increase the accuracy of such approaches. Another limitation of our model is that it has more than 15 million learned parameters and therefore the training of this model takes more than 8 h. The parameter of the proposed MSCD-UNet is three times larger than the standard 3D-UNet because the MSCD-UNet has three subnetworks for the T1, FLAIR, and T2 in parallel. We used T1, T2, and FLAIR as multi-channel input in the MSCD-UNet, and while the training time was substantially reduced, the performance of segmentation was not satisfactory. Therefore, we should focus on the relationship between this parallel architecture and the performance of segmentation. We believe that the performance of segmentation would be improved, even without this parallel architecture.

## 6. Conclusions

We propose a novel Multi-scale Spatial and Channel Dimension-based U-Net, referred to as MSCD-UNet, by integrating the multi-scale context information in spatial and channel dimensions for brain tissue segmentation. It contains three modules: MP, MDP, and multi-branch output. The MP is an extractor to capture spatial information during the encoding procedure, which consists of multiple max pooling with different kernel sizes in parallel. Extensive experiments indicate that the proposed information extractor MP can effectively enhance the representative ability by exploiting the multi-scale spatial information. The MDP and multi-branch output is a channel-based multi-scale feature extractor, which can recover the corresponding localization to a higher resolution layer in the decoding path. An ablation study demonstrates the effectiveness of the proposed MDP and multi-branch output. This reflects the importance of capturing multi-scale features in enhancing the learning ability in the encoding and decoding paths. We validated our proposed network on the MRBrainS13, IBSR18, and ISeg2017 datasets for brain tissue segmentation and achieved state-of-the-art results as compared to other existing approaches. The proposed method can promote the research on automated brain tissue segmentation as well as offer a useful and effective tool for assessing and diagnosing neurodegenerative diseases and disorders of human brain. In future work, we will explore the proposed network for other medical image challenges.

**Author Contributions:** Conceptualization: J.-S.L. and G.-Z.M.; funding acquisition: E.-M.S. and R.-C.J. investigation: J.-S.L.; methodology: J.-S.L.; software: E.-M.S.; formal analysis: G.-Z.M.; writing-original draft preparation: J.-S.L.; writing-review and editing: G.-Z.M.; All authors have read and agreed to the published version of the manuscript.

**Funding:** This research was financially supported by the National Natural Science Foundation of China (No. 81671768)

**Institutional Review Board Statement:** Not applicable.

**Informed Consent Statement:** Not applicable.

**Data Availability Statement:** Some publicly available datasets were used in this study. This data can be found here: https://mrbrains13.isi.uu.nl/data/ accessed on 7 September 2020, https://www.nitrc.org/frs/?group_id=48 accessed on 7 September 2020, and https://iseg2017.web.unc.edu/ accessed on 29 May 2021.

**Conflicts of Interest:** The authors declare no conflict of interest.

# References

1. Wright, R.; Kyriakopoulou, V.; Ledig, C.; Rutherford, M.; Hajnal, J.; Rueckert, D.; Aljabar, P. Automatic quantification of normal cortical folding patterns from fetal brain MRI. *Neuroimage* **2014**, *91*, 21–32. [CrossRef]
2. Wells, W.; Grimson, W.; Kikinis, R.; Jolesz, F. Adaptive segmentation of MRI data. *IEEE Trans. Med. Imaging* **1996**, *15*, 429–442. [CrossRef] [PubMed]
3. Assefa, D.; Keller, H.; Ménard, C.; Laperriere, N.; Ferrari, R.J.; Yeung, I. Robust texture features for response monitoring of glioblastoma multiforme on -weighted and -FLAIR MR images: A preliminary investigation in terms of identi-fication and segmentation. *Med. Phys.* **2010**, *37*, 1722–1736. [CrossRef] [PubMed]
4. Qin, C.; Guerrero, R.; Bowles, C.; Chen, L.; Dickie, D.A.; del Valdes-Hernandez, M.; Wardlaw, J.; Rueckert, D. A large margin algorithm for automated segmentation of white matter hy-perintensity. *Pattern Recognit.* **2018**, *77*, 150–159. [CrossRef]
5. Moeskops, P.; Benders, M.J.; Chiţă, S.M.; Kersbergen, K.J.; Groenendaal, F.; de Vries, L.S.; Viergever, M.A.; Išgum, I. Automatic segmentation of MR brain images of preterm infants using supervised classification. *Neuroimage* **2015**, *118*, 628–641. [CrossRef]
6. Maier, O.; Menze, B.H.; der Gablentz, J.; Häni, L.; Heinrich, M.P.; Liebrand, M.; Winzeck, S.; Basit, A.; Bentley, P.; Chen, L.; et al. ISLES 2015—A public evaluation benchmark for ischemic stroke lesion segmentation from multispectral MRI. *Med. Image Anal.* **2017**, *35*, 250–269. [CrossRef] [PubMed]
7. Janakasudha, G.; Jayashree, P. Early Detection of Alzheimer's Disease Using Multi-feature Fusion and an Ensemble of Classifiers. In *Advanced Computing and Intelligent Engineering*; Springer: Singapore, 2020; pp. 113–123.
8. Ashburner, J.; Friston, K.J. Unified segmentation. *Neuroimage* **2005**, *26*, 839–851. [CrossRef]
9. Rajchl, M.; Baxter, J.S.H.; McLeod, A.J.; Yuan, J.; Qiu, W.; Peters, T.M., Khan, A.R. A3eTb. MAP-based Brain Tissue Segmentation using Manifold Learning and Hierarchical Max-Flow regularization. In Proceedings of the MICCAI Grand Challenge on MR Brain Image Segmentation (MRBrainS'13), Nagoya, Japan, 26 September 2013.
10. Duchesne, S.; Pruessner, J.C.; Collins, D.L. Appearance-Based Segmentation of Medial Temporal Lobe Structures. *Neuroimage* **2002**, *17*, 515–531. [CrossRef] [PubMed]
11. Ballanger, B.; Tremblay, L.; Sgambato-Faure, V.; Beaudoin-Gobert, M.; Lavenne, F.; Le Bars, D.; Costes, N. A multi-atlas based method for automated anatom-ical Macaca fascicularis brain MRI segmentation and PET kinetic extraction. *Neuroimage* **2013**, *77*, 26–43. [CrossRef] [PubMed]
12. Heckemann, R.A.; Hajnal, J.V.; Aljabar, P.; Rueckert, D.; Hammers, A. Automatic anatomical brain MRI segmentation combining label propagation and decision fusion. *Neuroimage* **2006**, *33*, 115–126. [CrossRef] [PubMed]
13. Scherrer, B.F.F.; Garbay, C.; Dojat, M. Distributed local MRF models for tissue and structure brain segmentation. *IEEE Trans. Med. Imaging* **2009**, *28*, 1278–1295. [CrossRef] [PubMed]
14. Long, J.; Ma, G.; Liu, H.; Song, E.; Hung, C.-C.; Xu, X.; Jin, R.; Zhuang, Y.; Liu, D. Cascaded hybrid residual U-Net for glioma segmentation. *Multimed. Tools Appl.* **2020**, *79*, 24929–24947. [CrossRef]
15. Chen, H.; Qin, Z.; Ding, Y.; Tian, L.; Qin, Z. Brain tumor segmentation with deep convolutional symmetric neural network. *Neurocomputing* **2020**, *392*, 305–313. [CrossRef]
16. Havaei, M.; Davy, A.; Warde-Farley, D.; Biard, A.; Courville, A.; Bengio, Y.; Pal, C.; Jodoin, P.-M.; Larochelle, H. Brain tumor segmentation with Deep Neural Networks. *Med. Image Anal.* **2017**, *35*, 18–31. [CrossRef]
17. Coupé, P.; Mansencal, B.; Clément, M.; Giraud, R.; de Senneville, B.D.; Ta, V.; Lepetit, V.; Manjon, J.V. AssemblyNet: A large ensemble of CNNs for 3D whole brain MRI segmentation. *Neuroimage* **2020**, *219*, 117026. [CrossRef] [PubMed]
18. Qamar, S.; Jin, H.; Zheng, R.; Ahmad, P.; Usama, M. A variant form of 3D-UNet for infant brain segmenta-tion. *Future Gener. Comput. Syst.* **2020**, *108*, 613–623. [CrossRef]
19. Ronneberger, O.; Fischer, P.; Brox, T. U-Net: Convolutional Networks for Biomedical Image Segmentation. In *International Conference on Medical Image Computing and Computer-Assisted Intervention*; Springer: Cham, Switzerland, 2015. [CrossRef]

20. Zhao, H.; Shi, J.; Qi, X.; Wang, X.; Jia, J. Pyramid Scene Parsing Network. In Proceedings of the 2017 IEEE Conference on Computer Vision and Pattern Recognition (CVPR), Honolulu, HI, USA, 21–26 July 2017; pp. 1063–6919.

21. Chen, L.; Papandreou, G.; Schroff, F.; Adam, H. Rethinking Atrous Convolution for Semantic Image Segmentation. *arXiv* **2017**, arXiv:1706.05587.

22. Chen, L.; Zhu, Y.; Papandreou, G.; Schroff, F.; Adam, H. Encoder-Decoder with Atrous Separable Convolution for Semantic Image Segmentation. In Proceedings of the European Conference on Computer Vision, Munich, Germany, 8–14 September 2018.

23. Rajawat, A.S.; Jain, S. Fusion Deep Learning Based on Back Propagation Neural Network for Personali-zation. In Proceedings of the 2nd International Conference on Data, Engineering and Applications, Bhopal, India, 28–29 February 2020.

24. Zhang, Z.; Zhang, X.; Peng, C.; Xue, X.; Sun, J. ExFuse: Enhancing Feature Fusion for Semantic Segmenta-tion. In Proceedings of the European Conference on Computer Vision (ECCV), Munich, Germany, 8–14 September 2018; pp. 273–288.

25. Ma, W.; Gong, C.; Xu, S.; Zhang, X. Multi-scale spatial context-based semantic edge detection. *Inf. Fusion* **2020**, *64*, 238–251. [CrossRef]

26. González-Villà, S.; Oliver, A.; Valverde, S.; Wang, L.; Zwiggelaar, R.; Lladó, X. A review on brain structures segmentation in magnetic resonance imaging. *Artif. Intell. Med.* **2016**, *73*, 45–69. [CrossRef]

27. Makropoulos, A.; Counsell, S.J.; Rueckert, D. A review on automatic fetal and neonatal brain MRI seg-mentation. *Neuroimage* **2018**, *170*, 231–248. [CrossRef]

28. Weisenfeld, N.I.; Warfield, S.K. Automatic segmentation of newborn brain MRI. *Neuroimage* **2009**, *47*, 564–572. [CrossRef] [PubMed]

29. Anbeek, P.; Išgum, I.; Van Kooij, B.J.M.; Mol, C.P.; Kersbergen, K.J.; Groenendaal, F.; Viergever, M.A.; De Vries, L.S.; Benders, M.J.N.L. Automatic Segmentation of Eight Tissue Classes in Neonatal Brain MRI. *PLoS ONE* **2013**, *8*, e81895. [CrossRef]

30. Artaechevarria, X.; Munoz-Barrutia, A.; Ortiz-De-Solorzano, C. Combination Strategies in Multi-Atlas Image Segmentation: Application to Brain MR Data. *IEEE Trans. Med. Imaging* **2009**, *28*, 1266–1277. [CrossRef] [PubMed]

31. Wang, H.; Suh, J.W.; Das, S.R.; Pluta, J.B.; Craige, C.; Yushkevich, P.A. Multi-Atlas Segmentation with Joint Label Fusion. *IEEE Trans. Pattern Anal. Mach. Intell.* **2013**, *35*, 611–623. [CrossRef] [PubMed]

32. Wang, L.; Gao, Y.; Shi, F.; Li, G.; Gilmore, J.H.; Lin, W.; Shen, D. LINKS: Learning-based multi-source IntegratioN frameworK for Segmen-tation of infant brain images. *Neuroimage* **2015**, *108*, 160–172. [CrossRef] [PubMed]

33. Zhang, Y.; Brady, M.; Smith, S. Segmentation of brain MR images through a hidden Markov random field model and the expectation-maximization algorithm. *IEEE Trans. Med. Imaging* **2001**, *20*, 45–57. [CrossRef]

34. Mishro, P.K.; Agrawal, S.; Panda, R.; Abraham, A. A Novel Type-2 Fuzzy C-Means Clustering for Brain MR Image Segmentation. *IEEE Trans. Cybern.* **2020**, 1–12. [CrossRef]

35. Zhang, W.; Li, R.; Deng, H.; Wang, L.; Lin, W.; Ji, S.; Shen, D. Deep convolutional neural networks for multi-modality isointense infant brain image segmentation. *Neuroimage* **2015**, *108*, 214–224. [CrossRef]

36. Moeskops, P.; Viergever, M.A.; Mendrik, A.M.; De Vries, L.S.; Benders, M.J.N.L.; Isgum, I. Automatic Segmentation of MR Brain Images With a Convolutional Neural Network. *IEEE Trans. Med. Imaging* **2016**, *35*, 1252–1261. [CrossRef] [PubMed]

37. Bao, S.; Chung, A.C. Multi-scale structured CNN with label consistency for brain MR image seg-mentation. *Comput. Methods Biomech. Biomed. Eng. Imaging Vis.* **2018**, *6*, 113–117. [CrossRef]

38. Nie, D.; Wang, L.; Gao, Y.; Shen, D. Fully convolutional networks for multi-modality isointense infant brain image segmentation. In Proceedings of the 2016 IEEE 13th International Symposium on Biomedical Imaging (ISBI), Prague, Czech Republic, 13–16 April 2016; pp. 1342–1345.

39. Xu, Y.; Géraud, T.; Bloch, I. From neonatal to adult brain MR image segmentation in a few seconds using 3D-like fully convolutional network and transfer learning. In Proceedings of the 2017 IEEE International Conference on Image Processing (ICIP), Beijing, China, 17–20 September 2017; pp. 4417–4421.

40. Chen, H.; Dou, Q.; Yu, L.; Qin, J.; Heng, P.A. VoxResNet: Deep voxelwise residual networks for brain seg-mentation from 3D MR images. *Neuroimage* **2017**, *170*, 446–455. [CrossRef] [PubMed]

41. Dolz, J.; Gopinath, K.; Yuan, J.; Lombaert, H.; Desrosiers, C.; Ayed, I.B. HyperDense-Net: A Hyper-Densely Connected CNN for Multi-Modal Image Segmentation. *IEEE Trans. Med. Imaging* **2019**, *38*, 1116–1126. [CrossRef] [PubMed]

42. Li, J.; Yu, Z.L.; Gu, Z.; Liu, H.; Li, Y. MMAN: Multi-modality aggregation network for brain segmentation from MR images. *Neurocomputing* **2019**, *358*, 10–19. [CrossRef]

43. Chen, L.; Bentley, P.; Mori, K.; Misawa, K.; Fujiwara, M.; Rueckert, D. DRINet for Medical Image Segmentation. *IEEE Trans. Med. Imaging* **2018**, *37*, 2453–2462. [CrossRef]

44. Lei, Z.; Qi, L.; Wei, Y.; Zhou, Y.; Zhang, Y. Infant Brain MRI Segmentation with Dilated Convolution Pyra-Mid Downsampling and Self-Attention. *arXiv* **2019**, arXiv:1912.12570.

45. Yu, L.; Cheng, J.-Z.; Dou, Q.; Yang, X.; Chen, H.; Qin, J.; Heng, P.-A. Automatic 3D Cardiovascular MR Segmentation with Densely-Connected Volumetric ConvNets. In Proceedings of the International Conference on Medical Image Computing and Computer-Assisted Intervention, Quebec City, QC, Canada, 10–14 September 2017; Lecture Notes in Computer Science. Springer: Cham, Switzerland, 2017; Volume 2017, pp. 287–295.

46. Bui, T.D.; Shin, J.; Moon, T. 3D Densely Convolution Networks for Volumetric Segmentation. *arXiv* **2017**, arXiv:1709.03199v2.

47. Dolz, J.; Desrosiers, C.; Wang, L.; Yuan, J.; Shen, D.; Ben Ayed, I. Deep CNN ensembles and suggestive annotations for infant brain MRI segmentation. *Comput. Med. Imaging Graph.* **2020**, *79*, 101660. [CrossRef]

48. Sun, L.; Ma, W.; Ding, X.; Huang, Y.; Liang, D.; Paisley, J. A 3D Spatially Weighted Network for Segmentation of Brain Tissue from MRI. *IEEE Trans. Med. Imaging* **2019**, *39*, 898–909. [CrossRef] [PubMed]

49. Maas, A.L.; Hannun, A.Y.; Ng, A.Y. Rectifier nonlinearities improve neural network acoustic models. In Proceedings of the 30th International Conference on Ma-Chine Learning, Atlanta, GA, USA, 16–21 June 2013.

50. Çiçek, Ö.; Abdulkadir, A.; Lienkamp, S.S.; Brox, T.; Ronneberger, O. 3D U-Net: Learning Dense Volumetric Segmentation from Sparse Annotation. In Proceedings of the Medical Image Computing and Computer-Assisted Intervention—MICCAI 2016, Athens, Greece, 17–21 October 2016; Lecture Notes in Computer Science. Ourselin, S., Joskowicz, L., Sabuncu, M., Unal, G., Wells, W., Eds.; Springer: Cham, Switzerland, 2016; Volume 9901. [CrossRef]

51. Mendrik, A.M.; Vincken, K.L.; Kuijf, H.J.; Breeuwer, M.; Bouvy, W.H.; de Bresser, J.; Alansary, A.; de Bruijne, M.; Carass, A.; El-Baz, A.; et al. MRBrainS challenge: Online Evaluation Framework for Brain Image Segmentation in 3T MRI Scans. *Comput. Intell. Neuroence* **2015**, *2015*, 813696. [CrossRef]

52. Rohlfing, T.; Brandt, R.; Menzel, R.; Maurer, C.R. Evaluation of atlas selection strategies for atlas-based image segmentation with application to confocal microscopy images of bee brains. *Neuroimage* **2004**, *21*, 1428–1442. [CrossRef] [PubMed]

53. Wang, L.; Nie, D.; Li, G.; Puybareau, É.; Dolz, J.; Zhang, Q.; Wang, F.; Xia, J.; Wu, Z.; Chen, J.; et al. Benchmark on Automatic Six-Month-Old Infant Brain Segmentation Algo-rithms: The iSeg-2017 Challenge. *IEEE Trans. Med. Imaging* **2019**, *38*, 2219–2230. [CrossRef] [PubMed]

54. Rohlfing, T. Image Similarity and Tissue Overlaps as Surrogates for Image Registration Accuracy: Widely Used but Unreliable. *IEEE Trans. Med. Imaging* **2012**, *31*, 153–163. [CrossRef] [PubMed]

55. Andermatt, S.; Pezold, S.; Cattin, P. Multi-Dimensional Gated Recurrent Units for the Segmentation of Biomedical 3D-Data. In *Deep Learning and Data Labeling for Medical Applications*; Springer: Berlin, Germany, 2016; pp. 142–151.

56. Stollenga, M.F.; Byeon, W.; Liwicki, M.; Schmidhuber, J. Parallel Multi-Dimensional LSTM, with Appli-Cation to Fast Biomedical Volumetric Image Segmentation. *arXiv* **2015**, arXiv:1506.07452.

*Article*

# ResBCDU-Net: A Deep Learning Framework for Lung CT Image Segmentation

Yeganeh Jalali [1], Mansoor Fateh [1,*], Mohsen Rezvani [1], Vahid Abolghasemi [2,*] and Mohammad Hossein Anisi [2]

[1]  Faculty of Computer Engineering, Shahrood University of Technology, Shahrood 3619995161, Iran; jalali.yegane@gmail.com (Y.J.); mrezvani@shahroodut.ac.ir (M.R.)

[2]  School of Computer Science and Electronic Engineering, University of Essex, Colchester CO4 3SQ, UK; m.anisi@essex.ac.uk

*  Correspondence: Mansoor_fateh@shahroodut.ac.ir (M.F.); v.abolghasemi@essex.ac.uk (V.A.)

**Abstract:** Lung CT image segmentation is a key process in many applications such as lung cancer detection. It is considered a challenging problem due to existing similar image densities in the pulmonary structures, different types of scanners, and scanning protocols. Most of the current semi-automatic segmentation methods rely on human factors therefore it might suffer from lack of accuracy. Another shortcoming of these methods is their high false-positive rate. In recent years, several approaches, based on a deep learning framework, have been effectively applied in medical image segmentation. Among existing deep neural networks, the U-Net has provided great success in this field. In this paper, we propose a deep neural network architecture to perform an automatic lung CT image segmentation process. In the proposed method, several extensive preprocessing techniques are applied to raw CT images. Then, ground truths corresponding to these images are extracted via some morphological operations and manual reforms. Finally, all the prepared images with the corresponding ground truth are fed into a modified U-Net in which the encoder is replaced with a pre-trained ResNet-34 network (referred to as Res BCDU-Net). In the architecture, we employ BConvLSTM (Bidirectional Convolutional Long Short-term Memory)as an advanced integrator module instead of simple traditional concatenators. This is to merge the extracted feature maps of the corresponding contracting path into the previous expansion of the up-convolutional layer. Finally, a densely connected convolutional layer is utilized for the contracting path. The results of our extensive experiments on lung CT images (LIDC-IDRI database) confirm the effectiveness of the proposed method where a dice coefficient index of 97.31% is achieved.

**Keywords:** segmentation; lung; CT image; U-Net; ResNet-34; BConvLSTM

**Citation:** Jalali, Y.; Fateh, M.; Rezvani, M.; Abolghasemi, V.; Anisi, M.H. ResBCDU-Net: A Deep Learning Framework for Lung CT Image Segmentation. *Sensors* **2021**, *21*, 268. https://doi.org/10.3390/s21010268

Received: 1 December 2020
Accepted: 29 December 2020
Published: 3 January 2021



## 1. Introduction

Lung cancer is known as the second most prevalent type of cancers in both genders in the world [1]. According to the World Health Organization (WHO), lung cancer is responsible for 1.3 million deaths per year in the world [2]. It is estimated that around 228,820 new lung cancer cases (116,300 in men and 112,520 in women) and around 135,720 deaths from this disease (72,500 in men and 63,220 in women) are identified in the United States each year [3]. Lung cancer is known as a malignant tumor characterized by the unnatural growth of the cell in the lung tissue. Rapid diagnosis of this cancer can significantly decrease the death rate and enhance patient survival chances. This is very important in improving the clinical situation of patients. Thus, it is necessary to present an intelligent algorithm for the early diagnosing of lung cancer.

Recent advances in computer vision and image processing technologies have significantly helped the healthcare systems particularly in the analysis of medical images. In this regard, image segmentation is widely used as one of the most fundamental, useful, and well-studied topics in image analysis. Image segmentation can significantly improve the recognizability of parts of an image by assigning a label to each pixel in the image such that those pixels with the same labels have similar visual features characteristics.

Segmentation is a substantial process in medical image processing and can reveal very useful information concealed in the images. In some medical applications, the classification of image pixels into descriptive regions, such as bones and blood vessels, is of interest. While in other applications it is more appropriate to look for pathological regions, such as cancer or tissue deformities [4]. One of the most important segmentation tasks in medical images is to identify redundant pixels or unwanted regions located as background. This segmentation is considered as one of the most challenging steps, especially in CT (computed tomography) or MRI (magnetic resonance imaging), to provide critical information about the shapes and volume of body organs. In other words, the overall performance of automated cancer detection is highly dependent on the output of the segmentation stage [5].

In the lung segmentation stage, we seek to distinguish those pixels associated with the lung from every other pixel in the surrounding anatomy. Radiologists often use a CAD (computer-aided design) system to provide a secondary consideration for an accurate diagnosis. This method is useful for improving the efficacy of the cure. For many CAD systems, a precise segmentation process of the target organ is required, which is a fundamental step and a prerequisite for effective image analysis. The segmentation of lung fields is particularly challenging because the lung zone is highly inhomogeneous. In addition, pulmonary structures present similar congestions in different scanners and scanning protocols which make the segmentation difficult. It becomes even more challenging because of the presence of nodules attached closely to the lung wall. Figure 1 offers two examples of lung CT scans that show the exact location of the node attached to the lung wall. This figure also clearly represents the challenge of dividing the lungs despite these nodules.

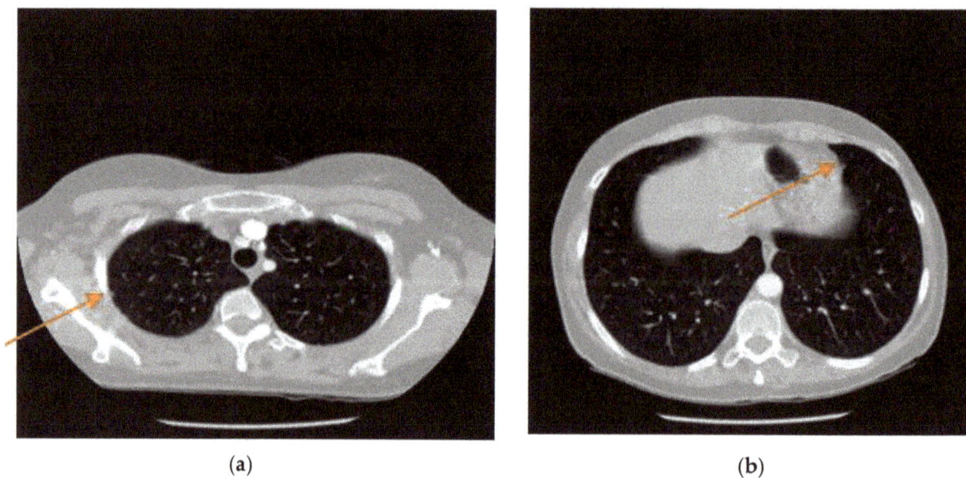

(a)        (b)

**Figure 1.** Two examples of nodules attached to the lung wall in CT-scan images. (**a**). represents one nodule attaching to the outer wall of the lung, (**b**). represents one nodule attaching to the outer wall of the lung (orange arrows).

Medical image segmentation is an important and inseparable step in the diagnosis process. For example, in the process of diagnosing lung cancer, the main steps are as follows: (1) image pre-processing; (2) image segmentation; (3) feature extraction; (4) lung cancer identification; (5) diagnosis of the disease [6,7]. It so happens that various algorithms directly use the segmentation step in their work [8–10]. For example, Wang et al. [10] conducted a study on differentiating COVID-19 from non-COVID-19 CT scans. In their proposed method, images of patients were first segmented during a single step using a deep neural network. Then, the images and tags were given to a network for classification. They could achieve a 0.959 ROC AUC score. Unlike the previous example, some methods extract the region of interest and do segmentation indirectly within the feature extraction stage [11–13]. For example, Pathak et al. [13] proposed a system for the detection of COVID-19 in CT scans that considered a prepressed transfer learning. The system used a neural network to extract the features from CT images, and a 2D convolutional neural network was considered for the classification. The proposed system was tested on 413 COVID-19 and 439 non-COVID19 images with 10-fold cross-validation, and it achieved 93.01% accuracy.

It is clear that medical image segmentation is always accompanied by disease detection algorithms. However, algorithms that specifically try to segment with high accuracy will ultimately perform better for the diagnostic model. For this reason, we will also present a robust system for accurate segmentation of the lung area in this article.

Generally, many techniques have been reported in the literature for the segmentation of medical images. The most important drawback of the existing methods is relying on the utilization of manual (hand-crafted) features to successfully segment the regions of interest. In addition, most techniques are unable to segment nodules attached to the lung wall. Recent advances in medical image processing by using deep learning-based methods have revealed great influences in clinical applications. These methods can appropriately learn important features of medical images and consequently overcome the limitation of hand-crafted features [14]. In this paper, we propose a deep learning-based method to accurately segment the lung tissue. In order to achieve a successful segmentation, we require the raw CT images with their associated ground truths. Unfortunately, current lung CT databases do not come with binary masks (ground truths). Hence, we propose a semi-automatic method to resolve this issue by producing the corresponding masks. Then, we apply appropriate pre-processing steps in order to enhance the quality of images used in the training phase. In the last phase, all these pre-processed images with corresponding binary masks are fed into a deep neural network. Our proposed deep model is a combination of the ResNet and BCDU-Net. In fact, the backbone and the basis of the deep learning network used in this paper are BCDU-Net. On the other hand, using pre-trained networks such as ResNet, which have been trained in the ImageNet data collection, increases the speed of training and the power of the network extension. So, the proposed method in this paper is a novel BCDU-Net architecture that takes the advantage of ResNet-34 instead of ordinary convolution layers in the encoding section.

The contributions of the current manuscript are:

- Applying novel extensive preprocessing techniques to improve quality of the raw images.
- Proposing a new method for extracting ground truths corresponding to the input images.
- Employing a new deep learning-based algorithm for proper segmentation of lungs.

The rest of this paper is organized as follows: Section 2 reviews some previous segmentation models. Section 3 introduces the proposed method in detail. Section 4 is devoted to evaluating the performance of our method through extensive experiments. Section 5 draws some conclusions. Section 6 highlights future works.

## 2. Related Works

There are several techniques that have been developed to address the segmentation task. Most of these approaches are mainly divided into five categories: threshold-based, edge-detection, region growing, deformable boundary, and learning-based methods. In what follows, we briefly review these categories.

### 2.1. Threshold-Based Methods

Since the lungs are filled with air during the CT scan, they are characterized by dark areas in the associated grayscale image. Therefore, threshold-based approaches rely on this principle that normal lung tissues have less density than the surrounding regions. On this basis, the lung regions are separated by specifying a suitable threshold on the images [15]. These approaches are of the most popular lung segmentation methods because of their simplicity in performance and computation. They can also be used in real-time applications.

However, these methods have some deficiencies in lung segmentation. (1) They are not able to effectively remove the trachea and main stem bronchi [16]. (2) Due to various conditions in different images like air volume and image acquisition protocol, a universal gray-level segmentation threshold would not be suitable [17]. (3) They are not often successful in cases where anomalies represent higher densities compared to those in natural lung tissues [18].

### 2.2. Edge-Detection Methods

Lung segmentation can be also performed by using edge detection techniques. Edge in image processing is defined as the boundary between the two regions with relatively distinct gray surface properties. Some of the well-established spatial edge detection techniques are Prewitt, Robert, Sobel, Prewitt, Laplacian, and Canny. In what follows, we refer to canny as the most effective edge detector algorithm.

Canny is a well-known conventional edge detection algorithm. It can find the edges of image regions by isolating noise from the image. The main advantage of this method is that it does not affect the properties of the image edges and find edges and critical thresholds. Canny is capable of achieving three important properties, i.e., great localization of edge points, small error rate, and one-to-one responses to every single edge. As a result, it normally performs well, thus, it is considered as one of the best methods to extract the edges compared to other existing methods [19]. Shin et al. [20] demonstrated the performance evaluation of different edge detectors and concluded that the Canny detector has the best performance and robustness compared to other edge detectors. In this regard, Campadelli et al. [21] detected edges from chest radiograph images and achieved an accuracy of 94.37%. Mendonca et al. [22] identified the image edges using a spatial detector for lung tissue segmentation in radiograph images. They used 47 radiograph images and achieved a sensitivity of 0.9225 and a positive predictive value of 0.968.

In brief, the benefits of edge-based methods are (1) performing well in discriminating between the background and the objects within an image, (2) high-level approach in image segmentation similar to the way human perception segments the images. The main deficiencies of these methods are: (1) sensitivity to noise, (2) working inappropriately on images with smooth transitions and low contrast.

### 2.3. Region Growing Methods

Segmentations based on image regions are called region growing techniques. The basic idea in this method is to collect pixels posing similar characteristics within a commonly formed area. In another word, this category of methods starts the segmentation process with a set of seeds. The seeds in any given image, can either be one single pixel or a group of several pixels. After forming the seeds, the next step is to determine whether the neighboring pixels must be added to the region or not. This is decided based on similarity criteria such as color, intensity, variance, texture, and motion. Gradually, these pixels begin to grow and form regions. Finally, when the image is completely divided by all the growing regions and all the textural stages of the image are obtained as the boundaries of the final regions, the algorithm is terminated. Region growing methods are utilized in many medical applications such as cavities segmentation in the cardiac images [23], blood vessel extraction in the angiographic data [24], renal segmentation [25], brain surface extraction [26], and lung CT image segmentation [27].

Region growing technique has some advantages including low computational complexity and high speed. However, its performance is highly dependent on the location of the seed points and the growing conditions. It can be stated that region growing methods are sensitive to noise or variation of intensity. This could result in holes or over-segmentation and also dependency performance on its initial seeds. Its particular disadvantage in lung CT images is that it cannot segment the nodules attached to the borders of the lung image [13].

### 2.4. Deformable Boundary Models

These models consider the entire object's boundary and can incorporate prior knowledge about the object's shape as a constraint toward a precise segmentation outcome. For example, in lung segmentation, the boundary of the lung is determined by the evolution of particular interior and exterior forces to fit the shape of the lung. Therefore, the parametric representations used in these models can provide a concise and analytical description of the lung. The most popular approach in deformable models is an active contour model or snake [28]. Itai et al. [29] segmented the lung region from a CT image using a 2D parametric deformable model, called the SNAKES algorithm, without considering any manual operations. Shi et al. [30] proposed an extraction technique for the lung region by using a new deformable model through radiograph images.

Also, there exist some active contour models with many privileges such as providing smooth and closed segmented contours and obtaining sub-pixel details of the object's boundaries [14]. However, one of the limitations of these models is that they often require human interaction within the construction of the initial contour. Therefore, they normally perform poorly in non-interactive applications, as the algorithm cannot be initialized close to the desired structure of interest. Another limitation of the SNAKE model is that they have weak convergence in the face of boundary concavities.

### 2.5. Learning-Based Models

Learning-based approaches are presented in the area of segmentation of medical images as well. In traditional learning-based methods, the segmentation process is addressed as engineered features. Pixel classification-based approach [31] is known as one of the most important categories in these techniques. However, it is very challenging to select sub-pixels and extract some features to train the classification of a greater number of pixels. To overcome this problem a super pixel learning-based method have used in [32] to prune the pixels and merge them with the confined regions of shape constraints to segment lung CT images. Generally, these methods have two shortcomings to extract the features. The first drawback is relying on using hand-crafted features to achieve the segmentation results. Another limitation is that designing the representative features for different applications is very difficult.

Segmentation techniques based on deep learning can be ranked as pixel-based learning techniques for classification. Unlike conventional pixel or super-pixel classification methods, which often use hand-crafted features, deep learning approaches can process natural data in its raw form as well as learning features and overcoming the limitations of hand-crafted features [19]. These approaches have predominately utilized for semantic segmentation of natural image scenes and have also found many applications in biomedical image segmentation tasks. They also contributed to decrease the manual manipulations needed for segmentation and improving the accuracy and speed of segmentation. One of the most important recent applications of segmentation is to accurately quantify the COVID-19 virus effects. In [33], a new deep-learning-based method is used for automatic screening of COVID-19 with limited samples in order to complete the screening of COVID-19 and prevent further spread of the virus.

Previous deep learning methods purposed for medical image segmentation are mostly based on the patches of images. Convolutional neural network (CNN) is the most successful and widely used approach among many deep learning architectures community for medical image analysis [34]. It is easy to use CNN to classify each pixel in the image separately by offering the extracted neighboring regions of a particular pixel. For example, the authors in [35] proposed a method based on light patches and sliding windows neuronal membranes segmentation in microscopic images. This method has two deficiencies: redundant computation caused from sliding window and huge overlap within input patches from neighbor pixels.

To overcome these problems, the use of a fully convolutional network (FCN) was introduced by Long et al. [36] in which the last fully connected layers of the CNN replaced by transpose convolutional layers. With emerging of the end-to-end FCN, Ronneberger et al. [37], using the idea of the FCN, proposed U-shape Net (U-Net) framework for biomedical image segmentation. U-Net is one of the most popular FCNs for segmentation of medical images. U-Net configuration (Figure 2) comprises two paths; a contracting path to capture context and a symmetric expanding path to obtain accurate localization. The contraction path includes consecutive convolutional layers and maxpooling layer. It is used to extract attributes while constraining the attributes map size. The expansion path achieves up-conversion and has the convolution layers to retrieve the size of the feature maps with the loss of localization knowledge. Also, the localization information is shared from the contraction layer to the expansion layer by applying skip connections. These connections are utilized in parallel and allows data to be transmitted directly from a network block to another with no extra computational cost. Ultimately, the convolution layer draws the attribute vector to the number of classes required at the final partitioning output. The U-Net model has some advantages compared to other patch-based segmentation approaches [38]: (1) It works well with very few training data. (2) It can utilize the global location and context information simultaneously. (3) It ensures maintenance of the complete texture of the input images.

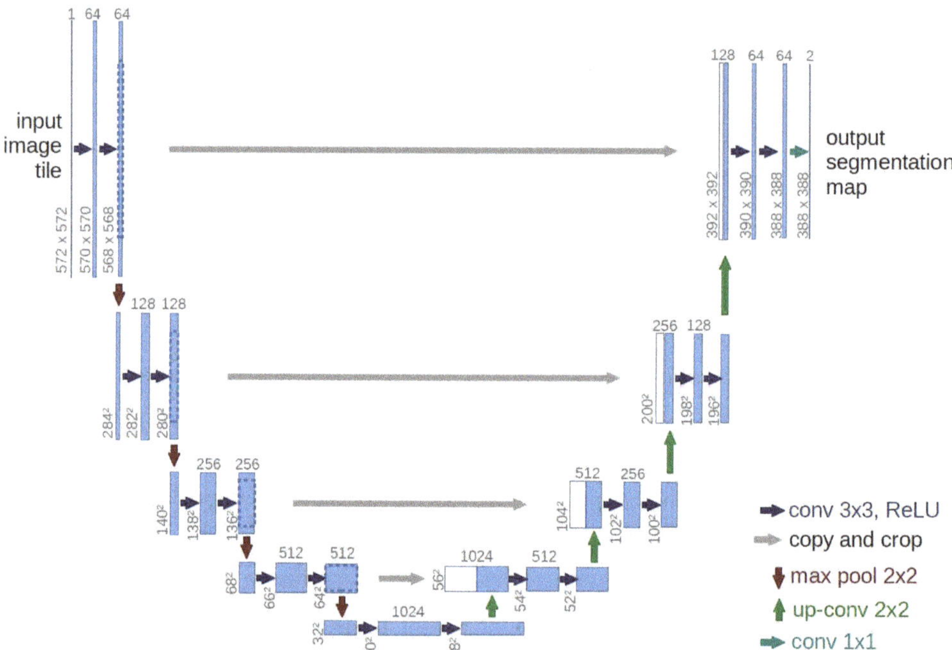


**Figure 2.** The U-Net architecture [37]. In the contraction path of this network, feature channels are doubled in each down-sampling. Conversely, the expansion path is responsible for decreasing feature channels. The skip connections are also displayed with gray arrows drawn to incorporate two feature maps.

U-Net has offered state-of-the-art performance in biomedical image segmentation. In recent years, different extensions of U-Net have been proposed [39–43]. For example, Milletari et al. [39] proposed V-Net as an extension of U-Net for 3D medical image segmentation. Furthermore, in an extended paper, Cicek [40] proposed a U-Net architecture for 3D images. Zhou et al. [41] developed a nested U-Net architecture. Other researchers have developed various extensions of the U-net. The most significant changes in these methods are mainly related to the skip connections. For example, in Attention U-Net [42], the extracted features at the skip connection are transferred to a processing stage first, and then they are concatenated to each other. One of the limitations of these networks is their two-stage process, i.e., first applying separate processing steps to each group feature map and then concatenating the feature maps together. In [43], a residual attention U-Net was proposed for automated segmentation of COVID-19 Chest CT images. This deep learning model is based on U-Net which uses the residual network and attention mechanism to enhance feature extraction and generate high-quality multi-class segmentation results. The use of this method has led to 10% improvement in the segmentation performance.

In order to improve the original U-Net network, instead of using the desired convolution layers, various other architectures can be used in the encoding part of this network. For example, a U-Net-based network is presented in [44] wherein the ResNet34 pre-training model is used in its contraction path (left U). The greatest advantage of this modification is increasing the speed of training and the power of the network extension.

In another work, U-Net has been extended to a network called BCDU-Net [45] and achieved better performance than modern alternatives for medical image segmentation. In this network, the encoding path includes four stages. Each stage is composed of two $3 \times 3$ convolutional filters on the image. After each convolution filter, there is a $2 \times 2$ max-pooling and a RELU activator. These three layers together form a down-sampling process. In each down-sampling, feature channels are doubled. The encoding path gradually extracts the representation of images and increases the dimensions of the representation layer by layer. This network offers two contributions. First, it uses densely connected convolutions to prevent the learning redundant features problem in successive convolutions in the last encoding path layer of general U-Net. Second, batch normalization is utilized in the decoding path after each up-sampling stage. Batch normalization helps to improve the performance, speed, and stability of neural networks. The resulting output from the batch normalization function is given to a bidirectional convolutional LSTM [46] (BConvLSTM). The feature maps are processed with BConvLSTM to integrate in a more complex way than simple concatenation in U-Net. BConvLSTM itself applies two ConvLSTMs on the input data in both forward and backward directions and then determines the data dependencies in both directions.

According to the above discussions and also the pre-trained ResNet framework [47] that makes the neural network wider, deeper, and faster, we propose an architecture that is mainly inspired by BCDU-Net and ResNet34 to automatically segment the lung CT images. In the next section, the proposed model will be described and presented with all the required details.

### 3. Proposed Method

The proposed model encompasses three major steps: (1) ground truth extraction, (2) image pre-processing and data preparation, and (3) deep learning-based segmentation. Moreover, our novel deep learning model is composed of BCDU-Net and ResNet34. The block-diagram of different steps of the proposed method is depicted in Figure 3. In what follows, we first introduce the database used in this study followed by a description of the process of semi-automatically re-producing database images. Then, we provide pre-processing operations to prepare data stepwise. Finally, we describe the method based on deep learning to segment these images and the corresponding masks.

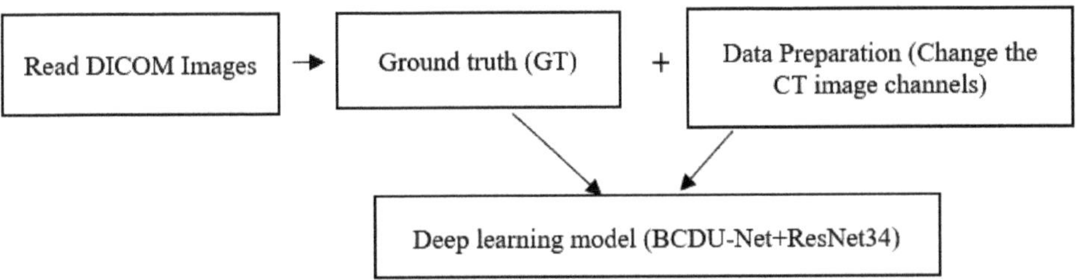


**Figure 3.** The pipeline of the proposed method.

### 3.1. DICOM Images Reading

In this paper, we used the LIDC-IDRI dataset which involves lung cancer CT scans with marked-up annotated lesions as well as diagnostic information [48]. It is an internationally available resource of development, training, and assessment of diagnostic methods used by the computer (CAD) to diagnose lung cancer. All CT scans are in DICOM format and measured in HU and they have three channels and a resolution of 512 × 512. The original DICOM images and their corresponding XML files are related to 1018 CT scans of 1010 patients registered in this data collection. These images consist of a chest CT scan and an XML file annotated by four professional medical experts. The first step is to read and import these DICOM images.

### 3.2. Ground Truth (GT) Extraction

Our deep learning architecture requires both input images and their corresponding ground truth for successful segmentation. This database lacks labels for lung images, thus, we need to manually extract every ground truth for CT images. Ground truth is in form of masks that could be used to extract ROI from images to be then fed to the deep learning model. Because the ground truth plays a vital role in the segmentation process, custom masks were created using a semi-automatic technique so that they could be verified to be 'correct'.

In the CT scans, the lungs are declared as dark zones, while lighter areas inside the lungs are considered to be blood vessels or air. The purpose of this step is to extract lung regions as accurately as possible from each CT scans slice. This step should be performed with extra care to avoid missing any region of interest particularly those attached to the lung wall. Seven steps are carried out to get the masked lungs. These are as follows [22]:

1. Conversion to binary image: In the first step, slices of DICOM images are converted into binary using the threshold method represented by Equation (1). A threshold of -604 HU was applied to extract lung parenchyma [23]. The transformed image to binary is shown in Figure 4b.

$$Binary\ (i,j) = \begin{cases} 1 & \text{if} \quad f(i,j) < T \\ 0 & \text{otherwise} \end{cases} \quad , \quad T = 604 \tag{1}$$

2. Removing the blobs connected to the CT image border: To classify the images correctly, the regions connected to the image border are removed, as shown in Figure 4c.

3. Labelling the image: Pixel neighbourhoods with the same intensity level can consider being a connected region. When this process is applied to the entire image some connected regions are formed. Figure 4a shows connected regions of integer array of the images that are labelled.

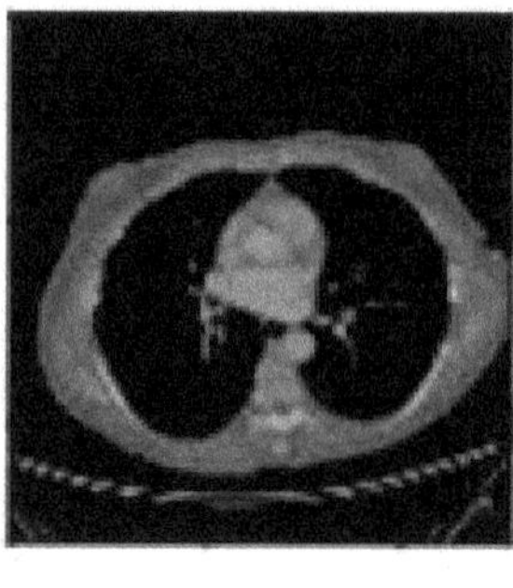 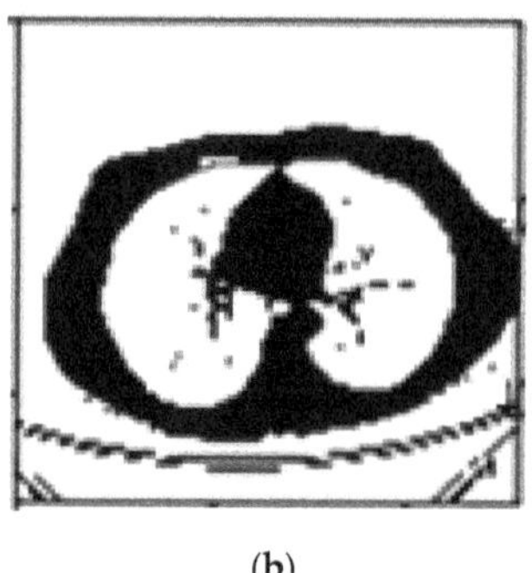 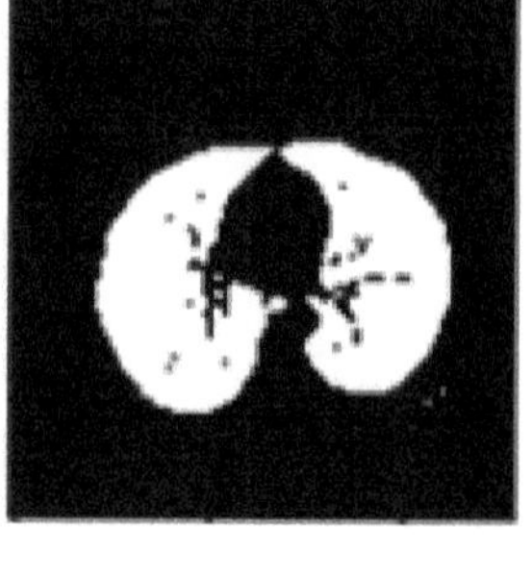

(a)      (b)      (c)

**Figure 4.** (a). Main CT image, (b). Binary image, (c). Image after eliminating border blobs.

4. Keeping the labels with two largest areas: As shown in Figure 5b, labels with the two largest areas (both lungs) are kept whereas the tissues with areas less than the expected lungs are removed.

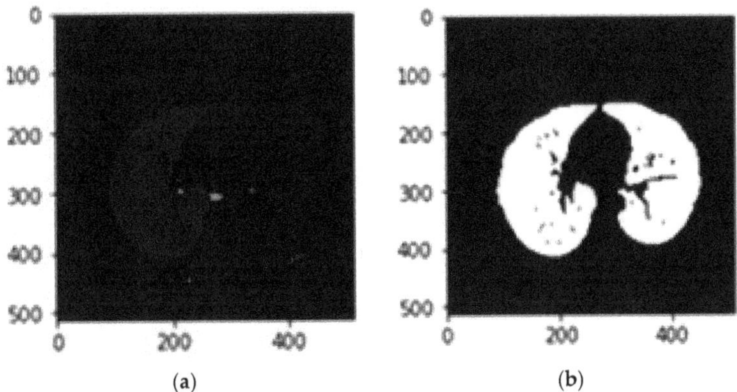

(a)  (b)

**Figure 5.** (a). Labeled image, (b). Image with the two largest labeled areas kept.

5. Applying erosion operation (with a disk of radius 2): This operation is applied on the image at this step to separate the pulmonary nodules attached to the lung wall from the blood vessels. The erosion operator reduces the bright areas of the image and makes the dark areas appear larger as shown in Figure 6a.
6. Applying closure operation (with a disk of radius 10) [15]: The aim of using this operator is to maintain the nodules connected to the lung wall. This operator can remove small dark spots from the image and connect small bright gaps. The image obtained by applying this operator is shown in Figure 6b.
7. Filling in the small holes within binary mask: In some cases, due to a breach in binary conversion using thresholding, a series of black pixels belong to the background appear in the binary image. These areas, known as holes, may be helpful. Therefore, we must obtain these areas by filling them as shown in Figure 6c.

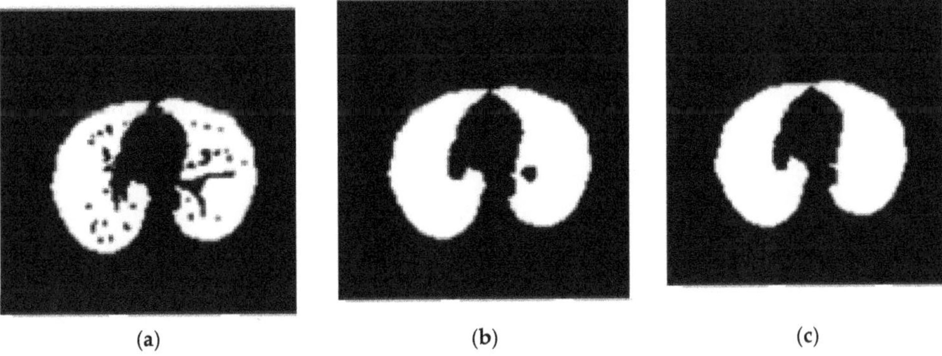

(a)  (b)  (c)

**Figure 6.** Results of applying (a). Erosion operation, (b). Closure operation, (c). Filling small holes (binary mask).

In the final step, binary masks are produced which are stored in '.bmp' format. The proposed steps sometimes fail and do not produce the correct binary mask due to two main reasons: (1) all the above steps may cause partial tissues, which could involve lung components, to be ignored in CT scan; (2) sometimes a closure operation, which connects small bright cracks, causes connection of two pixels that fill the non-pulmonary tissue, e.g., air instead of the lung. Figure 7 shows 2 samples of these problems.

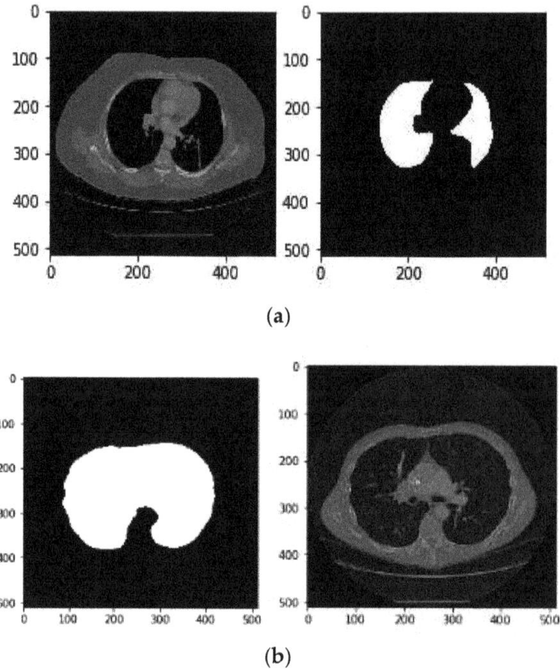

**Figure 7.** (**a**). Sample of missing a part of the lung in the generated mask, due to considering only the two largest areas, (**b**). Sample of misplaced pixels connecting that fills the non-pulmonary space with white pixels.

Motivated by the above discussions, we need to provide a manual segmentation after producing binary masks by the mentioned algorithm, if necessary. We extracted 1714 binary masks for 10 patients (averagely 170 samples for each patient) using this semi-automatic method. It takes hours to label each CT image by experts, while production of each mask takes on average around three minutes in our proposed method, considering the worst conditions and the need for manual reform. Therefore, the main advantage of this method is to save a lot of time. Also, we plan to publicize our produced masks soon to help other researchers using them in future researches.

### 3.3. Data Preparation

Following the GT extraction described above, we now aim to prepare input raw images to improve the training process of the deep learning network by applying a few preprocessing steps. Therefore, we use two stages including edge detection functions and dilation morphological operations.

According to the description of the LIDC-IDRI database in previous sections, all CT scans have $512 \times 512$ resolution and three channels. In this stage of the proposed method, we want to improve the overall segmentation performance. It seems that if we increase the focus of the network during training on a series of specific image features, it will help to improve the forecast. In this regard, we have changed the channels of each image. To do this, we convert these default channels for each CT image to three newly designed channels as follows. In this regard, we use several preprocessing operations such as edge detection functions and dilation morphological operations to generate new images. Then, these images are fed to our proposed network. The main advantage of this idea is that if these newly generated images are fed to a deep neural network, its training can be faster and more accurate. In other words, the proposed channels can provide focused information for the deep neural network which are compatible with the associated masks. This leads to more efficient training and ultimately reduction of false-positive measures. Details of the proposed image conversion are as follows:

(a)    Image binarization: In this process, a binary image is created with two values on the grey surface, i.e., black and white. The lung region poses a black colour with the value zero. Figure 8 shows the binarization process of a CT image.

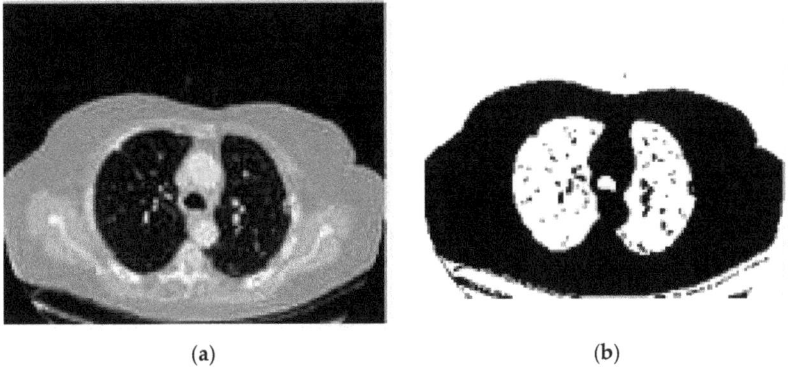

(**a**)                                        (**b**)

**Figure 8.** Image binarization process. (**a**). Original CT, (**b**). binarized.

(b)    Dilation morphological operation: Morphological operations, typically applied to binary images, are used to extract and describe the geometry of the object in the image [49,50]. As a result of the binarization process described before, there would still be remaining regions of white colour around the lungs regarded as unwanted noise. Thus, morphological operations can be used to remove these regions. Moreover, there could still be some small black holes in the lung's region, suspicious of noise caused by the binarization process. These holes should be also removed using morphological operations.

The morphological operation involves two basic operators: dilation and erosion. Dilation [51] is applied when the segmented object loses part of its target area. This operator increases the target area of the segmentation. It also increases the sensitivity but decreases the specificity. The dilation operation can be mathematically represented as Equation (2).

$$A \oplus B = \bigcup_{x \in B} A_x \tag{2}$$

where A is the image and B is the structuring element. In fact, Equation (1) means that the matrix A is transmitted by each of the points B and then the assembly of all the transferred matrices is calculated. We applied a dilation operation to remove redundant white regions around the lung and small black gaps inside it. Figure 9 shows the result of the dilation process. As can be seen, the orange arrow section (noise) in a binary image is removed in the dilation result.

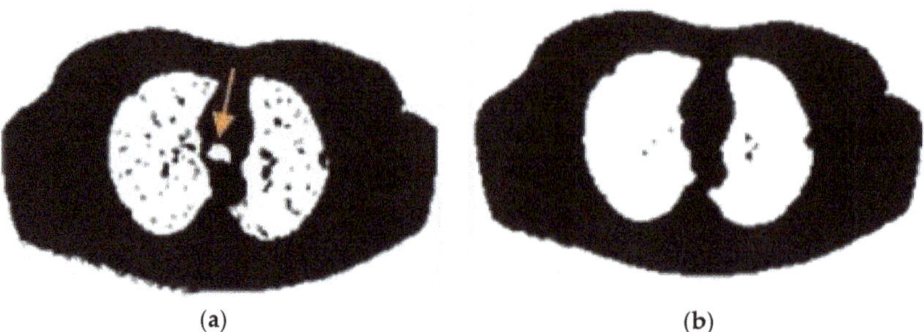

(a)                                                             (b)

**Figure 9.** Image after (**a**). binarization; (**b**). dilation.

(c)   Edge detection: As already stated, the edge detection filter determines the vertices of an object and the boundaries between objects and the background in the image. This process can also be used to improve the image and eliminate blur. An important advantage of the Canny technique is that it tries to remove the noise of an image before edge extraction and then applies the tendency to find the edges and the critical value of the threshold. Motivated by the advantages expressed so far, we also applied the Canny method to detect the edges in the source images. Figure 10 shows the result of the edge detection process.

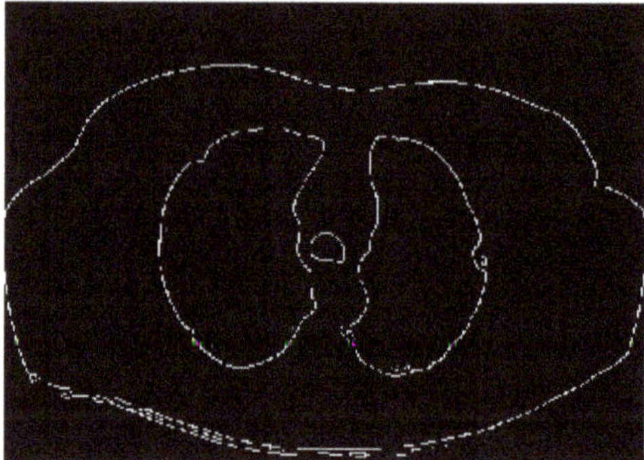

**Figure 10.** Edge detection using Canny.

As a result, it cuts down the data quantity and removes unwanted parts, while preserving the required structural features in the image. Next, we need to generate new images with proposed filled channels. The first image channel is filled with the original image (Figure 11a). The second channel of the output image would be an image containing an edge detection process (Figure 11b). In the end, the third new channel would be the image result of the dilation operation (Figure 11c). This helps to reduce the area around the object and also removes the noise. Figure 11 shows the result of the combination of channels. We generated 1714 new lung CT images for 10 patients using the above processing method.

As shown in Figure 11, the resulting image of the combination of the three channels is red. This is due to the arrangement of these channels. As mentioned earlier, the first channel of the new image contains the original image. The second and third channels have been replaced with edge detection processes and dilation operation, respectively. Since black pixels are dominant in the input image (including the edges and resulting image after applying the expansion operations) the final composite image receives the greatest effect from the first channel, leading to a dominant red color. However, if the main image is placed on the second channel, the output image will be green, and similarly blue for the third channel.

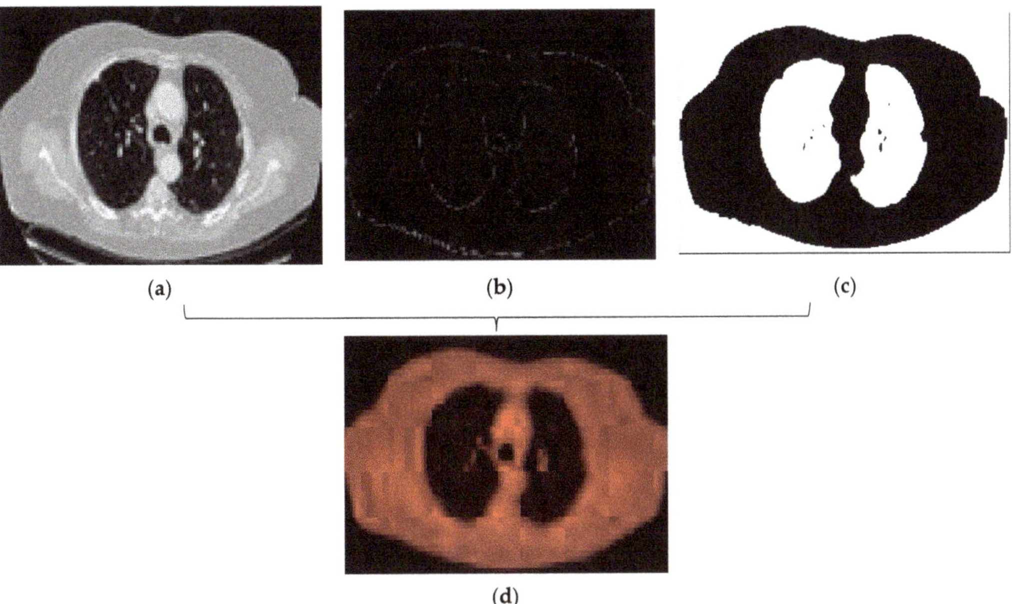

**Figure 11.** (**a**). First new channel, (**b**). Second new channel, (**c**). The second new channel (**d**). Result of a combination of new channels.

### *3.4. Lung Segmentation Using Deep Learning*

Since the main goal of this paper is to extract lungs from CT images, our proposed model must successfully address the semantic segmentation problem. U-Net is the most related available deep architecture in this regard. U-Net can learn from a relatively small-size training dataset. In addition, it vastly speeds up training time if a pre-trained model is used. Hence, a good starting point to train the network when dealing with image inputs is using a pre-trained ImageNet model along with its weights. On the other hand, ResBlocks architecture, which was proposed in [47,51], can facilitate the training process, while it offers a deeper network due to having all accumulated layers. Moreover, according to the experiments conducted in different networks and comparing their results, the use of the convolution layer instead of the pooling layer is preferred. This is because pooling layers generate huge semantic feature loss in the image. Thus, it seems ResNet architecture can be a more appropriate choice for the encoder part of the U-Net (the left half of the U). Figure 12 shows the block diagram of the ResNet-34 algorithm used in the encoder section of our proposed network. Our proposed model is mainly inspired by BCDU-Net and ResNet-34 [52] named as Res BCDU-Net. The backbone of this network is a ResNet-34 structure as the encoder which is shown in Figure 13. Details of different layers in the proposed model are described as follows.

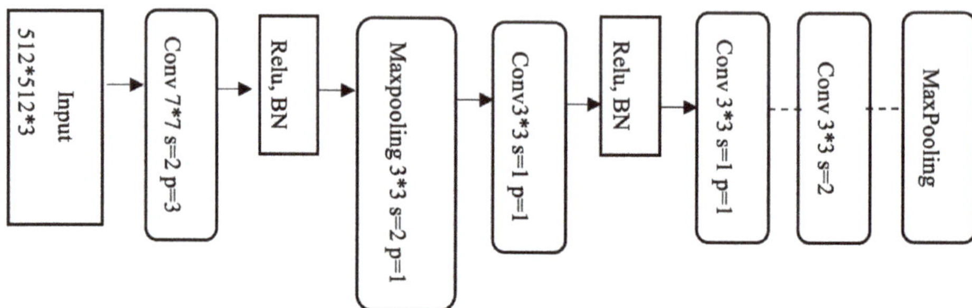

**Figure 12.** Block diagram of the ResNet-34 in the encoder of Res BCDU-Net.

- Encoding path: In Res BCDU-Net, the encoder is replaced with a pre-trained ResNet-34 network. The last layer of this path like BCDU-Net adopts a densely connected convolutions mechanism. So, the last layer, in contrast to all residual blocks in this path, never attempts to combine features through summation before being transferred to a layer; instead, it tries to concatenate the features. In other words, features that are learned per block are passed to the next block. This strategy can help the network to avoid learning redundant features. Figure 13 shows the difference between Res blocks and dense blocks.

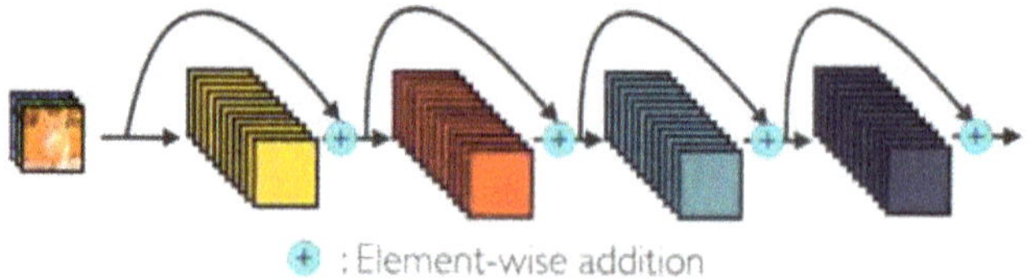


(a)

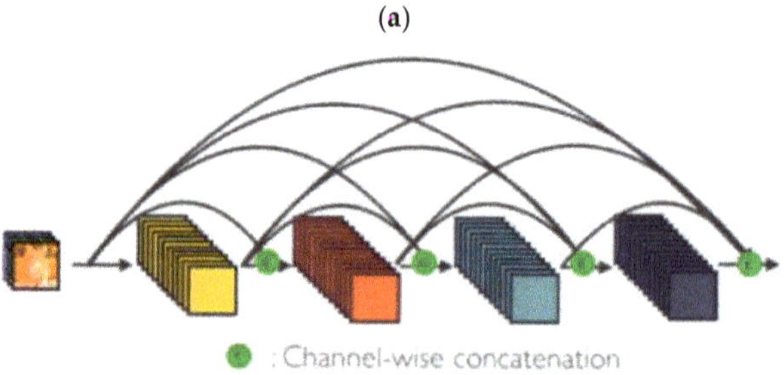


**Figure 13.** (a). ResNet Concept, (b). One Dense Block in Dense Net [53].

- Decoding path: In the decoding path, two feature maps should be concatenated: the feature maps corresponding to the same layer from the encoding path and those from the previous layer of the up-sampling function. In this Network, batch normalization was performed after the output of each up-sampling, before processing of two feature maps. Afterward, the resulting output is given to a BConvLSTM layer. In a standard ConvLSTM, only forward dependencies are processed. However, it is very important not to lose information concealed in any sequence. Therefore, the analysis of both forward and backward approaches has been proven to improve predictive network performance [54]. Both forward and backward ConvLSTMs are considered as standard processes. Therefore, two set parameters are considered as BConvLSTM. This layer can decide on the present input by verifying the data dependencies in both directions. Figure 14 illustrates our proposed network schematically.

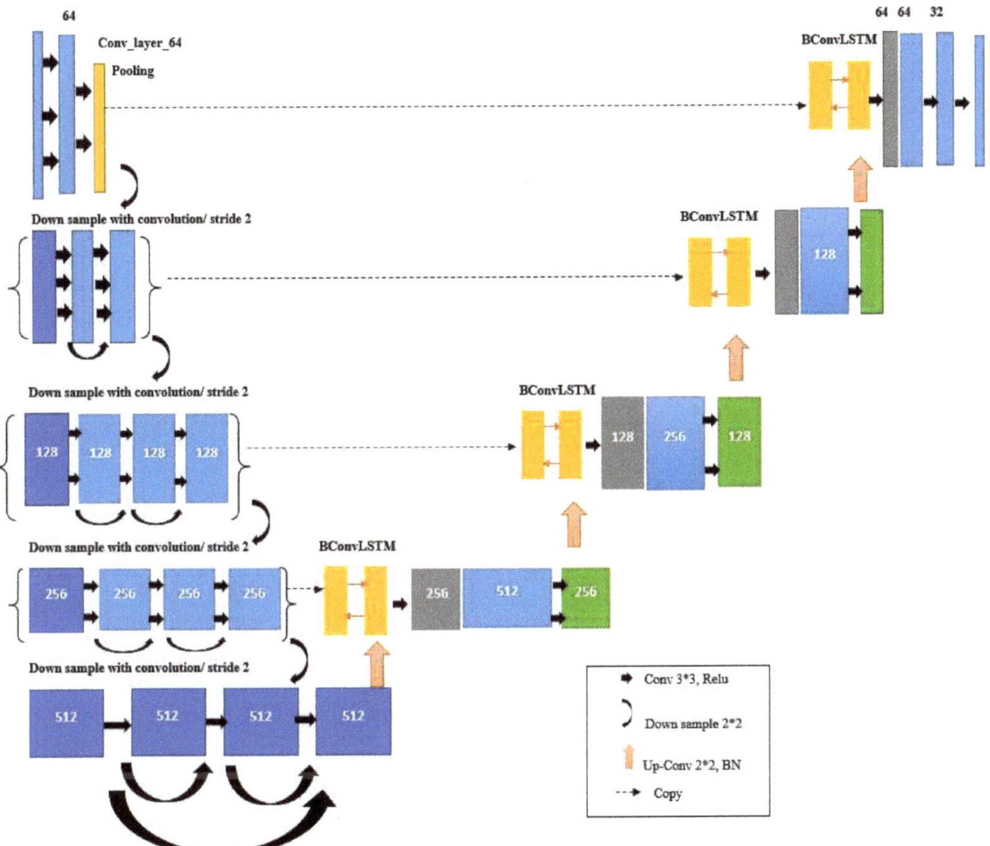

**Figure 14.** Res BCDU-Net architecture. The contraction path consists of Res blocks and a max-pooling layer. Such the U-Net, in each downsampling of encoding path, feature channels are doubled (64 to 128 to 256 to 512). In the last layer of the contracting path, we used 3 convolutional blocks with 2 dense connections. As seen, in the expansion path, the output of each batch normalized is given to a BConvLSTM layer.

## 4. Experimental Results

We evaluated the performance of our proposed neural network on 1714 CT images of the LIDC-IDRI dataset with the corresponding generated ground truth as described in the previous section. The experiments were implemented based on the Keras module with the TenserFlow backend. The network was trained for 50 epochs and batch size 32.

### 4.1. Evaluation Metrics

Several well-established criteria were used for performance evaluation of our proposed network, namely accuracy (AC), precision (Pr), recall (Re), and F1-score. We first calculated true positive (TP), false positive (FP), true negative (TN), and false negative (FN). These performance measures are mathematically expressed as follows:

$$\text{Accuracy} = \frac{TP + TN}{TP + TN + FP + FN} \tag{3}$$

$$\text{Precision} = \frac{TP}{TP + FP} \tag{4}$$

$$\text{Recall} = \frac{TP}{TP + FN} \tag{5}$$

$$F1 - \text{score} = 2 \times \frac{\text{Precision} \times \text{Recall}}{\text{Precision} + \text{Recall}} \tag{6}$$

To turn the results into a more reliable form, Dice's coefficient [55] is also used to evaluate our results. The Dice score is normally used to determine the performance of the segmentation step on the given images. This is a kind of similarity measure between two different objects. It is equal to the number of overlapping pixels between the two partitions divided by the size of the whole two objects. The Dice score is calculated as:

$$DSC = 2 \times \frac{|E \cap Q|}{|E| + |Q|} \tag{7}$$

where, $E$ is the segmented lung parenchyma area's pixels based on our network, $Q$ is the ground truth image's pixels and $|E \cap Q|$ represents the intersect pixels of two images. We also calculated the receiver operating characteristics (ROC) curve and the area under the curve (AUC). ROC curve is defined as a plot of TPR to FPR, with TPR placed on the $y$-axis and FPR on the $x$-axis. AUC is defined as the underlying area of the ROC curve. In other words, it measures the quality in which the network can segment the input data.

*4.2. Results*

We grouped randomly the dataset into training data (1200 images), validation data (257 images), and test data (257 images) in proportion 70%, 15%, and 15%. We also repeated our experiments 10 times and reported the obtained average performance across all run in this paper. All image sizes are $512 \times 512$. The input of the network consists of the CT images with three separate designed channels and corresponding ground truth annotations that we generated semi-automatically. Since the image segmentation process corresponds to a pixel-wise classification problem, the task of the neural network is to assign a label or class to all pixels of the input image. The output of the trained network is a pixel-wise mask of the image. Each pixel is given one of two categories:

Class 1: Pixels that fall within the lung area are labelled by '0'.
Class 2: Pixels related to the non-lung class are represented by the label '1'.

According to the above descriptions, first, we calculated the confusion matrix as shown in Figure 15.

According to Figure 14, we can see that the TP is very high, and also the point of attention achieved a very low FP. With respect to these values, calculated amounts for the accuracy, precision, recall, and F1-score measures are obtained as 97.83%, 99.93%, 97.45%, and 98.67%, respectively. Table 1 summarizes the results of the precision, recall, F1-score, accuracy, and dice score for another and our methods with LIDC dataset (The best-maintained metrics are highlighted in bold). We also provided some visual example results in Figure 15 to better compare U-Net and BCDU-Net.

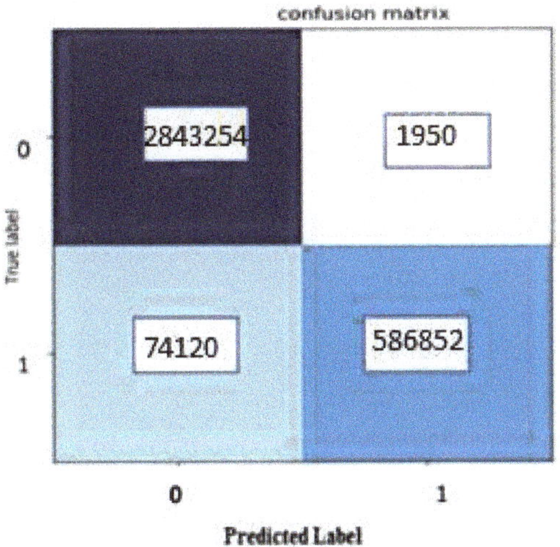


**Figure 15.** Confusion Matrix for the proposed method.

**Table 1.** Comparison of proposed network performance and the state-of-the-art alternatives on LIDC-IDRI dataset.

| Methods | Precision | Recall | F1-Score | Accuracy (%) | Dice Coefficient |
|---|---|---|---|---|---|
| U-Net [37] | 96.11 | 96.34 | 96.22 | 95.18 | 95.02 |
| RU-Net [38] | 95.52 | 97.21 | 96.35 | 97.15 | 94.93 |
| ResNet34-Unet [44] | 97.32 | 98.35 | 97.83 | 96.73 | 95.28 |
| BCDU-Net [45] | 99.02 | 98.03 | 98.52 | 97.21 | 96.32 |
| Proposed Method | 99.12 | 97.01 | 98.05 | 97.58 | 97.15 |

According to Table 1, we find that the performance of our proposed method performed better compared to related methods. According to this table, several results can be concluded as follow:

- Using the ResNet34 structure in the encoder section of the U-Net network has considerably improved the obtained results particularly in the quantity of recall.
- BCDU—Net model generally performs better than the ResNet structure in the contracting path of the U–Net.
- Using ResNet within BCDU-Net has achieved a better DSC similarity score compared to cases where these networks are used individually.
- Using images under our designed channels help to improve the quantitative results in all the evaluation criteria in comparison to using default channels.
- The high level of recall in our proposed model (with three new channels) arises from small FP as shown in the confusion matrix.

As shown in Figure 16, the U-Net model does not work well because of its deficiencies. The BCDU-Net model resolves much of the shortcomings in the image segmentation by U-Net but it sometimes appears a false-positive diagnosis mode (third column). In our proposed method, this problem has been resolved to a large extent and the final segmentation image is much similar to its corresponding mask (compare with U-Net and BCDU-Net in three last columns from right in Figure 16). It can be concluded that the combination of new channels to generate initial CT images and emphasis on components such as the edges and removal of additional items that are irrelevant in new filled channels greatly improves the adaptability power of the network.

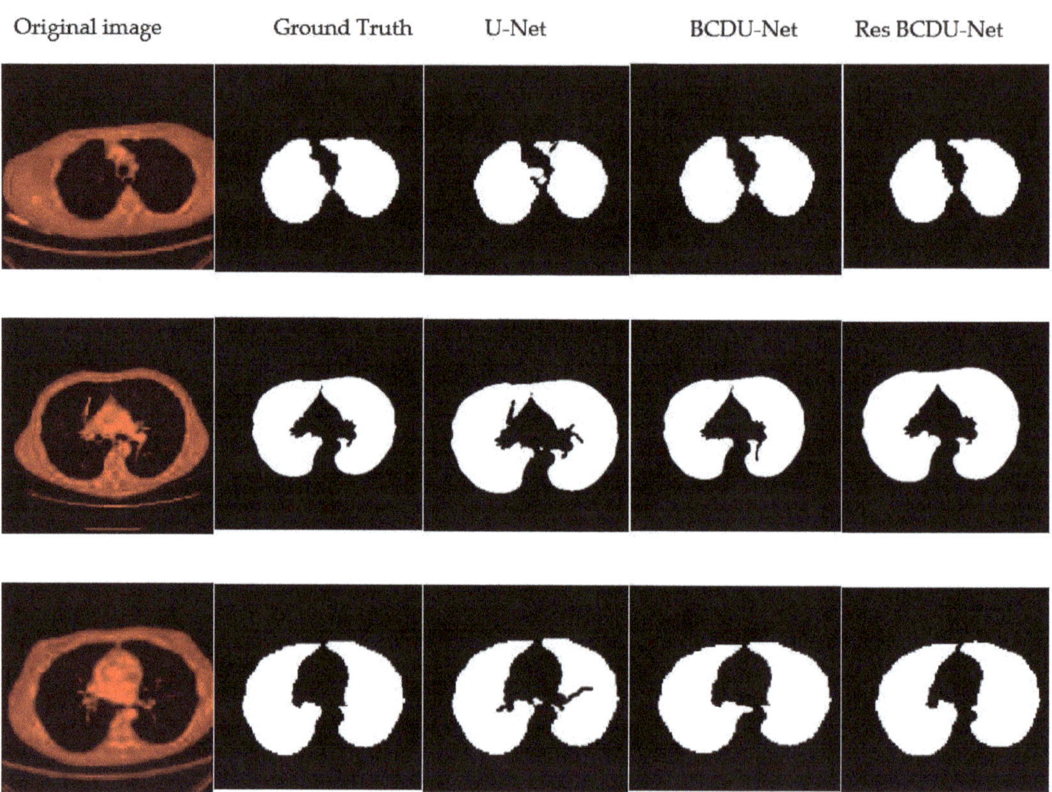

**Figure 16.** Sample results. From left to right: Original CT image, Ground Truth, U-Net, BCDU-Net, and Proposed method.

The proposed method has solved the high false-positive challenge as well. Also, losing the attached nodules to the lung wall challenge has been resolved by our proposed method (see and compare two last columns from right in Figure 16). It seems that the first challenge is resolved by the idea of combining three new channels in the CT images because it focuses on some components such as the edges and also removes the irrelevant objects and noise in the raw CT images. It can help the final segmentation network to be accurate. The second challenge is resolved by using the ResNet architecture in the first half of BCDU-Net because there is only one Pooling layer in the ResNet architecture and it causes less semantic information to be lost. In addition, the densely connected convolution mechanism in the last layer of the encoding path of the network plays an important role to prevent learning redundant features. To better represent the two above challenges and how the proposed method has resolved them, we have included these two challenges along with the components generated by our algorithm in Figure 17. It seems in this figure, the two challenges described, with the help of our proposed method, are solved using the new hybrid channels in the images and the use of ResNet34 architecture in the encoder section of the neural network.

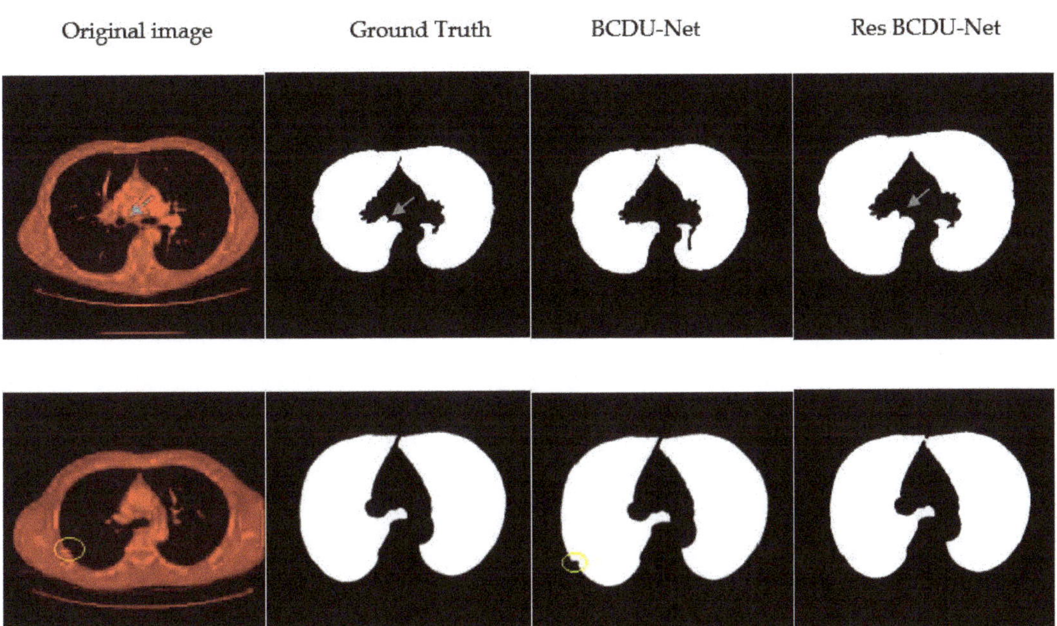

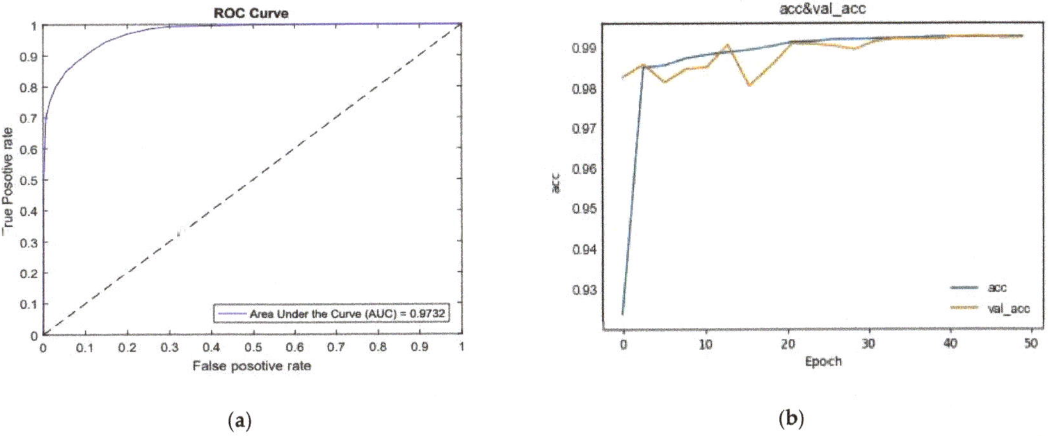

**Figure 17.** Visualizes the challenges for segmentation. First row presents the challenge of considering micro pulmonary tissues in the segmented image as the non-pulmonary region causing high false positive. Second row presents the challenge of losing attached nodules to the lung wall. (A yellow circle wrapped around the center of the nodule).

The overall performance of our proposed method, the ROC curve and also the accuracy of training and validation proposed network for LIDC-IDRI dataset are shown in Figure 18.

**Figure 18.** (**a**). ROC curve of Res BCDU-Net; (**b**). The accuracy of training and test for Res BCDU-Net.

According to Figure 18a, the AUC corresponds to 0.9732 which implies the effectiveness of the proposed model performance. Figure 18b shows that the network converges quickly; on the other hand, it converges after the 35th epoch. We also can see that the accuracy of training increases to over 99% after the 35th epochs. This is a good indicator of appropriate training of the network. In the validation phase, from epoch 0 to 30, it has a descending trend, which indicates inappropriate selection of weights, but the accuracy has been gradually increased from the 35th to 55th epochs. The training and validation accuracy will overlap between 35th and 50th.

*4.3. Ablation Study*

In this section, we conduct the ablation study to determine the effects of each component on the performance of the segmentation system. In detail, we intend to answer these questions in this section: (1) How does the use of images with the new three channels affect the overall performance of the system? (2) What is the effect of automatically producing binary labels for each of the images? (3) What is the effect on execution time and assisting the medical community? (4) What is the effect of using densely connected convolutions and BConvLSTM in the proposed deep neural network on the final performance of the system?

First, we discover the role of the new CT image channels in the segmentation performance. So, we did our experiments using images with their own default channels. The result can be found in Table 2. As we can see, the performance of our proposed method, where CT images are filled with newly designed channels, is higher than when they are filled with default channels.

**Table 2.** Impact of CT image channels on system performance.

| Channel Type in CT Images | Precision | Recall | F1-Score | Accuracy (%) | Dice Coefficient |
|---|---|---|---|---|---|
| Default | 99.12 | 97.01 | 98.05 | 97.58 | 97.15 |
| Proposed | 99.93 | 97.45 | 98.67 | 97.83 | 97.31 |

As the second work in this section, we look at the running time of the binary mask production algorithm. In this paper, we first used an automated algorithm to produce masks, and then, if necessary, we applied manual modification to each of the generated images. It takes hours to label each CT image taken by the Radiologists; whereas in our proposed method, without manual correction, all masks were produced within 10 min, on average. Considering the worst conditions and the need for manual correction and examination of each image produced by the algorithm, each mask requires 3 min to be made. Looking at Figure 19 the proposed method is capable of producing a similar number of images in a time of nearly 10 min. This figure shows the time of execution measured on the dimension of the data set from 50 to 1700 images. Furthermore, the execution time is reduced to 20% only with respect to the computation time without loading the image. As the number of images increases linearly, we can see that the execution time increases linearly, while the time required for the analysis of images by an expert will be greatly increased by increasing the number of images and parameters such as fatigue and so on.

Finally, we aim to examine the effect of densely connected convolution mechanism in the last layer of the encoding path of neural network and also the rule of using BConvLSTM on the skip connection. Table 3 shows these results. For this comparison, the CT images with new channels are assumed to be the network input, and the ResNet blocks are also used in the encoding section. Given the values in Table 3, we can observe the positive impact of using dense connection mechanism and BConvLSTM on system performance. (Please note that we have already discussed the role of ResNet blocks in the encoding path of the network in Table 1.)

**Table 3.** Impact of using densely connected convolutions and BConvLSTM on system performance.

| Method | Precision | Recall | F1-Score | Accuracy (%) | Dice Coefficient |
|---|---|---|---|---|---|
| Without Densely Connected Convolutions and BConvLSTM | 97.02 | 94.32 | 95.55 | 96.21 | 96.19 |
| Ours (With Densely Connected Convolutions and BConvLSTM) | 99.93 | 97.45 | 98.67 | 97.83 | 97.31 |

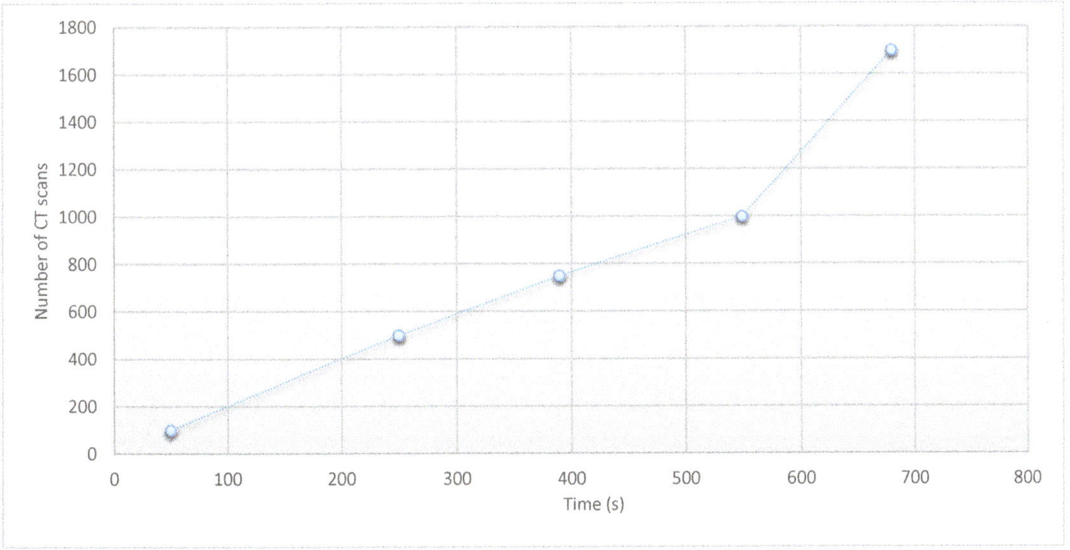

**Figure 19.** The execution time of the binary mask production algorithm.

## 5. Conclusions

In this paper, we proposed Res BCDU-Net to automatically and accurately segment the lung region from CT images. The proposed method consists of three main steps. First, we presented a semi-automatic technique to extract the ground truth for each lung. One of the great benefits of our method is that one can manage to produce all mask images, intelligently, without the need for the expertise of a radiologist and that saves a huge amount of time. Second, we proposed a novel three image channel generation and observed a significant decrease in the false positive rate and higher dice coefficients due to effective network input imagery. Finally, we designed the segmentation framework using a novel deep network architecture using a BCDU-Net with an encoder of pre-trained ResNet-34. This model was named Res BCDU-Net. It performed well, as verified through our extensive experiments on the large LIDC-IDRI dataset.

We have seen that combining ResNet and BCDU-Net networks as well as using CT images with newly designed channels in the proposed method has led to a few false positives as well as higher dice similarity scores. We have also seen that by using the automated algorithm used in the label production section for the dataset, the execution time is much less than the one used for producing masks and this is one of the most important advantages of this method.

The application of the proposed algorithm in daily work is being accepted. Because accurate and reliable segmentation of lung tissue is of particular importance in various clinical applications such as computer-assisted bronchoscopy, quantification of emphysema, and diagnosis of lung cancer. Therefore, the great potential goal of our work is applying it to clinical application to help the medical community in their daily work.

## 6. Future Works

One of the interesting research topics that could be pursued in the future is the adaptation and testing of the proposed method for 3D lung CT images. In this regard, a network such as V-Net can be used. Another idea for future works could involve using a combination of deep learning-based networks to segment medical images. It is also possible to examine the use of data enhancement methods and their impact on overall performance.

**Author Contributions:** Conceptualization, Y.J. and M.F.; data curation, Y.J.; formal analysis, M.R. and M.H.A.; investigation, M.F., M.R. and V.A.; methodology, Y.J.; project administration, M.F.; resources, V.A.; software, Y.J.; supervision, M.F.; validation, M.F. and M.R.; visualization, M.H.A.; writing—original draft, Y.J.; writing—review & editing, V.A. and M.H.A. All authors have read and agreed to the published version of the manuscript.

**Funding:** No external funding has received for conducting this research.

**Institutional Review Board Statement:** Not applicable.

**Informed Consent Statement:** Not applicable.

**Data Availability Statement:** Not applicable.

**Conflicts of Interest:** The authors of this manuscript declare no conflicts of interest.

## Abbreviations

Acronyms used in the paper.

| | |
|---|---|
| WHO | World Health Organization |
| CT | Computed Tomography |
| MRI | Magnetic Resonance Imaging |
| CAD | Computer-Aided Diagnosis |
| BCDU-Net | Bi-directional ConvLSTM U-Net with Densely connected convolutions |
| FCN | Fully Convolutional Neural Network |
| CNN | Convolutional Neural Network |
| BConvLSTM | Bidirectional Convolutional LSTM |
| LIDC | Lung Image Database Consortium |
| IDRI | Infectious Disease Research Institute |
| XML | Extensible Markup Language |
| DICOM | Digital Imaging and Communications in Medicine |
| HU | Hounsfield unit |
| ROC | Receiver Operating Characteristic |
| AUC | Area under the ROC Curve |

## References

1. Hossain, M.R.I.; Imran, A.; Kabir, M.H. Automatic lung tumor detection based on GLCM features. In *Asian Conference on Computer Vision*; Springer: Cham, Switzerland, 2014; pp. 109–121.
2. Sun, S.; Christian, B.; Reinhard, B. Automated 3-D segmentation of lungs with lung cancer in CT data using a novel robust active shape model approach. *IEEE Trans. Med. Imaging* **2011**, *31*, 449–460. [PubMed]
3. American Cancer Society's Publication, Cancer Facts & Figures 2020. Available online: https://www.cancer.org/research/cancer-facts-statistics/all-cancer-facts-figures/cancer-facts-figures-2020.html (accessed on 2 November 2020).
4. Wang, Y.; Guo, Q.; Zhu, Y. Medical image segmentation based on deformable models and its applications. In *Deformable Models*; Springer: New York, NY, USA, 2007; pp. 209–260.
5. Neeraj, S.; Aggarwal, L.M. Automated medical image segmentation techniques. *J. Med. Phys. Assoc. Med. Phys. India* **2010**, *35*, 3–14.
6. Asuntha, A.; Singh, N.; Srinivasan, A. PSO, genetic optimization and SVM algorithm used for lung cancer detection. *J. Chem. Pharm. Res.* **2016**, *8*, 351–359.
7. Jeyavathana, R.; Balasubramanian, D.; Pandian, A.A. A survey: Analysis on preprocessing and segmentation techniques for medical images. *Int. J. Res. Sci. Innov.* **2016**, *3*, 113–120.
8. Panwar, H.; Gupta, P.K.; Siddiqui, M.K.; Morales-Menendez, R.; Singh, V. Application of deep learning for fast detection of COVID-19 in X-Rays using nCOVnet. *Chaos. Solitons. Fractals* **2020**, *138*, 109944. [CrossRef]
9. Amine, A.; Modzelewski, R.; Li, H.; Su, R. Multi-task deep learning based CT imaging analysis for COVID-19 pneumonia: Classification and segmentation. *Comput. Biol. Med.* **2020**, *126*, 1–10.
10. Wang, X.; Deng, X.; Fu, Q.; Zhou, Q.; Feng, J.; Ma, H.; Liu, W.; Zheng, C. A Weakly-supervised Framework for COVID-19 Classification and Lesion Localization from Chest CT. *IEEE Trans. Med Imaging* **2020**, *39*, 2615–2625. [CrossRef]
11. Hira, S.; Bai, A.; Hira, S. An automatic approach based on CNN architecture to detect Covid-19 disease from chest X-ray images. *Appl. Intell.* **2020**. [CrossRef]
12. Cheng, J.; Chen, W.; Cao, Y.; Xu, Z.; Zhang, X.; Deng, L.; Zheng, C.; Zhou, J.; Shi, H.; Feng, J. Development and Evaluation of an AI System for COVID-19 Diagnosis. *medRxiv* **2020**. [CrossRef]

13. Pathak, Y.; Shukla, P.K.; Tiwari, A.; Stalin, S.; Singh, S.; Shukla, P.K. Deep Transfer Learning based Classification Model for COVID-19 Disease. *IRBM* **2020**. [CrossRef]

14. Rizwan, H.I.; Neubert, J. Deep learning approaches to biomedical image segmentation. *Inform. Med. Unlocked* **2020**, *18*, 1–12. [CrossRef]

15. Memon, N.A.; Mirza, A.M.; Gilani, S.A.M. Segmentation of lungs from CT scan images for early diagnosis of lung cancer. *Proc. World Acad. Sci. Eng. Technol.* **2006**, *14*, 228–233.

16. Omid, T.; Alirezaie, J.; Babyn, P. Lung segmentation in pulmonary CT images using wavelet transform. In Proceedings of the 2007 IEEE International Conference on Acoustics, Speech and Signal Processing-ICASSP'07, Honolulu, HI, USA, 15–20 April 2007; pp. 448–453.

17. Sasidhar, B.; Ramesh Babu, D.R.; Ravi Shankar, M.; Bhaskar Rao, N. Automated segmentation of lung regions using morphological operators in CT scan. *Int. J. Sci. Eng. Res.* **2013**, *4*, 114–118.

18. Keita, N.; Shimizu, A.; Kobatake, H.; Yakami, M.; Fujimoto, K.; Togashi, K. Multi-shape graph cuts with neighbor prior constraints and its application to lung segmentation from a chest CT volume. *Med. Image Anal.* **2013**, *17*, 62–77.

19. Geetanjali, J.; Kaur, S. A Review on Various Edge Detection Techniques in Distorted Images. *Int. J. Adv. Res. Comput. Sci. Softw. Eng.* **2017**, *7*, 942–945.

20. Shin, M.C.; Goldgof, D.B.; Bowyer, K.W.; Nikiforou, S. Comparison of edge detection algorithms using a structure from motion task. *IEEE Trans. Syst. Manand Cybern. Part B Cybern.* **2001**, *31*, 589–601. [CrossRef]

21. Paola, C.; Casiraghi, E.; Artioli, D. A fully automated method for lung nodule detection from postero-anterior chest radiographs. *IEEE Trans. Med Imaging* **2006**, *25*, 1588–1603.

22. Ana Maria, M.; da Silva, J.A.; Campilho, A. Automatic delimitation of lung fields on chest radiographs. In Proceedings of the 2004 2nd IEEE International Symposium on Biomedical Imaging: Nano to Macro (IEEE Cat No. 04EX821), Arlington, VA, USA, 18 April 2004; pp. 1287–1290.

23. Hu, X.; Alperin, N.; Levin, D.N.; Tan, K.K.; Mengeot, M. Visualization of MR angiographic data with segmentation and volume-rendering techniques. *J. Magn. Reson. Imaging* **1991**, *1*, 539–546. [CrossRef]

24. Tang, J.; Millington, S.; Acton, S.T.; Crandall, J.; Hurwitz, S. Surface extraction and thickness measurement of the articular cartilage from MR images using directional gradient vector flow snakes. *IEEE Trans. Biomed. Eng.* **2006**, *53*, 896–907. [CrossRef]

25. Cline, H.E.; Dumoulin, C.L.; Hart, H.R., Jr.; Lorensen, W.E.; Ludke, S. 3D reconstruction of the brain from magnetic resonance images using a connectivity algorithm. *Magn. Reson. Imaging* **1987**, *5*, 345–352. [CrossRef]

26. Nihad, M.; Grgic, M.; Huseinagic, H.; Males, M.; Skejic, E.; Smajlovic, M. Automatic CT image segmentation of the lungs with region growing algorithm. In Proceedings of the 18th International Conference on Systems, Signals and Image Processing-IWSSIP, Bratislava, Slovakia, 16–18 June 2011; pp. 395–400.

27. da Silva Felix, H.J.; Cortez, P.C.; Holanda, M.A.; Costa, R.C.S. Automatic Segmentation and Measurement of the Lungs in healthy persons and in patients with Chronic Obstructive Pulmonary Disease in CT Images. In Proceedings of the IV Latin American Congress on Biomedical Engineering 2007, Bioengineering Solutions for Latin America Health, Margarita Island, Venezuela, 24–28 September 2007; pp. 370–373.

28. Kass, M.; Witkin, A.; Terzopoulos, D. Snakes: Active contour models. *Int. J. Comput. Vis.* **1988**, *1*, 321–331. [CrossRef]

29. Yoshinori, I.; Kim, H.; Ishikawa, S.; Katsuragawa, S.; Ishida, T.; Nakamura, K.; Yamamoto, A. Automatic segmentation of lung areas based on SNAKES and extraction of abnormal areas. In Proceedings of the 17th IEEE International Conference on Tools with Artificial Intelligence (ICTAI'05), Hong Kong, China, 14–16 November 2005; pp. 5–10.

30. Shi, Y.; Qi, F.; Xue, Z.; Chen, L.; Ito, K.; Matsuo, H.; Shen, D. Segmenting lung fields in serial chest radiographs using both population-based and patient-specific shape statistics. *IEEE Trans. Med Imaging* **2008**, *27*, 481–494. [PubMed]

31. Cheng, J.; Liu, J.; Xu, Y.; Yin, F.; Wong, D.W.K.; Tan, N.-M.; Tao, D.; Cheng, C.-Y.; Aung, T.; Wong, T.Y. Superpixel classification based optic disc and optic cup segmentation for glaucoma screening. *IEEE Trans. Med Imaging* **2013**, *32*, 1019–1032. [CrossRef] [PubMed]

32. Titinunt, K.; Han, X.-H.; Chen, Y.-W. Liver segmentation using superpixel-based graph cuts and restricted regions of shape constrains. In Proceedings of the 2015 IEEE International Conference on Image Processing (ICIP), Quebec City, QC, Canada, 27–30 September 2015; pp. 3368–3371.

33. Chen, X.; Yao, L.; Zhou, T.; Dong, J.; Zhang, Y. Momentum contrastive learning for few-shot COVID-19 diagnosis from chest CT images. *arXiv* **2020**, arXiv:2006.13276.

34. Zhou, K.; Gu, Z.; Liu, W.; Luo, W.; Cheng, J.; Gao, S.; Liu, J. Multi-cell multi-task convolutional neural networks for diabetic retinopathy grading. In Proceedings of the 2018 40th Annual International Conference of the IEEE Engineering in Medicine and Biology Society (EMBC), Honolulu, HI, USA, 18–21 July 2018; pp. 2724–2727.

35. Dan, C.; Giusti, A.; Gambardella, L.M.; Schmidhuber, J. Deep neural networks segment neuronal membranes in electron microscopy images. In Proceedings of the advances in Neural Information Processing Systems, Lake Tahoe, NV, USA, 3–6 December 2012; pp. 2843–2851.

36. Jonathan, L.; Shelhamer, E.; Darrell, T. Fully convolutional networks for semantic segmentation. In Proceedings of the IEEE Conference on Computer Vision and Pattern Recognition, Boston, MA, USA, 7–12 June 2015; pp. 3431–3440.

37. Olaf, R.; Fischer, P.; Brox, T. U-net: Convolutional networks for biomedical image segmentation. In Proceedings of the International Conference on Medical Image Computing and Computer-Assisted Intervention, Munich, Germany, 5–9 October 2015; pp. 234–241.

38. Alom, M.Z.; Yakopcic, C.; Taha, T.M.; Asari, V.K. Nuclei Segmentation with Recurrent Residual Convolutional Neural Networks based U-Net (R2U-Net). In Proceedings of the NAECON 2018—IEEE National Aerospace and Electronics Conference, Dayton, OH, USA, 23–26 July 2018; pp. 228–233. [CrossRef]
39. Fausto, M.; Navab, N.; Seyed-Ahmad, A. V-net: Fully convolutional neural networks for volumetric medical image segmentation. In Proceedings of the 2016 Fourth International Conference on 3D Vision (3DV), Stanford, CA, USA, 25–28 October 2016; pp. 565–571.
40. Özgün, C.; Abdulkadir, A.; Lienkamp, S.S.; Brox, T.; Ronneberger, O. 3D U-Net: Learning dense volumetric segmentation from sparse annotation. In Proceedings of the International Conference on Medical Image Computing and Computer-Assisted Intervention, Athens, Greece, 17–21 October 2016; pp. 424–432.
41. Zhou, X.; Ito, T.; Takayama, R.; Wang, S.; Hara, T.; Fujita, H. Three-dimensional CT image segmentation by combining 2D fully convolutional network with 3D majority voting. In Proceedings of the Deep Learning and Data Labeling for Medical Applications, Athens, Greece, 21 October 2016; pp. 111–120.
42. Ozan, O.; Schlemper, J.; Folgoc, L.L.; Lee, M.; Heinrich, M.; Misawa, K.; Mori, K.; Mori, K.; McDonagh, S.; Hammerla, N.Y.; et al. Attention u-net: Learning where to look for the pancreas. In Proceedings of the 1st Conference on Medical Imaging with Deep Learning (MIDL 2018), Amsterdam, The Netherlands, 4–6 July 2018.
43. Ozsahin, I.; Sekeroglu, B.; Musa, M.S.; Mustapha, M.T.; Ozsahi, D.U. Review on Diagnosis of COVID-19 from Chest CT Images Using Artificial Intelligence. *Comput. Math. Methods Med.* **2020**. [CrossRef]
44. Stephen, L.; Chong, L.H.; Edwin, K.P.; Xu, T.; Wang, X. Automated Pavement Crack Segmentation Using U-Net-Based Convolutional Neural Network. *IEEE Access* **2020**, *8*, 114892–114899.
45. Reza, A.; Asadi-Aghbolaghi, M.; Fathy, M.; Escalera, S. Bi-directional ConvLSTM U-net with Densley connected convolutions. In Proceedings of the IEEE International Conference on Computer Vision Workshops, Seoul, Korea, 22 April 2019; pp. 1–10.
46. Song, H.; Wang, W.; Zhao, S.; Shen, J.; Lam, K.-M. Pyramid dilated deeper convlstm for video salient object detection. In Proceedings of the European Conference on Computer Vision (ECCV), Munich, Germany, 8–14 September 2018; pp. 715–731.
47. Christian, S.; Ioffe, S.; Vanhoucke, V.; Alemi, A.A. Inception-v4, inception-resnet and the impact of residual connections on learning. In Proceedings of the Thirty-First AAAI Conference on Artificial Intelligence, San Francisco, CA, USA, 4–9 February 2017; pp. 4278–4284.
48. Available online: https://wiki.cancerimagingarchive.net/display/Public/LIDC-IDRI (accessed on 10 September 2020).
49. Vanitha, U.; Prabhu Deepak, P.; PonNageswaran, N.; Sathappan, R. Tumor detection in brain using morphological image processing. *J. Appl. Sci. Eng. Methodol.* **2015**, *1*, 131–136.
50. Megha, G. Morphological image processing. *Int. J. Creat. Res. Thoughts* **2011**, *2*, 161–165.
51. He, K.; Zhang, X.; Ren, S.; Sun, J. Deep residual learning for image recognition. In Proceedings of the IEEE Conference on Computer Vision and Pattern Recognition, Las Vegas, NV, USA, 27–30 June 2016; pp. 770–778.
52. Gao, H.; Sun, Y.; Liu, Z.; Sedra, D.; Weinberger, K.O. Deep networks with stochastic depth. In Proceedings of the 14th European Conference on Computer Vision, Amsterdam, The Netherlands, 11–14 October 2016; pp. 646–661.
53. Gao, H.; Zhuang, L.; Van Der Maaten, L.; Weinberger, K.Q. Densely Connected Convolutional Networks. In Proceedings of the 2017 IEEE Conference on Computer Vision and Pattern Recognition (CVPR), Honolulu, HI, USA, 21–26 July 2017; pp. 2261–2269. [CrossRef]
54. Sayda, E. Deep Stacked Residual Neural Network and Bidirectional LSTM for Speed Prediction on Real-life Traffic Data. In Proceedings of the 24th European Conference on Artificial Intelligence—ECAI 2020, Santiago de Compostela, Spain, 12 June 2020.
55. Lee, D.R. Measures of the amount of ecologic association between species. *Ecology* **1945**, *26*, 297–302.

*Article*

# Imaging Tremor Quantification for Neurological Disease Diagnosis

**Yuichi Mitsui** [1], **Thi Thi Zin** [1,*], **Nobuyuki Ishii** [2] **and Hitoshi Mochizuki** [2]

1    Graduate School of Engineering, University of Miyazaki, Miyazaki 889-2192, Japan; mitsuiyuichi@outlook.jp
2    Department of Neurology, Faculty of Medicine, University of Miyazaki, Miyazaki 889-2192, Japan;
     nobuyuki_ishii@med.miyazaki-u.ac.jp (N.I.); mochizuki-h@umin.net (H.M.)
*    Correspondence: thithi@cc.miyazaki-u.ac.jp

Received: 16 October 2020; Accepted: 20 November 2020; Published: 22 November 2020

**Abstract:** In this paper, we introduce a simple method based on image analysis and deep learning that can be used in the objective assessment and measurement of tremors. A tremor is a neurological disorder that causes involuntary and rhythmic movements in a human body part or parts. There are many types of tremors, depending on their amplitude and frequency type. Appropriate treatment is only possible when there is an accurate diagnosis. Thus, a need exists for a technique to analyze tremors. In this paper, we propose a hybrid approach using imaging technology and machine learning techniques for quantification and extraction of the parameters associated with tremors. These extracted parameters are used to classify the tremor for subsequent identification of the disease. In particular, we focus on essential tremor and cerebellar disorders by monitoring the finger–nose–finger test. First of all, test results obtained from both patients and healthy individuals are analyzed using image processing techniques. Next, data were grouped in order to determine classes of typical responses. A machine learning method using a support vector machine is used to perform an unsupervised clustering. Experimental results showed the highest internal evaluation for distribution into three clusters, which could be used to differentiate the responses of healthy subjects, patients with essential tremor and patients with cerebellar disorders.

**Keywords:** tremor; essential tremor; ataxia; finger–nose–finger test

---

## 1. Introduction

A tremor is one of the most common involuntary movements seen in neurological disorders. It is characterized as a rhythmic, involuntary oscillation of a body part by muscle innervations that imply repetitive contractions [1–4]. Various types of tremors occur, depending on their causes. In general, tremors can be divided into two types: resting and action tremors. Action tremors can be further classified into postural tremors, kinetic tremors, task-specific tremors and intention tremors [5,6]. In clinical practice, a tremor is most commonly classified by its appearance and cause or origin. There are actually more than 20 types of tremors. Among them, the most common cause of resting tremors is Parkinson's disease (*PD*). The most common causes of postural and kinetic tremors are essential tremors (*ET*) and cerebellar disorders (*CD*). It is easy to distinguish *PD* resting tremors from other tremors because trembles occur at rest and weaken when the target muscles contract [4,5,7]. On the other hand, there are various causes of action tremor, and it is not easy to identify the cause.

The most common causes of action tremors are essential tremors (*ET*) and cerebellar disorders (*CD*). *ET* tremor behaves regularly, but *CD* tremor behaves irregularly and sometimes includes intention tremor [5,6]. Both *ET* and *CD* patients have several common features, such as increased tremor when mentally stressed and restricted fine movements. In clinical practice, clinicians try to distinguish these two tremors, *ET* and *CD*, by neurological examinations such as FNF (finger–nose–finger) test [1,6,7]. However, it is not easy to detect subtle irregularities of finger movement and observe where the finger tremor becomes stronger during the FNF test by usual observation. For this reason, distinguishing between *ET* and *CD* could be difficult even for a skilled neurologist. Therefore, we propose a non-contact method of distinguishing *ET* from *CD* featuring image processing technology and including measurement of tremor severity to confirm its effectiveness.

The rest of this paper is organized as follows: In Section 2, we review the literature relating to our research. In Section 3, the materials and methods are proposed. The method of disease diagnosis and analysis is described in Section 4. Then some experimental results are described in Section 5 using datasets collected by two neurologists who specialize in patients with tremors. In Section 6, we present discussions and plans for future research. Finally, we conclude the paper by giving remarks in Section 7.

## 2. Related Works

*ET* is a disease in which tremors only appear as a symptom and are not life-threatening. However, it is clinically important to make an early diagnosis, allowing treatment specific to *ET* when available and improving the patient's quality of life [8]. The prevalence of *ET* is approximately 2.5–10% of the population [9]. Although it can develop in any age group, it is mainly seen in the elderly: 4% of people over 40 years old and 5–14% of people over 65 years old have *ET* [10–12]. Although the cause of *ET* is not well understood, speculation exists that a hyperexcited state of the sympathetic nerve is involved because the symptoms increase with stress [13]. *CD* results from causes such as cerebellar infarction, inflammation, demyelination, autoimmunity, trauma, degeneration and tumors. Symptoms of *CD* include cerebellar ataxia, intention tremor and cerebellar sway of the upper limbs [5,7,12]. *ET* is rarely life-threatening, while *CD* can be. However, *ET* often disturbs a patient's quality of life. Therefore, an early diagnosis, differentiating *ET* from *CD*, is important. In addition, even in those who are highly skilled, it may be difficult to differentiate tremors when the patient first presents with the symptom.

Currently, the diagnosis of a patient's disease using the characteristics of tremor is performed subjectively based on the experience and skills of specialists. However, there are various problems with this. Doctors who are not specialists, such as family doctors and on-duty doctors, do not have a means of quantitatively evaluating a tremor, incurring the risk of misdiagnosis. Such quantification is important in determining the proper treatment.

Various tremor rating scales have been used to evaluate symptoms [14–16], but these are qualitative and subjective, and errors may occur depending on the person who assesses them [17,18]. Therefore, how to more accurately quantify tremor characteristics has become an urgent subject of research. For example, such research has included the acquisition of tremor signals using devices such as multipolar *EMGs*, electromagnetic tracking devices, accelerometers and gyroscopes, using the resulting data for evaluation and diagnosis [19]. However, since these methods require a large-scale dedicated device, it is unrealistic to use them in an examination room. In addition, these methods often require attaching a sensor or the like to the patient, resulting in different symptoms due to stress or the burden imposed during the examination. Recent tremor-related research has been done using magnetic resonance imaging (MRI) and a neurophysiological assessment [20]. Results indicate a significant association between severe tremors and malfunctions in specific areas of the brain. Moreover, the literature includes applicable developments in image processing in the framework of deep learning [21,22].

The severity assessment of *ET* or *CD* is determined by expert opinion and is likely to be subjective in nature. Several investigators have tried to quantify these symptoms. Analysis of FNF test, a classic neurological examination method, has been reported using an accelerometer or inertial sensor. Using inertial sensors, the changes of spatiotemporal parameters are related to the disability level in patients with multiple sclerosis [23] or cerebellar ataxia [24]. Using a three-dimensional motion capture system, analyses of body movements during FNF test in patients with poststroke could discriminate between patients with mild and moderate upper limb impairments [25]. These studies have been successful in quantifying the severity of symptoms. However, no research has ever tried to capture and distinguish between the characteristics of *ET* and *CD*. Furthermore, the fine movements of the fingertips during the FNF test has never been analyzed by monitoring them with a video over time.

## 3. Materials and Proposed Imaging Method for Tremor Quantification

In this section, we describe the architecture of our proposed system in which tremors are characterized based on visual data collected by performing the *FNF* (finger–nose–finger) test. Specifically, these visual data are collected using a smartphone, as in the actual diagnostic, for distinguishing *ET* from *CD*. The purpose of this analysis is to automatically and objectively diagnose the disease and measure its severity. In the *FNF* test, patients move their index finger back and forth between their nose and the examiner's finger to see whether tremors occur. As a result, non-specialist doctors such as family doctors or on-duty doctors can avoid misdiagnosis, detecting life-threatening problems, and averting *MRI* imaging and other unnecessary medical costs. In addition, without the need for sensors, no burden is placed on the patient, and the *FNF* test analysis can be performed easily. The system is composed of the following four components: the dataset collection system, image preprocessing, feature extraction, disease diagnosis and analysis.

### 3.1. Subjects and Design of Data Collection System

Data collection for the *FNF* test was conducted in an examination room at the Miyazaki University Hospital for *ET* patients ($N = 10$; female, $n = 4$; age, 71.5 +/− 8.1, mean +/− *SD*) and *CD* patients ($N = 18$; female, $n = 9$; age, 68.1 +/− 7.5), with images captured in a side view. Figure 1 illustrates the process of the data collection system. The smartphone camera is fixed at a distance of about 1~1.5 m from the doctor and the patient. The recorded video has a resolution of 480 × 640 pixels, and the frame rate is 30 frames per second (fps). The doctor places his index finger at various locations in front of the patient. The patient touches his/her index finger to the doctor's index finger and then touches his/her index finger to his/her own nose. Repeat several times with the doctors moving the target finger each time.

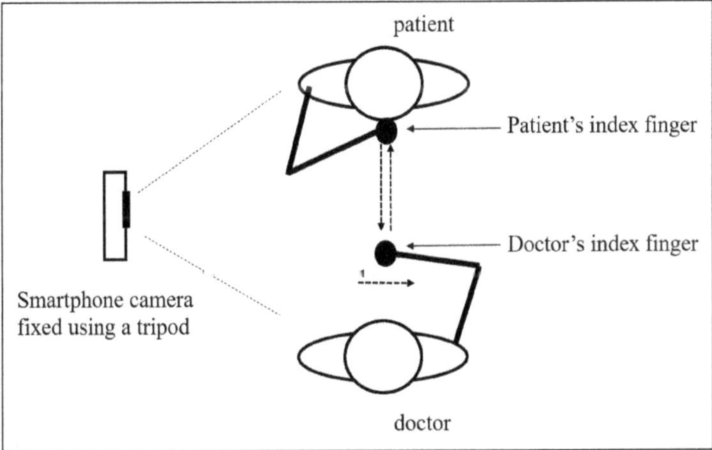


**Figure 1.** The illustration of the data collection system.

Two neurologists made diagnoses and evaluated the severity of the patients' tremors based on the essential tremor rating assessment scale [15] or the scale for assessing and rating ataxia [14]. According to these scales, patients were classified into two groups; mild (*ET*, upper limb tremor < 1 cm; *CD*, FNF test tremor < 2 cm) or severe (*ET*, upper limb tremor ≥ 1 cm; *CD*, FNF test tremor ≥ 2 cm). The video data were recorded by two neurologists using smartphones. The data or healthy subjects

($N$ = 8; female, $n$ = X; age, X +/− X) include data recorded with the cooperation of members of the laboratory, featuring a recorded video of the diagnostic test performed on two ordinary healthy people at the Miyazaki University Hospital. The *FNF* test was performed by winding red or green tape around the subject's finger. This protocol was approved by the Ethics Committee of the University of Miyazaki, with a waiver of written informed consent obtained from all participants.

*3.2. Image Preprocessing Component*

In this component, image preprocessing is performed in preparation for further analysis. In the first step, the patient's finger area must be extracted from the video image in each frame. Figure 2 shows the algorithm for extracting the finger area. First, a background image is detected as a noise source, and then the region of interest is set by removing the noise. Subsequent processes include inputting an image for extracting the finger area, threshold processing, noise processing and finger area estimation, finally obtaining the coordinates of the finger area.

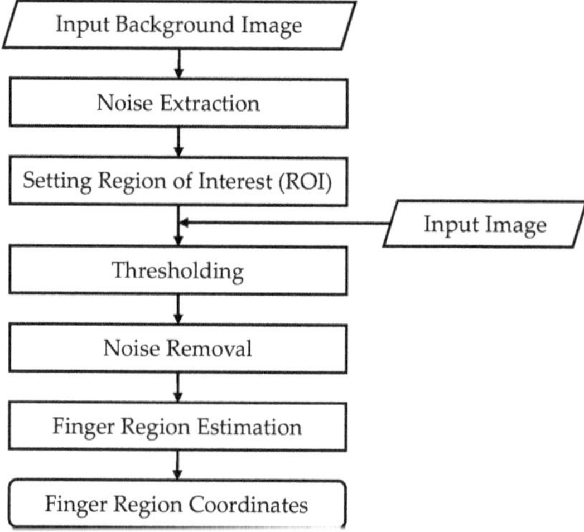


**Figure 2.** Finger area extraction algorithm.

Due to the location for recording video in the examination room, noise can occur for many reasons when extracting the finger area. In the presence of noises, we perform a noise removal process by using background modeling with an initial background as an image that does not include the target object. By converting the RGB image to HSV, we perform the defined thresholding of the hue information in order to obtain the finger object. The input image and converted HSV image are shown in Figure 3a,b, respectively [16]. Hue information representing the hue of the HSV image, Saturation information representing the saturation, and Value information representing the brightness are thresholded, and the finger is obtained by taking the logical product with the region of interest.

We also performed noise processing by calculating the aspect ratio of each area resulting from the labeling process. Since the finger area has a shape close to that of a square, threshold values are applied to the calculated aspect ratio to remove areas of the same size as finger areas that are elongated in vertical or horizontal orientations and could not be removed by noise processing using labels.

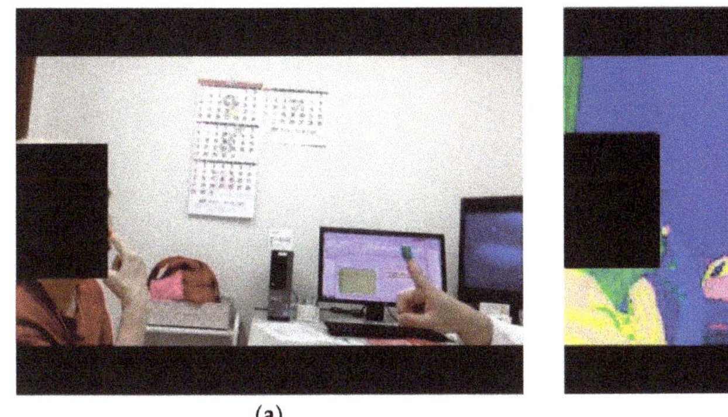
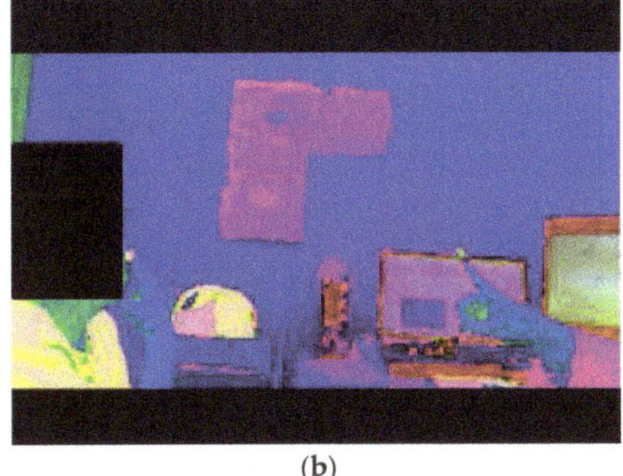

(a)         (b)

**Figure 3.** (**a**) Input image. (**b**) The converted HSV image for the input image.

### 3.2.1. Estimation of Finger Area when Multiple Labels Exist

Even if noise processing is performed, not all noise can be removed. In addition, noise processing sometimes removes the finger area. In such a case, the finger area is estimated as follows: If multiple labels exist, such as frame $t$, select the label closest to the coordinates of the finger area surrounded by the red circle detected in the previous frame. The coordinates of the finger area are obtained by calculating the center of gravity of the labeled object.

### 3.2.2. Estimation of Finger Area when No Label Remains

If the finger area is mistakenly removed during noise processing, all labels can be lost. In that case, a smoothing process is performed using finger area coordinates from preceding and subsequent frames to estimate the finger area. For example, as shown in Figure 4, when two frames with no label continue for two successive frames, the difference between $X$ coordinate and $Y$ coordinates is calculated from the preceding and subsequent frames, and the coordinate values are evenly calculated for the unlabeled frames. The finger area is estimated by substituting a value that changes.

| Frame | X | Y |
|-------|-----|-----|
| $t$ | 154 | 187 |
| $t+1$ | - | - |
| $t+2$ | - | - |
| $t+3$ | 286 | 199 |

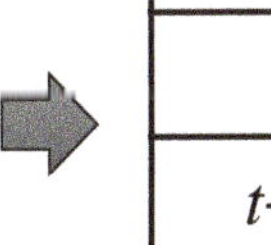

| Frame | X | Y |
|-------|-----|-----|
| $t$ | 154 | 187 |
| $t+1$ | 198 | 191 |
| $t+2$ | 242 | 195 |

**Figure 4.** Estimation of finger area when no label remains.

### 3.3. Feature Extraction Process

Now, we present the feature extraction process for the detected finger areas in an *FNF* test. In this process, we extract the following six measures.

### 3.3.1. Mean Square Deviation (RMSD: Root-Mean-Square Deviation)

This measure quantifies the vertical distance of the patient's up and down positions. In Figure 5, the finger region is shown as a graph in the rectangular coordinate plane. In order to do so, we employ a linear-quadratic function along with the least square method. Since the linear-quadratic function represents a parabola curve, we can estimate the vertical distance that the finger moves up and down by computing the root mean square deviation (RMSD) measure. The formula for calculating RMSD is shown below:

$$RMSD = \sqrt{\frac{\sum_{i=1}^{n}\left(y_i - \widehat{y}_i\right)^2}{n}},\tag{1}$$

where $n$ is the number of plotted data points, $y_i$ is the plotted value, and $\widehat{y}_i$ is the value of the approximated quadratic function.

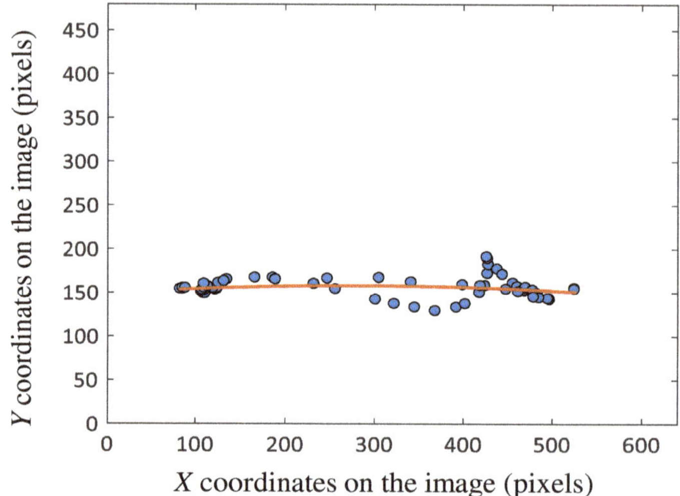


**Figure 5.** Plot of finger trajectory and approximate curve.

### 3.3.2. Dispersion of Acceleration

As the second measure in consideration, dispersion of acceleration digitizes variations in speed. This measure was selected because healthy people have constant finger movements. However, patients with tremor symptoms have varying rates of change. Therefore, acceleration is calculated for each frame using the coordinate data of the finger region. Next, by calculating the variance from the calculated acceleration, we can quantify the dispersion of acceleration. The formula for calculating this variance is shown below.

$$\text{Variance} = \frac{1}{n}\sum_{i=1}^{n}(x_i - \overline{x})^2,\tag{2}$$

where $n$ is the number of frames, $x_i$ is the acceleration in each frame, and $\overline{x}$ is the average value of the acceleration. Using the dispersion of acceleration that digitizes variation in speed change, we can quantify how the patient's finger is decelerating near the examiner's finger.

### 3.3.3. Histogram Feature

In the case of tremor patients, particularly *ET* patients, their fingers often slow down near the examiner's finger when performing the *FNF* test. Therefore, the following equation is used to determine whether the finger is moving back and forth with a constant rhythm.

$$\text{Histogram} = \frac{(h_{max} - h_{med})}{n},\tag{3}$$

where $n$ is the number of frames, $h_{max}$ the maximum value of the histogram, and $h_{med}$ is the median value of the histogram. The difference between the simple maximum value and the median value requires a different round-trip time (number of frames) depending on the moving image, so normalization is performed by dividing by the number of round-trip frames. In the histogram method, we first construct a histogram by taking the X coordinate on the image of the finger as the vertical axis and the frame number as the horizontal axis. We then divide the range between the maximum frequency of X coordinate and the minimum frequency of X coordinate into four equal parts [12]. Finger movement can be considered unstable if angle analysis indicates that the finger moves up and down an excessive number of times. Moreover, the ratio between the time required in the initial movement of the patient's finger from the examiner's finger to the patient's nose and the time required for the return trip is longer for tremor patients. In this case, the average moving distance is used for digitizing how much the finger is shaking.

### 3.3.4. Angular Feature

In the *FNF* test, the fingers are moved horizontally, but in patients with tremor symptoms, fine up and down vibrations occur. In order to detect this, the angle at which the finger has moved between frames is considered and calculated by using the following equation.

$$\theta = \tan^{-1}\frac{y_n - y_{n-1}}{x_n - x_{n-1}},\tag{4}$$

However, the range of $\theta$ is $-180° \le \theta \le 180°$. Here, $(x_n, y_n)$ is the coordinate data for the finger region of the *nth* frame, and $(x_{(n-1)}, y_{(n-1)})$ is the coordinate data for the $n - 1$ frame. The total number of frames satisfying $-150° < \theta < -30°$ or $30° < \theta < 150°$ is determined using the angle obtained from the above equation; the number of times the finger swings up and down is also obtained. An example of the finger movement angle is shown in Figure 6.

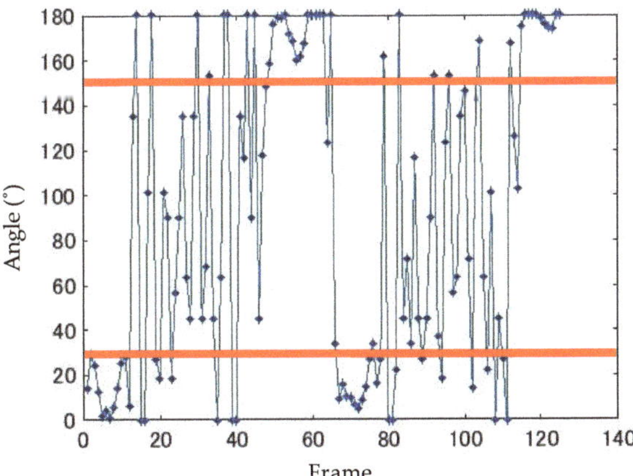

**Figure 6.** Angle of finger movement.

### 3.3.5. Measure of Round-Trip Time Ratio

In order to detect abnormality in the *FNF* test, the following equation is used to determine the ratio between the time required for initial and return paths in finger movement:

$$\text{Time Ratio} = \frac{\overline{f_0}}{\overline{f_r}}, \tag{5}$$

where $\overline{f_0}$ is the average number of frames on the initial path and $\overline{f_r}$ is the average number of frames on the return path. Here, a threshold value is set for the amount of finger movement in each frame, as shown in Figure 7. The frames in which finger movement exceeds the threshold value are extracted, and the average number of frames for the initial and the return path is obtained. Then, the ratio between the time required for initial and return trips is calculated.

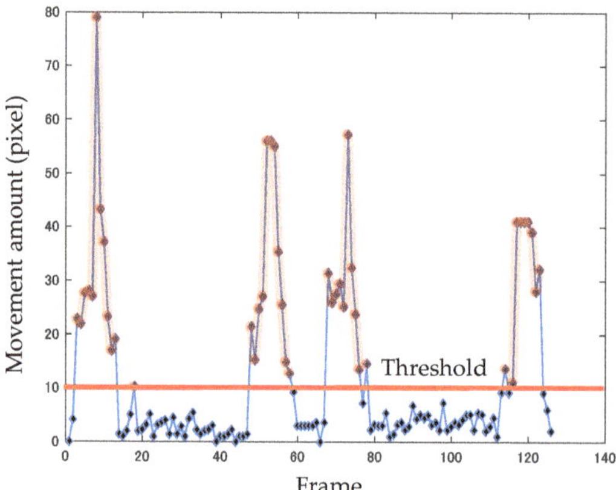


**Figure 7.** Amount of finger movement in each frame.

### 3.3.6. Measure of Average Travel Distance

The final feature to be extracted is the average moving distance digitizing how much the finger is shaking when first touching the examiner's finger. Tremor patients shake their fingers when touching the examiner's finger in the *FNF* test. This is especially remarkable in *ET* patients. Therefore, initial and return frames are extracted during the calculation process for the round-trip time ratio. Thus, the average moving distance of the fingers during that period is calculated using the following formula.

$$\overline{d} = \frac{1}{(m_2 - m_1) + (m_4 - m_3) + 2}\left(\sum_{i=m_1}^{m_2} d_i + \sum_{i=m_3}^{m_4} d_i\right) \tag{6}$$

where $\overline{d}$ represents the average moving distance, $d_i$ is the moving distance in the $i$ frame, $m_1$ is the first frame of the first touch, $m_2$ is the last frame of the first touch, and $m_3$ is the first frame of the second touch, $m_4$ is the frame at the end of the second touch.

## 4. Method of Disease Diagnosis and Analysis

A classifier is learned by using the feature values obtained in Section 3, and the disease is diagnosed by classifying the data using that classifier. In this study, we classify by supervised learning. Supervised learning is a method of learning a classifier that correctly outputs the relationship between data and class, using the information on the label for the data and the class of data provided in advance.

The methods used as classifiers are as follows: linear discriminant analysis, logistic regression analysis, support vector machine (*SVM*), and the *k*-nearest neighbor method (*k-NN* method). Verification of the classifier is performed by *k*-fold cross-validation. Briefly, we will describe these classifiers as follows:

### 4.1. Linear Discriminant Analysis

Linear discriminant analysis is a method of finding a straight line that can best classify which group to enter when new data are obtained using data provided in advance that was divided into different groups.

### 4.2. Logistic Regression Analysis

In medical statistics, logistic regression analysis is one of the statistical methods used in multivariate analysis. In this method, when the objective variable (class) is binary, the probability $P$ that an event occurs when one of the classes is an event is expressed by the equations in (7).

$$P' = \ln\left(\frac{P}{1-P}\right) = b_0 + b_1 x_1 + b_2 x_2 + \ldots + b_p x_p,$$
$$P = \frac{1}{1-e^{-P'}},$$

(7)

where $b_0$ is a constant, $b_p$ is a partial regression coefficient, and $x_p$ is a covariate (feature amount).

### 4.3. Severity Measurement in ET Patients

Figure 8 shows the algorithm for measuring severity in *ET* patients. Threshold processing is applied to the feature amount calculated by the histogram analysis, angle analysis, and the average moving distance proposed in Section 2, and if it is equal to or greater than the threshold, one point is added to each. If the total number of points finally scored is less than 2, the score is mild, and if the total score is 2 or more, the score is severe.

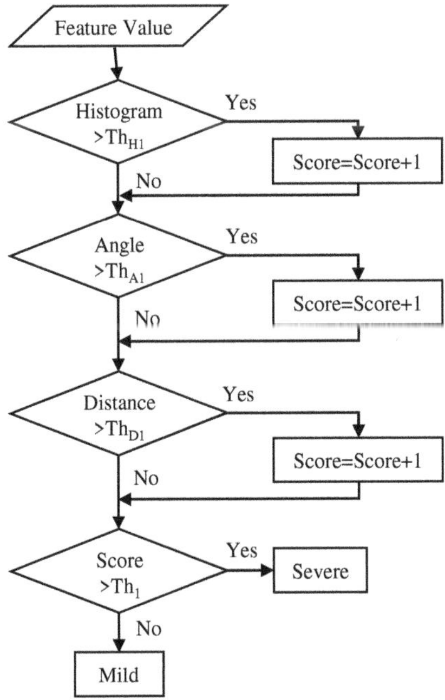


**Figure 8.** Severity measurement algorithm for essential tremors (*ET*) patients.

*4.4. Severity Measurement in CD Patients*

Figure 9 shows the algorithm for measuring severity in *CD* patients. Threshold processing is applied to the feature amount calculated by *RMSD*, histogram analysis, angle analysis, and average moving distance proposed in Section 2, and if it is above the threshold, one point is added to each. If the total score is less than 3 points, the severity score is mild, and if the total score is 3 points or more, the score is severe.

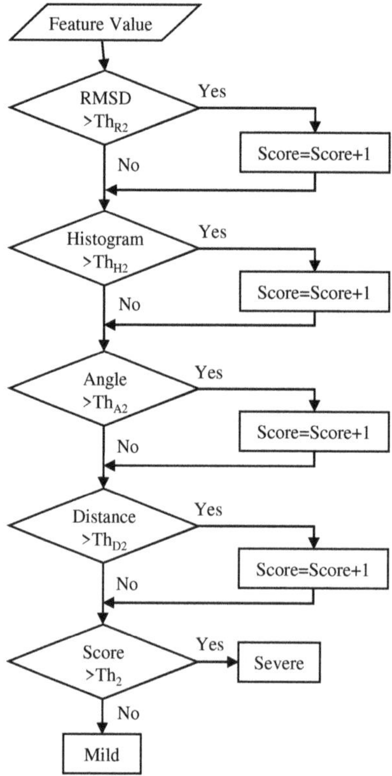


**Figure 9.** Algorithm for measuring severity in cerebellar disorders (*CD*) patients.

## 5. Experimental Results

This section describes the results of experiments using the proposed method. This time, we conducted an experiment using 8 sets of data for healthy people, 10 data for *ET* patients, and 18 data for *CD* patients. For training and testing data separation, we applied the $k$-fold cross-validation technique. In our system, we set $k = 5$ and therefore, the dataset is split into five folds. Machine learning is performed in experiments that each have multiple classes, as follows: (1) healthy subjects and tremor patients, (2) *ET* patients and *CD* patients and (3) healthy subjects, *ET* patients and *CD* patients. After learning is completed, $k$-fold cross-validation is performed. The results are shown in Tables 1–3, respectively.

**Table 1.** Classification results of healthy subjects and tremor patients.

| Classifier | Accuracy (%) |
| --- | --- |
| Linear discriminant | 83.9 |
| Logistic regression | 85.0 |
| *SVM* | 86.7 |
| *k-NN* | 83.4 |

**Table 2.** Classification results of *ET* patients and *CD* patients.

| Classifier | Accuracy (%) |
| --- | --- |
| Linear discrimination | 79.3 |
| Logistic regression | 72.2 |
| *SVM* | 83.6 |
| *k-NN* | 70.0 |

**Table 3.** Classification accuracy of healthy subjects, *ET* patients and *CD* patients.

| Classifier | Accuracy (%) |
| --- | --- |
| Linear discrimination | 68.9 |
| *SVM* | 76.1 |
| *k-NN* | 60.0 |

We have also conducted experiments measuring the tremor severity of *ET* and *CD* patients using the method proposed in Section 3. In these experiments, the threshold was determined by using a total of four training data in experiments featuring examinations by Doctor A and Doctor B, which had a low rate of accuracy in determining the severity of *ET* and *CD* patients. Tables 4 and 5 provide samples of the experiment results.

**Table 4.** Accuracy of severity measurement in *ET* patients.

| Severity Measurement | Total Number | Correct Number | Accuracy (%) |
| --- | --- | --- | --- |
| Mild | 4 | 3 | 75.0% |
| Severe | 4 | 3 | 75.0% |
| Total | 8 | 6 | 75.0% |

**Table 5.** Accuracy of severity measurement in *CD* patients.

| Severity Measurement | Total Number | Correct Number | Accuracy (%) |
| --- | --- | --- | --- |
| Mild | 7 | 6 | 85.7% |
| Severe | 9 | 6 | 66.7% |
| Total | 16 | 12 | 75.0% |

## 6. Discussion

As a result of training the classifier using the proposed feature quantity and *k*-division cross-validation, in the classification experiment featuring healthy subjects and tremor patients, Table 1 shows the best results using *SVM*, with an accuracy of 86.7%. In all three misdiagnoses, the examining doctors incorrectly assessed the symptoms to be mild and had difficulty in correctly assessing the symptoms even when reviewing the videos.

According to the classification of *ET* and *CD* patients, from Table 2, the best result was obtained with *SVM*, and its accuracy was 83.6%. Four misdiagnoses occurred, in three of which physicians incorrectly assessed the symptoms to be mild. The data for the single remaining *ET* patient presented *CD*-like characteristics, such as a high *RMSD* value due to severe symptoms with much shaking of the finger up and down and much time for the initial trip from examiner's finger to patient's nose. These severe symptoms seem to be the cause of the misclassification.

As shown in Table 3, a classification experiment featuring healthy subjects, *ET* patients and *CD* patients, the best result was obtained using *SVM*, with an accuracy of 76.1%. The accuracy was slightly less than with the above two classification experiments. In order to improve the accuracy, new features that better differentiate classes must be studied, and the dataset should be enlarged.

### 6.1. About Severity Measurement

Improving the measurement of severity first involves adapting the threshold to the amount of feature data, calculating the score, and then taking the measurement. As a result of doing so, the measurement accuracy for both *ET* and *CD* patients could be improved to 75.0%. In one example, the inaccuracy resulted from repeating training for the *FNF* test. In this case, the score was low, and the doctor incorrectly assessed the symptoms as severe, though the experimental results indicate that the symptoms were actually mild. In addition, since the purpose of this study is to facilitate measurements in the examination room, the conditions for recording video, such as camera placement, have not yet been optimized. For this reason, some erroneous results were obtained because the feature amount for the average moving distance of the finger increases when the camera is close and decreases when the camera is far. In order to solve this problem, normalization processing could be added so that the feature amount does not change depending on the conditions of video recording.

### 6.2. Future Outlook

In the future, the main issues will be the examination of new features and the use of new methods. The *ET* tremor is regular, and its frequency is generally 4–12 Hz [4,26,27]. In order to focus on the frequency component of tremors in future research, it is expected that diagnostic accuracy will be improved by performing analysis using the fast-Fourier transform.

Due to the fact that severity is difficult to define, we only differentiated mild and severe symptoms rather than attempting a more granular assessment. As a future challenge, we will quantify the degree of severity. Doing so will allow understanding the effect of therapy and enables doctors to modify prescriptions when symptoms do not improve.

## 7. Conclusions

In this paper, we proposed a non-contact method of discriminating *ET* and *CD* and a method of measuring tremor severity by analyzing the *FNF* test using image processing technology. We proposed feature quantities to quantify what a doctor actually focuses on in the *FNF* test and trained a classifier using these feature quantities. As a result of performing *k*-fold cross-validation on the classifier, *SVM* obtained an accuracy of 83.6% in classifying *ET* and *CD* patients. In addition, threshold processing was applied to the amount of feature data in each dataset, the score was calculated, and the severity was evaluated. As a result, the severity of symptoms for both *ET* and *CD* patients could be evaluated with an accuracy of 75.0%.

In the future, we expect to improve diagnostic accuracy by examining new features and using new methods, including some analysis of frames per second (fps) increase and Eigen background models focusing on tremor frequency and on detecting the nose of the patient and the finger of the examiner. Since the Eigen background model is based on the method of principal component analysis, we expect that a more clear foreground image (in our case, the finger area) would be extracted. It is also necessary to consider various approaches to quantify severity. In addition, we will increase the amount of data collected and aim to build a more reliable system. Moreover, in our future work, we would like to explore and analyze the raw recorded data by using a machine learning approach, such as using recurrent convolutional neural networks to extract prominent features from the data.

**Author Contributions:** This paper is organized and written by the second author T.T.Z. The second author also laid down the conceptual model and supervised experiments by the first author Y.M. who was her Master's student. The third author N.I. and the fourth author H.M. provided the experimental environments to obtain real-life data and gave medical interpretations. All authors have read and agreed to the published version of the manuscript.

**Funding:** This research received no external funding.

**Conflicts of Interest:** The authors declare no conflict of interest.

# References

1.  Miskin, C.; Carvalho, K.S. Tremors: Essential Tremor and Beyond. In *Seminars in Pediatric Neurology*; WB Saunders: Philadelphia, PA, USA, 2018; Volume 25, pp. 34–41.

2.  Mansur, P.H.G.; Cury, L.K.P.; Andrade, A.O.; Pereira, A.A.; Miotto, G.A.A.; Soares, A.B.; Naves, E.L. A review on techniques for tremor recording and quantification. *Crit. Rev. Biomed. Eng.* **2007**, *35*, 343–362. [CrossRef] [PubMed]

3.  Jombík, P.; Spodniak, P.; Bahýľ, V.; Necpál, J. Visualization of Parkinsonian, Essential and Physiological Tremor Planes in 3D Space. *Physiol. Res* **2020**, *69*, 331–337. [CrossRef] [PubMed]

4.  Wang, X.X.; Feng, Y.; Li, X.; Zhu, X.Y.; Truong, D.; Ondo, W.G.; Wu, Y.C. Prodromal Markers of Parkinson's Disease in Patients With Essential Tremor. *Front. Neurol.* **2020**, *11*, 874. [CrossRef] [PubMed]

5.  Buijink, A.W.; Contarino, M.F.; Koelman, J.H.; Speelman, J.D.; Van Rootselaar, A.F. How to tackle tremor -systematic review of the literature and diagnostic work-up. *Front. Neurol.* **2012**, *3*, 146. [CrossRef]

6.  Kamble, N.; Pal, P.K. Tremor syndromes: A review. *Neurol. India* **2018**, *66*, 36–47. [CrossRef]

7.  Crawford, P.; Zimmerman, E. Differentiation and diagnosis of tremor. *Am. Fam. Phys.* **2011**, *83*, 697–702.

8.  Lee, S.; Chung, S.J.; Shin, H.W. Neuropsychiatric Symptoms and Quality of Life in Patients with Adult-Onset Idiopathic Focal Dystonia and Essential Tremor. *Front. Neurol.* **2020**, 11. [CrossRef]

9.  Geraghty, J.J.; Jankovic, J.; Zetusky, W.J. Association between essential tremor and Parkinson's disease. *Ann. Neurol.* **1985**, *17*, 329–333. [CrossRef]

10. Dogu, O.; Sevim, S.; Camdeviren, H.; Un, S.; Louis, E.D. Prevalence of essential tremor: Door-to-door neurologic exams in Mersin Province, Turkey. *Neurology* **2003**, *61*, 1804–1806. [CrossRef]

11. Louis, E.D.; Marder, K.; Cote, L.; Wilder, D.; Tang, M.X.; Lantigua, R.; Gurland, B.; Mayeux, R. Prevalence of a history of shaking in persons 65 years of age and older: Diagnostic and functional correlates. *Mov. Disord.* **1996**, *11*, 63–69. [CrossRef]

12. Louis, E.D.; Faust, P.L. Essential tremor: The most common form of cerebellar degeneration? *Cerebellum Ataxias* **2020**, *7*, 1–10. [CrossRef] [PubMed]

13. Handforth, A.; Parker, G.A. Conditions associated with essential tremor in veterans: A potential role for chronic stress. *Tremor Other Hyperkinetic Mov.* **2018**, *8*, 517. [CrossRef]

14. Schmitz-Hübsch, T.; Du Montcel, S.T.; Baliko, L.; Berciano, J.; Boesch, S.; Depondt, C.; Giunti, P.; Globas, C.; Infante, J.; Kang, J.S.; et al. Scale for the assessment and rating of ataxia: Development of a new clinical scale. *Neurology* **2006**, *66*, 17–20. [CrossRef] [PubMed]

15. Elble, R.; Comella, C.; Fahn, S.; Hallett, M.; Jankovic, J.; Juncos, J.L.; LeWitt, P.; Lyons, K.; Ondo, W.; Pahwa, R.; et al. Reliability of a new scale for essential tremor. *Mov. Disord.* **2012**, *27*, 1567–1569. [CrossRef]

16. Mitsui, Y.; Ishii, N.; Mochizuki, H.; Zin, T.T. A Study on Disease Diagnosis by Tremor Analysis. *Int. Multi Conf. Eng. Comput. Sci.* **2018**, *1*, 14–16.

17. Bilge, S.; Jenq-Neng, H.; Su-In, L.; Linda, S. Tremor Detection Using Motion Filtering and SVM. In Proceedings of the 21st International Conference on Pattern Recognition (ICPR 2012), Tsukuba, Japan, 11–15 November 2012; pp. 178–181.

18. Ishii, N.; Mochizuki, Y.; Shiomi, K.; Nakazato, M.; Mochizuki, H. Spiral drawing: Quantitative analysis and artificial-intelligence-based diagnosis using a smartphone. *J. Neurol. Sci.* **2020**, *411*, 116723. [CrossRef]

19. Zdenka, U.; Otakar, S.; Martina, H.; Arnost, K.; Olga, U.; Václav, H.; Chris, D.N.; Evzen, R. Validation of a new tool for automatic assessment of tremor frequency from video recordings. *J. Neurosci. Methods* **2011**, *198*, 110–113. [CrossRef]

20. Benito-León, J.; Serrano, J.I.; Louis, E.D.; Holobar, A.; Romero, J.P.; Povalej-Bržan, P.; Kranjec, J.; Bermejo-Pareja, F.; Del Castillo, M.D.; Posada, I.J.; et al. Essential tremor severity and anatomical changes in brain areas controlling movement sequencing. *Ann. Clin. Transl. Neurol.* **2019**, *6*, 83–97.
21. Yuan, X.; Liu, Q.; Long, J.; Hu, L.; Wang, Y. Deep Image Similarity Measurement Based on the Improved Triplet Network with Spatial Pyramid Pooling. *Information* **2019**, *10*, 129. [CrossRef]
22. Figueroa-Mata, G.; Mata-Montero, E. Using a Convolutional Siamese Network for Image-Based Plant Species Identification with Small Datasets. *Biomimetics* **2020**, *5*, 8. [CrossRef]
23. Daunoraviciene, K.; Ziziene, J.; Griskevicius, J.; Pauk, J.; Ovcinikova, A.; Kizlaitiene, R.; Kaubrys, G. Quantitative assessment of upper extremities 3 motor function in multiple sclerosis. *Technol. Health Care* **2018**, *26*, 647–653. [CrossRef] [PubMed]
24. Krishna, R.; Pathirana, P.N.; Horne, M.; Power, L.; Szmulewicz, D.J. Quantitative assessment of cerebellar ataxia, through automated limb functional tests. *J. Neuroeng. Rehabil.* **2019**, *16*, 31. [CrossRef] [PubMed]
25. Johansson, G.M.; Grip, H.; Levin, M.F.; Häger, C.K. The added value of kinematic evaluation of the timed finger-to-nose test in persons post-stroke. *J. Neuroeng. Rehabil.* **2017**, *14*, 11. [CrossRef] [PubMed]
26. Sharma, S.; Pandey, S. Approach to a tremor patient. *Ann. Indian Acad. Neurol.* **2016**, *19*, 433–443. [CrossRef]
27. Martuscello, R.T.; Kerridge, C.A.; Chatterjee, D.; Hartstone, W.G.; Kuo, S.H.; Sims, P.A.; Louis, E.D.; Faust, P.L. Gene expression analysis of the cerebellar cortex in essential tremor. *Neurosci. Lett.* **2020**, *721*, 134540. [CrossRef]

**Publisher's Note:** MDPI stays neutral with regard to jurisdictional claims in published maps and institutional affiliations.

MDPI
St. Alban-Anlage 66
4052 Basel
Switzerland
Tel. +41 61 683 77 34
Fax +41 61 302 89 18
www.mdpi.com

*Sensors* Editorial Office
E-mail: sensors@mdpi.com
www.mdpi.com/journal/sensors

www.ingramcontent.com/pod-product-compliance
Lightning Source LLC
LaVergne TN
LVHW070108170726
843515LV00007B/1635